The Philip E. Lilienthal imprint
honors special books
in commemoration of a man whose work
at University of California Press from 1954 to 1979
was marked by dedication to young authors
and to high standards in the field of Asian Studies.
Friends, family, authors, and foundations have together
endowed the Lilienthal Fund, which enables UC Press
to publish under this imprint selected books
in a way that reflects the taste and judgment
of a great and beloved editor.

*The publisher and the University of California Press Foundation gratefully acknowledge the generous support of the Philip E. Lilienthal Imprint in Asian Studies, established by a major gift from Sally Lilienthal.*

*Nuclear Ghost*

CALIFORNIA SERIES IN PUBLIC ANTHROPOLOGY

The California Series in Public Anthropology emphasizes the anthropologist's role as an engaged intellectual. It continues anthropology's commitment to being an ethnographic witness, to describing, in human terms, how life is lived beyond the borders of many readers' experiences. But it also adds a commitment, through ethnography, to reframing the terms of public debate—transforming received, accepted understandings of social issues with new insights, new framings.

*Series Editor: Ieva Jusionyte (Brown University)*

*Founding Editor: Robert Borofsky (Hawaii Pacific University)*

*Advisory Board: Catherine Besteman (Colby College), Philippe Bourgois (UCLA), Jason De León (UCLA), Laurence Ralph (Princeton University), and Nancy Scheper-Hughes (UC Berkeley)*

# Nuclear Ghost

ATOMIC LIVELIHOODS IN
FUKUSHIMA'S GRAY ZONE

*Ryo Morimoto*

UNIVERSITY OF CALIFORNIA PRESS

University of California Press
Oakland, California

Library of Congress Cataloging-in-Publication Data

Names: Morimoto, Ryo, author.
Title: Nuclear ghost : atomic livelihoods in Fukushima's gray zone / Ryo
    Morimoto.
Description: Oakland, California : University of California Press, [2023] |
    Includes bibliographical references and index.
Identifiers: LCCN 2022047687 | ISBN 9780520394100 (hardback) |
    ISBN 9780520394117 (paperback) | ISBN 9780520394124 (ebook)
Subjects: LCSH: Radioactive pollution—Japan—Fukushima-ken. |
    Fukushima Nuclear Disaster, Japan, 2011. | Fukushima-ken (Japan)—
    Social conditions—21st century.
Classification: LCC HN730.F873 M675 2023 | DDC 363.17/990952117—dc23/
    eng/20221117
LC record available at https://lccn.loc.gov/2022047687

32  31  30  29  28  27  26  25  24  23
10  9  8  7  6  5  4  3  2  1

CONTENTS

# ILLUSTRATIONS

## MAPS

## FIGURES

TABLE

# *Introduction*

There is a nuclear ghost [*houshanō obake*] in Minamisōma.

HATSUMI

"A CITY WITH NUCLEAR GHOSTS" was how Hatsumi, a woman in her sixties, described the state of Minamisōma city, Fukushima Prefecture. "Do you believe in ghosts?" she asked me. Noticing my dumbfounded face, she offered me a chance to respond. I could not reply right away and, to earn some time, reached my hand to a glass of cold barley tea she had served me.

It was late July 2013, during a hot, humid summer. I had just moved to Minamisōma from Massachusetts for my dissertation fieldwork. Talking to residents like Hatsumi, I wanted to understand why many people lived on the edges of nuclear evacuation zones despite the elevated risk of radiation exposure that the media, social media, and scientific reports made undeniably visible. As an outsider, I struggled to understand the polarized discourses concerning postfallout Fukushima. On the one hand, it was argued that the state and the electric company had acted inhumanely to "force" people to reside in the irradiated environment.[1] On the other hand, the local and national government spent so much money and so many resources to make it possible for people to "stay in" and "return to" the region. The same tension still exists at the time of this writing, in 2022, more than eleven years after the disasters.

On March 11, 2011, when the magnitude 9.0 earthquake and the tsunami hit the Tōhoku (northeastern) region of Japan, I was in Massachusetts, more than 6,500 miles away and fourteen hours behind. As I woke up that morning, I witnessed the chaotic unfolding of the combined disasters (*Fukugō Saigai*/複合災害), or what is called 3.11 (*san ten ichi ichi*), in the recorded images of the tsunami overcoming the coasts of Tōhoku and eastern Japan. The images of destruction bombarded my senses, and I could barely follow the constantly accumulating numbers of people confirmed dead and missing. Now we know that the earthquake and tsunami killed 15,900 people in twelve prefectures, 2,523 are still missing, and the physical and material damages have cost the country over $1.4 trillion.

The situation became even direr as the tsunami devastated what was believed to be the robust assemblage of the Tokyo Electric Power Company (TEPCO) Fukushima Dai-ichi Nuclear Power Plant (colloquially referred to as 1F [*ichi efu*]) in the coastal region of Fukushima and disabled its backup power generators. As a result, the reactors' cooling system was incapacitated, and hydrogen explosions occurred at three of the six reactors between March 12 and 15, causing the haphazard distribution of radioactive debris throughout the planet and mass evacuation in the surrounding region. "There is no immediate danger," the chief secretary of the cabinet, Yukio Edano, repeated like a broken record. Focusing on containing the fear among the citizens instead of disseminating information, the state acted on what Clarke and Chess (2008) call "elite panic."[2] The natural hazards, technological accident, and the subsequent elite panics later became known as "the Fukushima nuclear disaster." In this book, however, I refer to the nuclear accident as "the TEPCO accident." In calling it the *TEPCO* accident, I want to make it evident that the accident occurred at the power plant in *Fukushima* owned and operated by *Tokyo* Electric to generate electricity exclusively for the people of central Japan. As I will show, this shift in the naming convention for the English-speaking audience signals the core of my ethnographic project, which aims to decenter the radiation-centered narrative to instead explore the local, more granular conditions surrounding 3.11.

Unlike the Chornobyl disaster in 1986, which remained secret until the neighboring countries traced spiked radiation-monitoring data back to the city of Pripyat, the globally circulated live images of hydrogen explosions and the ensuing efforts to contain the crippled reactors made the TEPCO accident in Fukushima a global "media event" (Beck 1987).[3] In a day, Fukushima became known to the world as the land of contamination. At the same time, while these Fukushima nuclear spectacles brought Fukushima to global attention, they frequently erased the losses the residents experienced from the earthquake and tsunami. Fukushima Prefecture alone lost 1,614 people, including two individuals in their twenties who were surveying the earthquake damage inside reactor four at 1F, and 196 people are still nowhere to be found.[4] The city of Minamisōma, where Hatsumi lived, experienced the highest death tolls in the prefecture, losing 636 people, and 111 people were still missing as of March 2022. The TEPCO accident and the subsequent evacuation order made the losses even more traumatic for those who had to give up searching for their missing friends and families.[5]

Even though I was terribly disturbed by what I saw from a distance, I could not keep my eyes off my computer screen, news reports, and social media.[6] I kept wondering if it was the end of Japan as I knew it. My sense of loss was surreal. I was not familiar with most of the places mentioned or depicted in the news. Growing up in western Japan, I knew no one in Tōhoku. My family, who lived far away and experienced only the aftershocks of the rattling earth and the incessant media spectacles, did not help me make sense of the disasters. They described 3.11 as a "big deal" and compared it to the magnitude 7.0 Hanshin-Awaji (Kobe) earthquake, which killed over 6,300 people in 1995, which we had experienced more intimately.

The overwhelming sense of uncertainty and fear of the unknown in Fukushima, however, suggested that something unusual was creeping up (Inose 2014). Sociologist Kai Erikson (1994) calls invisible threats like radiation and its lingering dread a "new species of trouble." It unsettles our taken-for-granted idea about the boundedness of an event—a plot with a clear beginning and end—and our assumptions about the safety and security of being in the world (Parkes 1967). For my family and me, what was happening in Fukushima felt closer to the chilling sensation caused by the sudden awareness of the invisible and unknown we had confronted after the Tokyo subway sarin attack, an act of chemical and religious terrorism by Aum Shinrikyo on March 20, 1995, following the Kobe earthquake on January 17.[7] Although 1995 was a dark year for Japan, 3.11 posed a different kind of existential challenge, and we were all seeking some reference for it in the past.[8] For making sense of this "unprecedented/*soutei gai*" disaster (Bestor 2013), our historical and cultural pockets were empty.

In the summer of 2013, when Hatsumi told me about the nuclear ghost of Minamisōma, I was still haunted by my exposure to the Fukushima spectacles. As a result, I could not help but interpret the "nuclear ghost" as the ghostly presence of radiation in the city, which can only be experienced with technoscientific instruments like a Geiger counter. By interpretating Hatsumi's nuclear ghost this way, I revealed the fundamental assumption I had brought with me to Minamisōma rather than the city's actual state. I went there to confirm my belief that it is an unsafe place to live and residents are in denial, just like the media and academic depictions of Fukushima suggested.[9] I had imagined that my research would explore the unarticulated danger, people's profound fear of imperceptible radiation, corporate and state secrecy about the scale and extent of contamination, and visible health defects among the residents, just as in the cases of Hiroshima, Nagasaki, Chornobyl, Hanford, the Four Corners, the Marshall Islands, French Polynesia, and other sites of nuclear fallout. After all, isn't a nuclear accident all about radiation exposure and its detrimental biological and environmental consequences? If the nuclear ghost is not radiation, what could it be?

Approaching Fukushima from this radiation-centered angle made individuals and their experiences less crucial; radiation impacts people equally, and if people think otherwise, it must be the result of manipulation. Following these presumptions, I failed to record any information about Hatsumi in that meeting, such as who she was, why she stayed there, what kind of life she had lived, and how she imagined her future in Minamisōma. In contrast, I duly documented the readings of my Geiger counter, which I thought indicated the world's objective state—that X amount of radiation is present in a specific locale regardless of who measures it—as if that information defined the place where Hatsumi resided and the life she lived.

I was wrong. It took me a long time, many mistakes, and many more interactions with residents like Hatsumi and others to come to learn otherwise. Believing that I was studying a disaster rather than individuals in a disaster, I initially searched for "the victim" of the TEPCO accident, those individuals who would fit in the category of "the sufferer" (*higaisha*) and "the exposed" (*hibakusha*), yet I could not find many; in my initial twenty or so semistructured interviews, people frequently ended the conversation by referring me to someone else who they thought "suffered" more.[10] Some people lost their family members, while others lost their homes from the tsunami or from contamination. This referral process eventually led to individuals who often appeared in the media reports of a "disaster-affected area/*hisaichi*." Those individuals had remarkable and elaborate stories of suffering and loss to share.[11] They also knew what they were expected to say to meet the distant others' gaze so that others could be spared from a similar scrutiny.

The failure of my initial approach made it apparent that the search for suffering only fulfilled outsiders' expectations and reproduced the hierarchy of suffering in the local community. This early experience in the field made me ask whether the goal of disaster ethnography is to locate and represent the experience of the people who suffered most? In the field I have often wondered about the role an anthropologist plays in the postdisaster context. Sometimes, I was unsure if ethnography was any different from Naomi Klein's influential idea of "disaster capitalism," in which individuals, institutions, corporations, and so on benefit from a disaster and its victims, or "disaster pornography," in which the depictions of sufferings become the mode of consumption, entertainment, and the reality. Jean Baudrillard ([1981] 1994) calls such a constructed reality "hyperreality," where repeated representations come to shape the reality through the process of self-fulfilling prophecy.

Although I have found it challenging to grapple with the question of my position as an unintentional extractivist in regard to locals, and I will keep coming back to this dilemma throughout the book, one thing was undeniable: the way that public discourse figured the tsunami and TEPCO accident did not match neatly with how each resident experienced, narrated, and remembered them differently on the ground.[12] One tsunami survivor in Minamisōma I met in 2013 hesitantly shared that "while I feel lucky to have survived the tsunami unlike some people, sometimes I cannot be confident that I am in a better state because of the nuclear accident afterward. I never thought being lucky is bad luck."

Suffering comes in many shapes and different tempos. Minamisōma's residents all experienced the same event, the so-called 3.11, but where they happened to be situated mattered to how they came to experience, live with, and process the aftermaths (Hastrup 2011). Residents often disagreed with each other about their situated experience of 3.11, and, more importantly, their relationship to it—their memory, interpretation, and experience—changed over time.

This interpretive struggle between individual residents, the public, and the state and experts about 3.11 and the ensuing social and political fragmentation reminded me of industrial disasters like the Minamata disease caused by mercury poisoning in the 1950s. Environmental scientist and activist Ui Jun ([1971] 2006, ch. 9), who laid the foundation of environmental studies in Japan, characterizes environmental disasters as extrascientific and surreal (*cho-genjitsuteki*) experiences that manifest themselves in society through social discrimination and divisions. I believe 3.11 is a similar species of trouble full of ambiguities, absurdities, and ambivalences.

To tell this convoluted story of 3.11, throughout this book I borrow words from novelist Haruki Murakami. His writings guide me to explore the gray zone between what is considered real or surreal, and scientific or not scientific, in how people experienced, remembered, and narrated 3.11. Patching together archives, memories, words, and narratives that illuminate diverse livelihoods despite radiation, I attempt to offer a horizon of social science of the surreal. My goal is to abduct, as Murakami (2001, 226–27) puts it in his writing about the Tokyo Gas Attack, "words coming from another direction, new words for a new narrative. Another narrative to purify this [radiation-centered] narrative."

*The Nuclear Ghost* is an ethnographic monograph about my chance encounters with various livelihoods in radioactive landscapes of coastal Fukushima. Here, residents' sustained efforts have helped recover and reconstruct the past tsunami damage done to physical structures and reopen former evacuation zones, while the damaged nuclear power plant continues to release contaminants each and every day. There I met many individuals who decided to stay or have returned to the region for various reasons, despite the risk of radiation exposure and sometimes in the face of others' harsh judgment of their character for doing so.

A retelling of the lives of those who did not leave, have returned, and have moved into coastal Fukushima, my stories might appear to some to be underplaying the decision of those who left the region and the potential adverse effects of radiation exposure and thus spreading a radiation-tolerant, pronuclear perspective. That is not my intention. Instead, I invite readers to witness the residents who, for one reason or another, felt compelled to stay or to return despite the risks and often incompetent, inflexible, and conservative state authority (Kainuma 2012; Samuels 2013). Whether coastal Fukushima is irreversibly contaminated or not, I met and spent significant time with people who still called it their home and desired to live with their ancestors and pass their land, cultures, and histories to future generations. I wanted to understand why, by hearing their stories. And now I am passing them on to you.

## PERIPHERY OF PERIPHERIES (*SYŪ-EN NO SYŪ-EN*)

A Minamisōma native, Hatsumi lived there her entire life. Like many other residents, however, she was not particularly fond of Minamisōma. "There is nothing here," she would say, "so I watch the travel channel and plan for my next trip abroad!" She often traveled outside the country to see places and experience things that she felt the rural city could not offer. From traveling abroad, Hatsumi was keenly aware of how outside people perceived postfallout Fukushima. One time, Hatsumi proudly shared the story of unintentionally scaring a worker at a boutique in Paris. According to Hatsumi, when she asked a salesclerk to send her purchased goods to Fukushima, she was met with an adverse reaction. "The store lady stepped a few inches back to take some distance from me! Fukushima is now famous, you know." She laughed and continued, "The fact is it is universal. Many Japanese people in the same tour reacted in a similar way."

Despite these negative experiences, she did not lose her desire to see the world. Instead, she said, "I just learned not to tell people I am from Fukushima. Sometimes it is hard to keep my story straight or not to speak with a strong regional dialect, especially to other people in the same tour," and she tried a few words in an off-sounding Tokyo dialect to illustrate her technique. "What is more frustrating to me is how inconvenient it is for us to get to places," she often complained to me. "Minamisōma is far from everything, and nothing is close enough. The nuclear accident made it worse than before!" This general sense of remoteness was something I heard repeatedly in the city.

More than two hundred kilometers north of Tokyo, Fukushima Prefecture is the third-largest prefecture in Japan, consisting of three distinct regions—Aizu, central Fukushima (Nakadōri), and coastal Fukushima (Hamadōri), which correspond with the mountainous area of western Fukushima, the middle (the most populous area), and the coastal side—each with a unique history and culture. Eight percent of Fukushima Prefecture, an area about the size of San Antonio, Texas, fell under the evacuation order in 2011.[13] Minamisōma and 1F belong to Hamadōri, where locals described the region as a *rikuno kotō* (an inaccessible corner of the land). Miri Yu's novel *Tokyo Ueno Station* captures the complex center-periphery relations through the story of a migrant laborer from Minamisōma who travels to Tokyo in search of seasonal employment. As I will detail in chapter 6, the nuclear power plant came to Hamadōri to revitalize the region so that, the state officials told locals, they could remain there and be with their families throughout the year. About fifty years after the plant was built, in 2011, the TEPCO accident ironically resulted in displacing families and sometimes separating their members.[14]

Although Minamisōma is the second most populous city in Hamadōri, with a population of more than seventy thousand people in 2010, its residents have always felt that it is a provincial city with respect to Fukushima Prefecture, let alone with respect to the rest of the country. The former mayor of the city, Katsunobu Sakurai, who became known globally for his SOS message on YouTube on March 26, 2011,[15] lamented the historical underpinning for the peripheralization of the city: "Unlike Futaba and Ōkuma towns, which host the nuclear plant, Minamisōma has never depended directly on the nuclear power plant" and, therefore, "what we [Minamisōma] have received from TEPCO is nothing but lopsided harm."[16] The TEPCO accident and its aftermaths only exacerbated this general sentiment. The mass evacuation (both forced and volunteer) caused a reduction of Minamisōma's population and further aged its population, and as of 2022 more than 37 percent of its residents are older than sixty-five. That percentage is even higher in the former evacuation zone, at almost 50 percent. The state reterritorialization of evacuation zones cut the infrastructure (roads and railways) between different parts of Fukushima and the outside for a long time and caused the disintegration of existing community by differently compensating residents for their damages. As I will illustrate in chapters 1 through 4, despite all the challenges, many individuals, whether willingly or unwillingly, lived with residual radiation on the edges of evacuation zones.

This type of local geographical and demographic complexity is one of many facets of Fukushima that the media reports and academic discourses did not capture fully. Instead, most viewers of the nuclear catastrophe unfamiliar with the region, including myself, rendered this catastrophe visible with the spectacular imaginary of contaminated Fukushima, which overshadowed the ongoing lives within it. In Minamisōma, as of the beginning of 2022, around 57,600 new and old residents lived in the city, and more than 6,600 people were back in city's former evacuation zone, less than twenty kilometers from ground zero of the TEPCO accident.

During my fieldwork between 2013 and 2019, I witnessed the gradual return of the area's residents following a series of disaster-recovery and reconstruction efforts led by the national government, the local government, and residents themselves.[17] In 2016, upon a public viewing of a documentary film about Minamisōma filmed between 2012 and 2014, a couple of residents commented to me how people looked very stern in the first few years after the accident: "Then, we were all looked deadly serious. Each day, we were trying to survive. Your face gets stiff when you forget how to smile or laugh, you know."[18] As many of us witnessed from afar, 3.11 was truly catastrophic to those who experienced it firsthand. Yet, contrary to popular expectation and persistent imagination, the actual primary health effects have not come from radiation itself. Instead, the residents suffered from various secondary health issues such as diabetes, cardiovascular issues, and obesity, as well as mental health issues.[19]

By stating the importance of secondary health issues, my goal is not to downplay the harm from the routinized low-dose radiation exposure and its slow, though accumulative, effect over a long span of time. Nor am I denying incidences of cancers (e.g., thyroid cancer among minors) and other injuries experienced by people who had higher-dose exposures immediately following the fallout or from serving to get crippled reactors under control.[20] Even though TEPCO should always remain accountable for the nuclear accident, I acknowledge invisible work by local and nonlocal employees of TEPCO and its related companies who have been laboring and exposing themselves to contain and decommission 1F.[21] The same goes for the workers at the Chornobyl nuclear power plant in Ukraine and other nuclear laborers throughout the world.

It is inevitable and crucial that the nuclear accidents in Fukushima and Chornobyl would be compared, since Chornobyl shaped our orientation to the TEPCO accident.[22] On April 15, 2011, the national government used the International Nuclear and Radiological Event Scale by the International Atomic Energy Agency (IAEA) and gave the nuclear accident in Fukushima the highest-level designation—seven, major accident—which was equivalent to the Chornobyl accident.[23] However, radioactive materials released from the TEPCO accident turned out to be one-tenth of those in the Chornobyl accident, which released 5,300 Peta (quadrillion) becquerels of radionuclides into the environment (Onda et al. 2020).[24]

The most significant difference between the two accidents, aside from social, cultural, and political particularities, is the following: after the Chornobyl accident, a zone with a thirty-kilometer radius from the plant was declared off-limits, and it remains so thirty-six years later. In contrast, in coastal Fukushima various zones have gradually been cleaned up and officially reopened, and many residents have gone back. In the last ten years the size of the evacuation zone has been reduced from 8 percent (1,150 km$^2$) to less than 2.5 percent (370 km$^2$) of the prefecture, and the number of evacuees in Fukushima has shrunk, from 164,865 at its peak in May 2012 to around 35,000 (reported) as of April 2022.[25]

Nevertheless, the presence of radiation in Fukushima has remained a central focus because of environmental monitoring and remediation, as well as lay public, academic, and media discussions of its potential negative impacts. In 2019, the Mitsubishi Research Institute surveyed one thousand Tokyo residents about their knowledge of Fukushima's reconstruction and the potential adverse health effects of radiation exposure.[26] The results revealed that urban people are persistently concerned about radiation's intergenerational impacts in particular.

From these types of radiation surveillance came the coexistence of two distinct radiation awarenesses. On the one hand, people became aware that radiation is omnipresent, as a universal phenomenon. On the other hand, the media, as well as medical and academic discourses, represented radiation hyperlocally as a specific coastal Fukushima phenomenon and bounded event. Caught between the universal and hyperlocal presence of radiation, residents struggled to live simultaneously with the invisible phenomenon, its techno-sensory representations, the experts' optimism, and outsiders' cynicism.

I do not intend to suggest that Fukushima's residents are no longer concerned about the long-term risks from routine radiation exposure and persistent environmental contamination. In fact, the most recent health survey in the prefecture revealed that more than 20 percent of Fukushima residents were still concerned about the potential adverse intergenerational impacts from past exposure.[27] My argument, rather, is that the technoscientifically determinable presence of radiation does not necessarily correspond to the *experience* of it.[28] The distinction is particularly critical in most parts of coastal Fukushima, where the postfallout radiation level turned out to not be high enough to cause immediate, visible, and scientifically significant harm.

Unlike the liquidators and nuclear laborers that Adriana Petryna (2013) worked with in her ethnography of postaccident Chornobyl, the individuals I interacted with did not use their irradiated bodies—their "biological citizenship"—as a way to legitimize their victimhood. Individuals I met at times resisted the idea of using their bodies as a tool against the state and TEPCO. They intentionally avoided getting their bodily contamination levels tested since they interpreted the monitoring as a way for the state to experiment with their bodies. Instead, remaining residents have demonstrated their sustained livelihood in Hamadōri by maintaining legal land ownership, genealogical continuity of their household, and cultural and historical continuity.

Long-term evacuation caused a radical change to the local landscape. Since 2013, hundreds of solar panels managed by nonlocal companies have covered the depopulated former evacuation zone. Not knowing when they could return and restart farming safely, many elderly residents felt they had no choice but to give up their farmlands. Observing this trend, one Minamisōma resident who has continued farming in the city said, "They [the state and TEPCO] have already messed up our lands, and we are not going to leave and let them steal our ancestral lands." For the residents, one major lesson from the fallout was the importance of not handing over their land to the state and private sector, who then could use that land to destroy their homes, cultures, histories, and spiritual ties to ancestors. Without those invisible ties and intangible things that coexist in their land, many residents felt the meaning of their life was halved. How did the residents act when radiation and its sociocultural, political, scientific, and economic significance suddenly came to delimit what it means to live as fully as they desired?

In this book I explore what happened to people's ongoing livelihoods when the state, experts, and public approached the TEPCO accident narrowly as a technical issue with potential medical consequences to individual bodies. As I will show, such technocratic and biomedical framing of the accident is one understudied harm that has caused significant damage to the local ecology, the residents' sense of belonging, and their relationship to the land, the dead, nonhuman others, and the ecosystem. I will delve into the broader impacts of the accident in chapters 6 through 9.

If, as architect Kishō Kurokawa (1977) imagines, peripheries (*shū-en*) are where ambiguities coexist and dissimilar entities find their interconnections, Minamisōma is one such *shū-en* where I have witnessed people apprehending the TEPCO accident and its sheer complexities.[29] In this highly politicized though provincialized place, I met various individuals who taught me how to attend to a category of experience Hatsumi identified as "the nuclear ghost" and its message.

## *EN* (縁) AND THE ETHNOGRAPHY OF FALLOUT

My interaction with Hatsumi in July 2013 was where my ethnography of postfallout coastal Fukushima began. As did many others, Hatsumi often described our meeting as the result of *en*. A term and idea originally drawn from the Buddhist concept of *pratyaya, en* means an indirect cause or a hidden condition of possibility for webs of interconnections. Paired with another term, *in,* or a direct cause, they make up the Buddhist theory of everything, or *innen* (see Hashimoto 2013; Jensen, Ishii, and Swift 2016; Minakata 2015; Rowe 2011). Colloquially, Japanese people use the term *en* to mean some form of connectedness ranging from a biological kinship (*ketsu-en*/血縁), to an inseverable relationship (*kusare-en*/腐れ縁), to a mere chance encounter (*kaikou*/邂逅).[30] Anthropologist Shunsuke Nozawa (2015) discusses *en* as a floating signifier or "relationship as such," which researchers of Japan inevitably observe in the Japanese people's interpretation of their chance encounter as ineffably fated (*Nanikano en*). From this Japanese folk perspective, ethnography is about both submitting to and weaving invisible threads, or *en*, by not questioning the cause and meaning of each encounter and instead letting each encounter be the guide. *En* itself is not meaningful, but one's belief in its potential as the invisible infrastructure for meaningfulness is.

My *en* with Hatsumi was a pure chance encounter mediated by a foreign film director whom I accompanied as a translator. Hatsumi was the wife of a Buddhist monk, the mother of two adult daughters, and the grandmother of their two children. She was part of the cast of the documentary film the German director had been shooting since 2012. On the way to Hatsumi's place for my first visit, the director told me about her critical role in the film as an angry, ordinary, and elderly citizen who had been dissatisfied with the dishonest state and the electric company. "She is a very tiny lady but a strong-willed and vocal woman," the director added, "unlike many other shy people I met during my time in coastal Fukushima." Privately, he frequently expressed frustration with locals who were somewhat reserved and hesitant to be on camera and speak up for themselves. "I am trying to give a voice to people who have been betrayed by the authority, but why aren't they angry like all of us?" was the question he asked me whenever he had a chance.

My first interaction with Hatsumi was disastrous; she shut the door in our faces while yelling that she did not want to be filmed ever again. We convinced her to give us a chance, but even as we sat across a table from her, she did not look us in the eye. Noticing the negative energy in the air, the director pushed me to persuade Hatsumi that her sustained involvement was essential for the completion of the film, and, quite intimidated by two strangers—the frustrated director and unhappy Hatsumi—I did what I could to mediate the bilingual negotiation. Before we had the chance to drink the glass of iced barley tea Hatsumi served, we left Hatsumi's place without securing a promise that we could continue filming her.

After my first failed attempt, however, I returned once again to Hatsumi's place. This time I was alone. A few days earlier, at the end of July, I had left the film project after a huge fight with the director.[31] (That's another story!) As I timidly approached her house, adjoined to the wooden temple where monks would practice Buddhism in front of Buddha statues of various sizes, she was feeding the sparrows that were sunbathing by her residence. "They know I come out this time of the day to feed them," she told me. "I've named them, but I cannot tell the difference between them." She laughed and gestured to me to follow her inside the temple.

During this meeting she revealed to me that she was upset about how we had lacked basic manners and respect in our approach to her. Hatsumi claimed that we did not bring a gift like people usually do in Japan, and we did not call her in advance for a visit. However, as we became closer, she eventually admitted to me the real reason she had turned us away: she had become increasingly dissatisfied with the director's view of Minamisōma. "When I first met the director [in 2012]," she told me, "I too thought the city was doomed. I too was angry, confused, and scared. I was scared for my older daughter and her young sons. I wanted to tell the world that! But you know, even the radioactive environment changes and we cannot be the victim forever. Like it or not, life goes on."

Hatsumi and I talked in her well-maintained Buddhist temple, which had been in the family since the early nineteenth century. Situated approximately twenty-five kilometers (15.5 miles) north of 1F, the temple served as one of the destinations for the cremated but unidentified tsunami victims no one had yet to claim (*muen botoke*) after 3.11. The presence of *muen botoke,* or the deceased without any familial ties, was the main reason the director had chosen Hatsumi's family for his film.[32] The director found this foster caring of the unidentified physical remains to be exotic and picturesque, and Hatsumi explained that she stayed in Minamisōma despite the fallout in order to care for the alienated remains with her husband for the community.

At the Buddhist temple, where the boundary between the dead and the living felt blurrier than in other places, Hatsumi's reference to the nuclear ghost carried an aura of credibility. According to the Japanese conceptualization, a ghost/*obake* (or *yūrei*) indicates a state of shapeshifting—one thing transforming into another thing. *Obake* is neither the dead nor the living. It belongs to the preternatural state. "Ghost," I remember from growing up in Japan, refers to the spirits that linger after bodies' death when the dead have some remaining concern or regret about the world of living. The ghost, as sociologist Avery Gordon (1997, xvi) puts it, "has a real presence and demands its due, your attention." In Minamisōma, the nuclear ghost emerged when the collective attention was directed to and fixated by radiation and its manifold representations. The nuclear ghost—the shapeshifting figure—has been lingering to tell the story of the absence in coastal Fukushima when radiation, the contaminated environment, and its potential detrimental health effects became the dominant, expected, and selectively curated story.

In Japan, ghosts are not an uncommon topic of conversation, especially after an unfortunate event. Authors such as Richard Lloyd Parry (2017) and Yuka Kudō (2016) have written about ghosts encountered in northeastern Japan by the survivors of the 2011 tsunami.[33] However ordinary ghosts and the supernatural are in the Japanese collective experience, this knowledge did not help me decipher Hatsumi's nuclear ghost reference. I continued to wonder, how can the nuclear or the radioactive thing become a ghost? If it has shifted its shape, what concerns does it have about the living? Or is the ghost merely a metaphor for describing the imperceptibility and the ambiguous, ever-present, and haunting psychosocial effects of radioactivity on human subjectivity, which literary scholar Gabriele Schwab (2020) metaphorically calls "radioactive ghosts"?

That day in late July 2013, I asked Hatsumi to speak more about how and why there is a nuclear ghost in Minamisōma. This is what she told me:

> Ghost is something some people see and believe in, but others don't see or believe at all. You might not see it ever and forget that it is there or could be there, but some other people can see what you do not see and tell you it is there. Then you doubt if there is something wrong with you for not seeing what others see or seeing what others don't. Once the ghost reveals itself to you or is revealed through others' determination, there's no turning back. It changes how you think about your past, present, and future. That's the nuclear ghost in Minamisōma.

As my eyes drifted to a Buddhist motif covered with gold leaf on the temple's high ceiling (specific to the Sōtō school), I told her that I was unsure if I believed in ghosts, nuclear or otherwise. She told me it was okay if I did not know, but she immediately pointed out that I carried a Geiger counter, so I must have been suspicious of imperceptible phenomena in Minamisōma.

"A Geiger counter is one way to ascertain the presence and absence of phenomena that you cannot sense, you know," Hatsumi said, winking to signal, as I read it, that she had caught my inconsistency. She was right. Before my first brief visit to Minamisōma a year before in 2012, I had unexpectedly found and bought a Geiger counter at a home-improvement center in Kawamata town, Fukushima Prefecture, expecting to "see" and "feel" something in the region that is beyond the threshold of the human experience.

I was taking a trip with a Tokyo-based journalist who was researching evacuees in Fukushima. We took a bullet train from Tokyo and rented a car in the Koriyama city station in central Fukushima. We drove through

Nihonmatsu city in the north to talk to a few evacuees from Namie town and then went east toward Kawamata town and through Iitate village to go as far east as possible. Our goal was to see how close we could get to 1F. The trip ended abruptly in Minamisōma when we were stopped by police officers who were guarding the barricade at the twenty-kilometer boundary from the power plant on Route 6. Back then, Minamisōma was still the final northern frontier of the nuclear disaster. As I will show in this book, the geography of nuclear disaster has radically changed since.

During this trip the Geiger counter mesmerized me. It enabled me to sense something beyond my senses. I tirelessly checked my Geiger counter both inside and outside the car as it made distinct clicking sounds as radioisotopes in the air made the inert gas inside conductive. Each click was a reminder of the presence of ionizing radiation in the vicinity. From the deep woods to barren pig farms and to abandoned houses, the technosensory device often controlled where I went. The higher the reading and the louder the clicking sound, the more real I felt about being in Fukushima. I fully believed that "true Fukushima" was hidden and that the truth was only accessible through my Geiger counter.

My faith in the Geiger counter exemplified a cognitive disorientation that anthropologist Joseph Masco (2006) calls the "nuclear uncanny."[34] In his study of the psychosocial effects on citizens of America's long-term engagement with the nuclear weapons project, Masco explains how radiation's imperceptibility abducts the human sensibility by "blurring the distinction between the animate and the inanimate, and between the natural and supernatural" (2006, 30; see also Beck 1987). In the early stage of my fieldwork, I had to believe that the numerical translation of radiation's presence on the technoscientific instrument's screen provided me a means of visualizing the TEPCO accident and its aftermaths that would help me orient myself to postfallout Fukushima.

When I returned to Minamisōma in 2013 for the second time, first to accompany the film director and then to conduct fieldwork, I made sure to bring the Geiger counter. I carried it to Hatsumi's because I was curious to learn if a traditional structure like a temple surrounded by many tall trees might be highly contaminated. From my previous experience, I knew that contaminants tend to be concentrated around forests and other bushy areas.[35]

My possession of the Geiger counter, as Hatsumi identified, gestured to my subconscious belief that radiation is the central issue in Minamisōma. My fieldwork, however, taught me that understanding and talking about postfallout coastal Fukushima by foregrounding what is invisible over what is visible is what locals like Hatsumi found problematic and even dehumanizing. As historian of nuclear technologies Robert Jacobs (2014) puts it, "Radiation makes people invisible" by denying all of their previous identities, histories, and lives. The selective attention to radiation, I also learned, elides the very reason radiation is the critical object of knowledge in the first place: human well-being. Radiation, therefore, makes not just people but everything else invisible. Unlike the Japanese conceptualization of *obake* or ghost, radiation has no message of its own, no regret about the living, unless humans assign meaning to its presence or absence and its polyvalent representations.

Not knowing what to say to her about the nuclear ghost, I offered to tell Hatsumi the reading of my Geiger counter. Hatsumi refused to hear it and raised her voice:

> What do you do with the knowledge of it? Are you going to use it to tell me that I should not be living here or judge me that my family is not smart to have young children living with us here? You might be fine doing what you think is right because you do not have to live here. If you want to, you can go elsewhere and not deal with any of it. But I have to continue living here. I have nowhere to go. After all, how do you know even what your gadget tells you is correct, and I am ignorant more than you are?

She took my gesture as a threat to the integrity of her senses and her judgment that an increased level of ambient radiation in the region is—and has to be, due to social and domestic circumstances—a tolerable threat. Her perspective taught me that the biological risk of low-dose exposure is not the absolute objective metric of life in Minamisōma.

Since 3.11, the Japanese lay public has used science and technology to exercise their rights as citizens in relation to postdisaster toxicity in food. Their efforts, anthropologist Nicolas Sternsdorff-Cisterna argues, challenged the tolerance level (*anzen*) set forth by state policy by illustrating the equal, if not higher, significance of obtaining a subjective sense of safety (*anshin*). Hatsumi's comment suggests an alternative formulation of toxicity and contamination, one that hinges not only on technoscience or personal perception of a given threat but instead on a metric that relies on ethical and social imperatives as well as local circumstances. For Hatsumi, accompanying her husband to care for *muen botoke* and people in the community who relied on their spiritual and religious services was personally more meaningful for her sense of well-being. Her personal values as a community member and a care-taker of the dead served to translate the risk into a tolerable threat.

In Minamisōma, Hatsumi was implying, seeing or not seeing—or speak-ing or not speaking of—radiation has an ethical and social consequence. Locals are careful about interpreting the meaning of radiation and speaking about radiation not because they want to disavow its presence but because many people, especially outsiders, cared only about radiation. Hidden in Hatsumi's skepticism about the accuracy of my Geiger counter, I now believe, was her critique of the stance that the only modes of narrating the aftermaths of the TEPCO accident were through technoscientific mediation and its visualizations of the invisible. Disaster does not happen in a nowhere land. It happens to specific groups of people and in a place with particular histories, cultures, and relations.

Some people stayed to maintain their connection to the land, while others gave up, each for their own reasons. The encounter reminded me of what Tewa Pueblo anthropologist Alfonzo Ortiz (1977, 19) advises Western schol-ars to consider when approaching traditional native North American cul-tures. If the places (where events happened) matter more than the time (when something happened) to how people tell stories and experience events, we must learn to "spatialize time" instead of turning "space into time." In the beginning of my fieldwork, I lacked adequate knowledge about the place where radiation and Hatsumi's nuclear ghost dwelled.

It took me a while to understand that radiation has a social and sometimes surreal life beyond its objectifiable presence in the environment, and beyond how and why it matters to the locals' everyday lives and their sense of self and place. In order to experience the messy life of radiation in society, I needed to subject myself—my own body and my personality—to myriads of local con-tingencies. I lived in Minamisōma to allow myself to be vulnerable, just like the residents have been, in the face of the sudden awareness of the invisible threat and its uncertain effects, to understand wholeheartedly, as sociologist Erving Goffman (1989, 125) would put it, "what [radiation] does to the life" of the locals and the broader ecology of coastal Fukushima.

Only through long-term participant observation—interacting with the residents of coastal Fukushima; driving around the region; exploring local historic sites; sitting through city council meetings; attending resident-organized workshops; farming and harvesting vegetables, fruit, cotton, and indigo plants; taking a tour at the crippled nuclear power plant multiple times; doing archival research at local libraries; and chasing wildlife—was

I able to tune into Hatsumi's riddle of the nuclear ghost.[36] Following Hatsumi's lead, *Nuclear Ghost* explores the social, historical, and cultural circumstances of individuals in order to understand how and why the residents have chosen to live with the incessantly visualized presence of radiation in Minamisōma, the city the residents experienced as the "gray zone" of the TEPCO accident.

## GRAY ZONE

Since July 2013 I have been fortunate enough to discover and cultivate my invisible *en* with many individuals like Hatsumi in coastal Fukushima. They resisted the idea that a nuclear catastrophe's significance lies only in the quantifiable contamination levels in bodies, things, and environments and in technoscientific explanations. Many residents did, in fact, tell me that their initial reaction to the fallout was fear—fear of the unknown—and that they desperately wanted to find some way to "sense" and give a shape to the invisible. The same fear still roams somewhere in the back of their minds every once in a while. However, they also attributed their everyday struggles to the hypervisualized presence of radiation in Fukushima, how radiation exposure has been represented and interpreted in society against their subjective experiences. This concern is deeply cultural, since taking the subject position of "the exposed" is tightly emmeshed with the collective memory of the stigmatized victims and survivors of atomic bombs in Hiroshima and Nagasaki from 1945.

In Japanese nomenclature, radiation exposure / *hibaku* could mean two distinct categories of experience in the written Japanese (被爆 and 被曝), although it is homophonic in speech. The difference is subtly placed on the radical of the second Chinese character that accompanies 被(*hi*), or to receive. The former term, historically highly charged, indicates any exposure to radiation from an explosion 爆 (*baku*), as in the victim of an atomic bomb, while the latter means any other radiation exposure 曝 (*baku*), including medical exposure from a CT scan or X-ray.[37] This inherent historicity of radiation exposure complicates the Japanese relationship to the fallout and the local experience of chronic low-dose exposure resulting from the TEPCO accident. In other words, the experience of radiation exposure evoked by the TEPCO accident is layered with additional temporality: the radioactive legacy of Hiroshima, Nagasaki, and now Fukushima.[38] In this book, I make no historical claim about exposure (Saito 2018). Fukushima is unlike Hiroshima and Nagasaki. I use *radiation exposure* to mean low-dose exposure that is far below the level that causes immediate biophysiological threat. The chronic presence of radiation, even if medically negligible, makes its effects ambiguous, allows the continuation of postfallout livelihoods, and invites manifold interpretations and social, cultural, and moral controversies, disintegrations, and atomization.

A prominent Japanese sociologist known for his deep engagement with coastal Fukushima, Hiroshi Kainuma, told me in 2013 how he was hesitant to write about Minamisōma. His reluctance came from Minamisōma's sheer complexity and messiness, which contested any attempt to capture it holistically. Situated between ten and thirty-four kilometers north of 1F, Minamisōma has been exposed to all aspects of 3.11 and its aftermaths, ranging from the severe tsunami damage to mass evacuation, the constant reterritorialization of zones, disaster recovery and reconstruction, environmental contamination and remediation, chronic low-dose radiation exposure, lifestyle-related health diseases, disaster compensations, community disintegration, suicides, wildlife issues, population decline, and an aging society. Nonetheless, Minamisōma has been placed rather at the periphery of the academic discourse, especially outside of Japan.[39]

Minamisōma is where all postfallout issues converge and coexist without any resolution. From the state and TEPCO's perspective, Minamisōma is neither severely contaminated nor historically embedded enough in the political economy of the TEPCO nuclear power plant to require thorough attention, unlike the municipalities in the south. Mindful of coastal Fukushima's local geopolitics, a handful of Minamisōma residents referred to the city as the fallout's "gray zone." They used the color gray to apprehend their lived experience of being in an ambiguous, ever-changing, and bewildering place, where nothing appeared describable with the simple binarisms of white or black, safe or unsafe. As anthropologist Hiroko Kumaki (2021, 1) explains, the residents in coastal Fukushima are living a paradox for which "there is no correct answer."

In *The Culture of Gray,* architect Kishō Kurokawa (1977) explains that gray is an ambiguous and indefinable color, composed when all the colors get mixed and cancel each other.[40] He perceives gray as the color that stands for the continuous state of transition, intermediariness, and shapeshifting. Kurokawa's fascination with this color comes from its perspectival and contextual quality; it concurrently encompasses everything and nothing, depending on where people perceive the color, what time of the day they see it, and where the color is located in the environment.

Linking the opaque quality of gray to the quotidian aesthetic sense of the Japanese, Kurokawa explores its sensibility in the Buddhist philosophy of emptiness (*kū*) and cosmology (*innen*) (Minakata 2015).[41] In particular, he sees the affinity of the idea of *en*—the conditions of possibility for all relations or an indirect cause of interconnection—with the Japanese architectural significance of roads. Kurokawa identifies roads as the kernel of Japanese architecture, peripherally situated (*shū-en*) and typically hidden but central in interconnecting dissimilar entities or flattening spatial boundaries. As he notes, "Unlike the Western rationalism, I see the core of the Japanese way of thinking in the Buddhist concept of *en*. This concept, call it a gray zone or in-betweenness, embraces ambiguities while providing a tie to different and conflicting things and ideas. The coexistence of ambiguities without any apparent resolution is what enables the movement forward in this system of thinking" (28).

Although Kurokawa's thinking is rather ethnocentric and eccentric, the simultaneity of multiple things and the transient quality of gray are lenses through which I have come to perceive Minamisōma and its residents' relationship to 3.11. The gray zone of the TEPCO accident is where Hatsumi's nuclear ghost lurks. As such, the nuclear ghost emerges out of ambiguities and resists ready-made analytical binarisms, like safe and unsafe, "purity and danger" (Douglas [1966] 2002), that help us curve the world at its joints, as it were. "Once the ghost reveals itself to you or is revealed through others' determination, there's no turning back," was Hatsumi's forewarning about the gray zone.

Working within this gray zone, *Nuclear Ghost* is the record of constant shapeshifting between raw experience, technosensory representations, and their lacuna where "'everything is happening . . . as though the future could no longer be imagined except as the memory of a disaster which we only have a foreboding of right now'" (Marc Augé cited in Virilio 2010, 7). In this book, I invite readers to explore the grayness not as a riddle to be solved but as a discriminatory condition in which only a certain group of people, but not others, was believed to live in a radioactive world. Detailing the lives lived under the hypervisualized regime of radiation in coastal Fukushima serves as an entryway into discovering *en* with unfamiliar people and their surreal lives with radiation.

While, medically, radiation and nuclear things are a force and agent that can sever molecular relations by disintegrating cells and damaging DNA, Laguna Pueblo novelist Leslie Marmon Silko (1977, 246) ponders the undeniable, embodied interconnections that nuclear things have produced for humanity. Linking the spatiotemporally dislocated events—the destruction of Japan by atomic bombs and the nuclear colonialism of the American southwest with the material residues from uranium mining—she writes in her novel *Ceremony,* "Human beings were one clan again, united by the fate the destroyers planned for all of them, for all living things."[42] Thinking with *en* and the ways that the residents entertain these invisible and ineffable connections as real amid the chronic threat of their disintegration allows us to reframe radiological contaminants not in terms of their essential property—that they are biomedically harmful—but in terms of how they have been and are distributed technopolitically across time, space, and human/nonhuman bodies as a condition of possibility for making a particular kind of relationship (im)possible.[43]

I contend that if we continue engaging with the TEPCO accident by focusing only on radiation's biomedical damage to humans and selectively curating narratives that confirm our expectations, we become complicit in ignoring the agency of the residents, eliding their situated experiences, and denying their willingness to sustain their livelihoods in coastal Fukushima. *Nuclear Ghost* invites you, the reader, to bear witness to the less spectacular, though very (sur)real, lives lived with radiation in the gray zone. For me, someone who was initially haunted by the Fukushima spectacles, witnessing such lives required an act of remembering other ways of being in the world that challenged my own assumptions about what radiation does to life.

My hope is that those yet-to-be imagined postfallout lives and livelihoods will become uncannily familiar to you. Being exposed to radioactive materials is not an extraordinary event or phenomenon that has happened and will happen only in peripheries and the gray zone. Since the mining of uranium in Belgian Congo, in the First Nations in Canada, and in the Navajo Nation as well as the first detonation of the atomic bomb in New Mexico's desert in 1945 (Bannerman 2018), radiation exposure has been part and parcel of what it means to live on a planet with "blasted landscapes" (Tsing 2015). We tend to displace the contaminated and exposed to the peripheries and forget them unless their bodies demand medical attention. The nuclear ghost emerges.

### ATOMIC LIVELIHOODS

In my ethnography of fallout, I present what the residents did when their physical well-being was simultaneously determined by the state and experts to be unharmed and imagined by the public as irreversibly damaged. In this ambiguous gray zone, what are the "constraints, causes, hopes, and possibilities—the practicalities of life" (Geertz 1992, 133) that the residents improvise, experiment with, and demonstrate?

In choosing to focus on the people who have stayed in or returned to coastal Fukushima since the fallout, I am attempting to resist engaging in what Indigenous scholar Eve Tuck (2009) calls damaged-centered research and thus occupying the role of those in power who define what a disaster is, what exposure is, what it should be, and who suffers. Tuck defines damage-centered research as "research that operates, even benevolently, from a theory of change that establishes harm or injury in order to achieve reparation" (413). Such research fixates its subjects as "broken" (409) and further harms already dispossessed subjects. I write this book not to embellish further nor to consume the suffering people have already experienced but to attend to their ongoing lives with radiation that have been overshadowed by our collective enchantment with radiation exposure's expected dehumanizing effects. *Nuclear Ghost* is a call for us to suspend our fixation with radiological damage in order to narrate the ongoing livelihoods under threats of chronic low-dose radiation exposure.[44] It focuses not on radiation but on the residents' hopes and desires to live and die well in the city, where the state and public's reliance on technoscientific definitions of radiation exposure as the only legible metrics of the nuclear accident has silenced and atomized their lived experiences.

Residents' and evacuees' concern about not being able to die at home and in Minamisōma and not being able to live in ways familiar to them provided a powerful counternarrative against the common imagination of Fukushima as a land of contamination where the state has coerced residents into staying. One of the main reasons for the residents' long-term evacuation and the heightened radiation surveillance, as I understood it before my fieldwork, was to protect locals from irradiation so that they could remain alive and healthy. However, for many locals displaced and living within the city in the aftermaths of the TEPCO accident, the idea of home—as an enduring physical structure and the nexus of intergenerational succession still standing despite the disaster and contamination—has become inseparable from their idea of dying well (*ii shinikata*).

How can people die well in the postfallout world? Alternatively, how do people live and live well in the protracted aftermath of radiological danger with the awareness of their undeniable copresence with contaminants? Scholarship on the postfallout contexts in Japan has tended to emphasize what sociologist

Ulrich Beck (1987, 158) calls the "'emergency-scientization' of everyday life."[45] While focusing mainly on languages and techniques for regaining a sense of safety and for reclaiming rights as citizens, those works have not been equally critical of the cultural and social consequences of nuclear things at home as the ground of the political and the intergenerational.[46]

While these interventions are certainly helpful for describing the initial attitude I had in Fukushima, the lopsided emphasis on the political (and biopolitical) has had the impact of exacerbating or even reproducing polarizing discourses, reflective not of cultural elaborations but of preexisting political associations and ideologies, such as anti- versus pronuclear and them versus us (Cleveland 2014). Observing the corrosion of the social fabric after the Chornobyl accident, Beck (1987, 159) diagnoses this postfallout condition accordingly: "Everyday and political rigidifications and fanaticizations proliferate in the contradiction between survival and the perception of dangers. The most extreme positions are adopted: some refuse to perceive the dangers at all, while others energetically insist on blanket condemnations in the name of 'self-protection' or the preservation of 'life on this earth.'"

The residents and their postfallout livelihoods have taught me how such a narrative treats exposure and contamination as an ahistorical, onetime event that causes irreversible changes to individuals' relationship to their own senses, to themselves, and to the world. It also fails to attend to what historian of science and technology Michelle Murphy (2017) calls "alterlife," where contaminants have already altered humans and nonhumans alike. The narrow focus on the here-and-nowness of contaminants and framing the accident as a bounded event is what I call the *half-life politics* of nuclear things—a technosensory politics that foregrounds the presence and absence of radiation in the environment and its potential harm on humans above all other things, like the residents' quality of life, cultural and intergenerational continuity, personal belongings, sustainability of livelihood, and biodiversity.[47]

As Masco (2015) argues eloquently, the critical theory of fallout can aid those impacted in transcending the ideological divide by drawing attention to the planetary consequences of nuclear fallout and the flow of contaminants beyond ecological, geographical, political, and sociocultural boundaries. Nevertheless, it is also vital to grapple with the possibility of a critical ethnography of fallout that explores an intermediary way between the self and the planetary in the scientized everyday life, a scale that breaks open the persistent, dichotomous narrative of containment and contamination.

As I will show in the book, staying and residing in coastal Fukushima was sometimes as hard as being displaced. The issue at hand is not merely about individual decisions to stay or leave but also about the situation that forced them to make life-changing decisions—namely, the TEPCO accident in coastal Fukushima and the lack of accountability for this accident. Most of my interlocutors decided to keep their physical, social, psychological, and spiritual ties to the region. They neither sought approval from outsiders like me nor did they ask for pity. In her ethnographic study of Black homeless girls in Detroit, anthropologist Aimee Cox (2015, 8) puts it aptly that "their [the interlocutors'] lives do not need sanitizing, normalizing, rectifying or translating so they can be deemed worthy of care and serious consideration." Instead, residents' various ongoing livelihoods call for a deeper reflection on our global engagement with nuclear things, the asymmetrical distribution of their risks, and the structural inequalities they (re)produce.

In fact, numerous local accounts of the TEPCO accident demand our attention beyond the initial "moral panic" (Cleveland 2014) associated with the fear of radiation exposure and require us to focus on the accident's ever-evolving aftermaths.[48] Incorporating those vernacular accounts, this book provides the longue durée analysis of broader local contexts of the TEPCO accident to articulate the embeddedness of risks in the region and approaches locals as individuals caught in historical, cultural, geopolitical, and intergenerational webs. I am interested in how Hatsumi's nuclear ghost emerged in Minamisōma and what its message is, rather than what the ghost is or whether or not it exists. All I know is that my ethnography would have been impossible without believing in this surreal shapeshifting figure.

I contend that the failure to encounter people as individuals with unique histories, perspectives, and positions victimizes them. This failure can be as harmful as, if not more harmful than, the average radiation exposure from living in Minamisōma today. Anthropologist Susanna Hoffman (2003, 68) describes this process as "secondary victimization," where uncritically applying categories like "victim" to approach people elides their unique individual experiences and make them unknowable. Instead, *Nuclear Ghost* focuses on residents' everyday struggles with the accident and its prolonged aftermath as well as their hopes and desires *despite* the accident and the technoscientific gaze and its spectacles. It is an investigation not of "atomic victimhood" but of what I call *atomic livelihoods*—how people live and what they do, feel, think, remember, and forget in the postfallout environment and the prolonged state of uncertainty, and what the social and historical conditions of possibility are for their present livelihoods.[49]

These livelihoods are, as I see it, "atomic" in two senses. First, it refers to the fact that people have been forced to live in an objectively more contaminated environment near the damaged power plant.[50] Second, their livelihoods are atomic in the sociological sense, meaning that someone or something has been individualized, isolated, and alienated. In designating individual livelihoods in coastal Fukushima as atomic, I am following Ui's contention ([1971] 2006) that environmental disaster is, at heart, about discrimination that triggers divisions and corrodes a community. Against the constant threat of disintegrations, *Nuclear Ghost* tells stories of connections that the residents gave up or protected and discovered anew when their postfallout experiences became increasingly atomized and almost incommensurable with "the story of exposure." By attuning myself to residents' atomic livelihoods, which have been largely ignored in the name of disaster reconstruction and radiation safety, my goal is not to make a normative claim about a livelihood in a contaminated region; different individuals came to confront, perceive, and experience their now-contaminated environment differently. There is no one right way in the gray zone. Nonetheless, as I will show, the residents' postfallout lives became the subjects of scrutiny and discrimination within and across societies.

Shifting from the study of "nuclear victimhood" to "atomic livelihoods" makes it possible to account for people's often-surreal postfallout experiences in which individual, social, political, and scientific determinations of the threshold of exposure are not only "impartial knowledge" (Petryna 2013) but are also subject to elaborations and manipulations (Kuchinskaya 2014). As I will show, many residents have found that in the time of radiological danger, science provided only "probable security" (Beck 1987, 157). At the same time, science helped produce the belief that measurable technosensory representations are the only valid mode of talking about an irradiated, contaminated environment (Button 2010; Das 2000), serving as a technique of state-sanctioned social control (Carroll 2006). Against this push toward the scientization of everyday life through half-life politics, *Nuclear Ghost* explores situated knowledge among the residents to help broaden our understanding of coastal Fukushima in particular and, more broadly, the more-than-biological effects of radiation and environmental contamination in society, which configure locally specific relations of "agency, causation, and uncertainty" (Jasanoff 1999, 135).

As much as this book is about everyday life with radiation in coastal Fukushima, it also is a cultural archive of the shifting Japanese relationship to disaster and nuclear things.[51] Putting this book together, I have benefited from the growing reservoir of scholarship in Japanese. When possible, I have attempted to expose readers to the wealth of knowledge and conversations that have emerged over the past eleven years and are still being produced in Japan.[52] Whether written by local actors, journalists, academics, or others, each source has contributed to the collective efforts to make sense of 3.11 and its significance beyond one country.

While using those texts to guide my ethnography, I have also interacted and shared space and sometimes food with not only ordinary residents but also local and national government officials, radiation safety experts, nuclear engineers, activists, volunteers, decontamination workers, entrepreneurs, biologists, medical doctors, ecologists, and TEPCO employees. They all have been engaging in their own projects to remediate coastal Fukushima and develop more transparent representations of radiation and its toxicity. More often than not, they are also residents of coastal Fukushima and have a personal stake in the accident and its aftermaths. As their thinking changed with time, so did mine.

Through my sustained engagements with changing people, growing texts, and a still-ongoing accident, the central questions I explore in the book are these: Who lives in the region and why? What does it mean to live with radiation and its real and surreal or imagined presence? And what do they hope to leave behind for the next generations? To answer these questions, each chapter of *Nuclear Ghost* tells stories of the social life of radiation and the unexpected ways that nuclear things have impacted and woven *en* among people, nonhuman others, histories, cultures, and the environment in coastal Fukushima. As such, the book experiments with and takes a chance on the potential of nuclear things to generate novel *en,* however ineffable, ghostly, surreal, and less scientific such mode of relationality between humans, nonhumans, land, ideas, and the environment might be. Put differently, because of the accident, I met people in coastal Fukushima and am writing about their lives, places, histories, and environments for you to know them and hopefully probe further. As you will see, the story of how I explore invisible threads and their interconnections amid the threat of physical, social, political, and moral disintegrations is a story of learning to accept and to be challenged, judged, and vulnerable when ethnographically subjecting oneself to others' worlds and worldviews.

The first part of the book, chapters 1 through 4, takes the readers through multiple "drives" in coastal Fukushima. Each chapter exposes readers to messy aftermaths of the TEPCO accident and differently situated residents' experiences of the fallout's social, political, economic, and psychological afterlives. Together, these chapters situate Minamisōma as a place that the residents stayed in, returned to, and decided to live and die in, to help readers imagine the postfallout condition of living with the presence of radiation. Chapter 5 moves the discussion from the radiation-centered discourse of the TEPCO disaster—the science of half-lives—to its impacts on resident's livelihoods. This analytical move redirects our attention to the more expansive effects of the accident and its remediation on the interconnected webs of things, the living, and the dead, which I explore in the second part, chapters 6 through 9. Chapters 6 and 7 delve into the local history and culture and the challenge of selective remembering and forgetting of the accident; chapter 8 explores the copresence of various invisible things; and chapter 9 investigates the multispecies ecopolitics of contamination and containment.

Together these nine chapters explore the nuclear ghost of Minamisōma, which has fallen through the cracks of the public imagination, academic discourses, local and national recovery and reconstruction, statecraft, scientific objectivity, and individual moral compasses. I contend that the inchoate experience of the nuclear catastrophe, which began in March 2011, will continue to stay with the residents and with the rest of the world for a long, long time. As such, my goal is not to ethnographically abduct nor analytically exonerate the ghost but to learn to live with its ongoing presence and shapeshifting message.

In this sense, I follow Kim Fortun's approach to disaster (2001, 350) by using this book as a form of advocacy, "a performance of ethics in anticipation for the future." The future of coastal Fukushima is rather grim, with the still highly contaminated 1F, its accumulating tritiated water, the newly built Interim Storage Facility for decontaminated waste, untouchable radioactive debris, and population decline and an aging society. The TEPCO accident has, quite literally and already, "colonized a future" (Masco 2006, 38; cf. van Wyck 2005) of one segment of coastal Fukushima and probably more. Despite the risks of nuclear energy highlighted in Japan in 2011, Chornobyl in 1986, and Three Mile Island in 1979, there are 445 commercial nuclear power reactors in thirty-two countries providing around 10 percent of the world's energy, many of which are aging. Meanwhile, fifty-five reactors are under construction.[53]

In anticipation of our collective radioactive future, it is critical that we hail the attention of readers who might be less familiar with coastal Fukushima or nuclear things. Such "strangers" have a special capacity. As sociologist Georg Simmel (1950, 402) asserts, "[the stranger] is not radically committed to the unique ingredients and peculiar tendencies of the group, and therefore approaches them with the specific attitude of 'objectivity.' But objectivity does not simply involve passivity and detachment; it is a particular structure composed of distance and nearness, indifference and involvement."

I share my experience in Minamisōma and the surrounding areas in coastal Fukushima in the hope that my book will serve as a window into the diverse livelihoods that people have chosen and constructed despite the fall-out and its protracted aftermaths. I offer my stories of the nuclear ghost to unearth the radioactive *en* of our collective nuclear legacies and hopefully to make you interested in coastal Fukushima so that maybe one day you will visit. As Haruki Murakami (2005, 22) puts it, "Chance encounters are what keep us going." A nuclear ghost still exists in the region, waiting to narrate its stories to those who are open to discounting their Geiger counters and attending instead to them. Hatsumi was right. The nuclear ghost of Minamisōma—the shapeshifting figure—has changed how I think about my past, present, and future. I hope it will have lasting impacts on yours as well.

———

# ONE

# Naming the Nuclear Ghosts

Please remember: things are not what they seem.

HARUKI MURAKAMI, *1Q84*

## HYPERVISIBLE RADIATION

Hatsumi did not want to speak much about the "nuclear ghost" of Minamisōma. Throughout my fieldwork between 2013 and 2019, I heard her elaborate on the nuclear ghost only once, in July 2013. After that I asked Hatsumi about the ghost every opportunity I had, but she resisted expanding on it, as if she had mentioned it to me by mistake, like accidentally sharing a secret she was not supposed to tell an outsider. On one occasion, she brushed off my inquiry about the ghost as "irrelevant" for understanding Minamisōma. On another, she accused me of being "lazy." If I were a true researcher, she said, it was my job to figure it out.

After living in Minamisōma for a while, I came to realize that Hatsumi's act of articulating her experience of alterity as the nuclear ghost was meant to express the demoralizing effects the radiation-centered discourses had on her postfallout life. Zooming in on contamination, these half-life politics selectively rendered the presence of radiation visible and put everything else out of focus. Various globally circulating visualizations of contamination in the region represented coastal Fukushima through multiple colors of radiation's intensity: purple, yellow, green, dark blue, and sky blue (see, for example, map 1, where the colors are represented with shades of gray). At the same time, those visualizations left the remaining residents' everyday lives without much color or resolution. Minamisōma became a strange place where what was not immediately perceptible became central to how outsiders imagined residents' lives. Here, residents experienced their lives in gray. In the gray zone, nothing was certain, even the much-discussed radiation and its anticipated effects on the body. Despite her reluctance to speak of the nuclear ghost

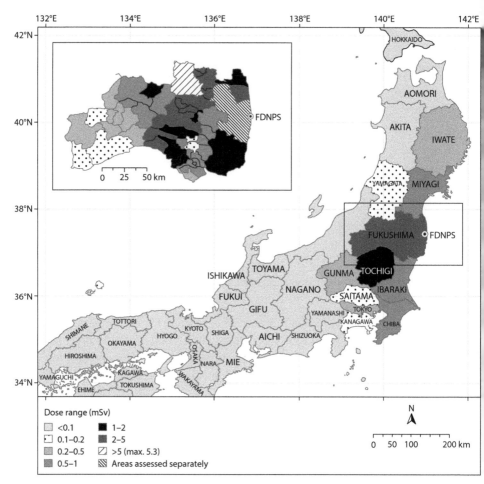

MAP I. Estimated prefecture-average effective doses to infants in the first year following the accident, in millisieverts (mSv), taken from the United Nations Scientific Committee on the Effects of Atomic Radiation (UNSCEAR) (2021, 62), https://web.archive.org/web/20220521154755/https://www.unscear.org/unscear/uploads/documents/unscear-reports/UNSCEAR_2020_21_Report_Vol.II.pdf.

of Minamisōma, Hatsumi made one thing clear to me: "You will know what I mean when you learn about Minamisōma." The nuclear ghost was not something to talk about, but something to experience. Whether I believed it existed or not, it was there.

Linking his experience in postfallout Minamisōma with the COVID-19 pandemic, journalist Makoto Baba (2021) expressed the general atmosphere of the city as "*usura samui*" (slightly chilly). This visceral feeling of a chill strangely resonates with the Japanese folk usage of ghost stories (*Kaidan*) to chill people down so that they can endure the humid summer.[1] Commenting on Baba's diagnosis of city's chilly atmosphere, one resident described to me in 2021 that "there was something creeping up in the city, especially until 2018. It is hard to name what it was. Something was just not right then." Although Hatsumi was the only person in coastal Fukushima to articulate her postfallout experience through the nuclear ghost, she was not the only Japanese person to name the existence of such a ghost in postfallout Japan. In 2015 I unexpectedly encountered the same expression on the internet from someone who lived far away.

Masahiro Ono, an immunologist at the Imperial College London, published a series of opinion pieces on Yahoo! Japan News between January and April 2015 titled "Radiophobia: The Poison of Democracy." In the series, Ono introduces the term *houshanō obake,* or nuclear ghost, to name the baseless fear of low-dose radiation observed in public discourses on Fukushima.[2] Speaking as a medical expert, he argues that the general lack of knowledge surrounding nuclear things made it impossible for Japanese society to have productive discussions about the TEPCO accident and its aftermaths. "In contemporary Japan," he observes, "the ghosts are openly walking around in daylight. They are the ghosts of a nuclear kind. These ghosts indoctrinate fear in people. The fear becomes the poison that circulates throughout society, not only making it hard to solve the issues of radiation but also paralyzing democracy."[3]

For Ono, science cannot be the sole authority determining what is safe during the postfallout period since there is no scientific consensus on the effects of chronic low-dose exposure. The only thing science knows for certain is the threshold, the level of exposure that leads to observable molecular damage. Therefore, the scientific position on radiation exposure is rather vague and general: less exposure is better. Due to the grayness of radiation exposure in science, its discussion necessitates a social and political negotiation about what is permissible.[4] Nonetheless, he argues, the ghost—the fear that prevents logical thinking—has incapacitated people from engaging in discussion. Ono proposes a solution: "If we can unmask the nature of the nuclear ghost, we can cause this haunting existence to stay in its appropriate place. Only then can we regain our ability to discuss radiation rationally and seek a social contract." Ono contends that naming the nuclear ghost could alleviate the collective failure to discuss what is permissible—

that is, how to live *with* low-dose radiation exposure and the awareness of its presence.[5]

Ono's discussion of the nuclear ghost illuminates the limits of scientific authority and sociopolitical constructions of fear and uncertainty in a time of uncertainty. However, his belief that it is possible to contain such fear by naming it does not fully account for Hatsumi's experience. His argument, which focuses on the (in)visibility of nuclear things, represents the radiation-centered, half-life politics that Hatsumi rejects with her reference to the nuclear ghost. For Hatsumi, the ghost is not merely a symptom of "irrational" fear creeping up in society or a by-product of the societal failure to trust science or engage in a democracy based on science. Naming the ghost would not get rid of it, nor did she seek others' permission to stay in Minamisōma and live with radiation. Instead, whatever the ghost stood for, it had palpable and long-lasting effects on Hatsumi. As she put it, "Once the ghost reveals itself to you or is revealed through others' determination, there's no turning back. It changes how you think about your past, present, and future." Unlike Ono's nuclear ghost, which roams discursively across spaces and inflicts a baseless fear in people, Hatsumi's ghost was a situated experience.[6]

In this chapter I take readers to Minamisōma during the aftermaths of 3.11, an unsettling social, political, and scientific gray zone of the TEPCO accident. We will explore the places people live and the way we might approach the residents who continue to live with radiation and its hypervisualized presence. In Minamisōma, where the debris fell and the state emplaced multiple evacuation zones, the residents' postfallout lives began to assert the possibility of living in close proximity to radiation and its sources. Their sustained livelihoods have been challenging the existing social contract related to and the reluctance to discuss what is safe and unsafe, as well as the scientific boundary between what is uncertain and certain about radiation and its effects on humans, nonhumans, and the environment.

In Minamisōma, radiation has certainly been an important issue, but it is not the chief one. Only by suspending our judgment about life with radiation and giving the benefit of the doubt to the residents' expertise in this matter can we begin experiencing the nuclear ghost. The nuclear ghost is not an irrational fear. It is a yet-to-be named experience, hidden underneath the hypervisualized presence of contaminants, waiting to be seen and heard. However, before we enter the gray zone, a few cautionary tales on the cultural imagination of distance are required to approach coastal Fukushima and its residents' livelihoods.

"What country are you from?" a resident asked me when I encountered her on the main street of the Kashima district in July 2013. The first place I lived in Minamisōma was in Kashima, a makeshift base camp for outside volunteers. Sachiko, a local woman in her late fifties, told me that she had witnessed me walking across the two-lane main road, Route 6, earlier, and I had caught her eye. When I told her that I was Japanese, that did not appear to be a satisfactory answer, since "no one would be walking to cross Route 6! You must be an outsider!" My unusual behavior, Sachiko claimed, was a warning to her because "there have been many outsiders since the disasters. And we are watching because we do not want outsiders coming in and doing bad things." In the brief interaction, Sachiko signaled the general local skepticism toward outsiders. Walking across Route 6 was evidently something only an outsider would do. I had also experienced such suspicion in the tsunami-stricken cities in Iwate and Miyagi, where I volunteered between 2011 and 2012. In Minamisōma and elsewhere, 3.11 introduced a flux of outside people (*yosomono*) to rural regions in northeastern Japan, where many residents described their towns and cities as relatively secluded places.

While those places needed all the outside help they could muster to recover from the havoc 3.11 produced, many residents struggled with outsiders coming in and sometimes staying on. Sachiko explained how people in Minamisōma tend to be closed off to outsiders. This, she suggested, was not a postfallout phenomenon but something that had long existed in rural cities like Minamisōma. "We are a tight community here. We know each other by face [*kaono mieru kankei*]," she said. In addition to radiological danger, 3.11 posed a new challenge to their tight-knit communities; since 3.11, many outsiders had come to Minamisōma, including evacuees from the south, developers, construction workers, politicians, researchers, journalists, TV and film crews, artists, religious groups, volunteers, and disaster tourists. The residents' concerns about the outsiders' increased presence manifested in rumors about crimes and other wrongdoing committed by outsiders, such as sexual harassment, burglaries, and attempted kidnappings. Some young evacuees with children claimed that Minamisōma had become an "unsafe place" to return to because of radiation and the presence of outsiders. For them, *yosomono* were a threat to be held at a distance, just like the invisible radiation, although the state carefully monitored only the latter.

In the first couple of years after the accident, residents were particularly on alert for the presence of outsiders. Many locals felt that *yosomono* came to the city to "uncover" the hidden truth of the TEPCO accident. Observing patterns in the questions being thrown at them, locals surmised that outsiders were after narratives that affirmed their imagination of Minamisōma: the spectacular life of postfallout. "How contaminated is your place?" "Are you compensated?" "What lies have the state and TEPCO told you?" These were familiar questions and perhaps fair ones, given that most outsiders learned about Minamisōma after the TEPCO accident and through various visualizations of radiation within its geography.

At community meetings, at local festivals, and during casual chats at a grocery store, residents shared tips with each other for entertaining outsiders. "Just complain about the state and electric company!" Others lamented how sharing their honest feelings did not help them in any way, saying that people who had the option to leave coastal Fukushima did not care how people lived. "They only care about contamination and nothing else," another said. What both groups agreed on was the implicit judgment *yosomono* had about the city and its residents. As one resident put it, "I can sense that people are judging us, thinking that living with radiation is an irrational decision. What is annoying is that even if I know more about where I live, their self-claimed knowledge about radiation and its danger makes me worry that Minamisōma is no longer the place I thought I knew." For the residents, the TEPCO accident and the national government's initial designation of evacuation zones produced two Minamisōmas: the contaminated space on the map and the place where people lived despite contamination. *Yosomono* encountered Minamisōma only as the former.

Despite the general suspicion of outsiders, residents were usually willing to talk, especially if you showed genuine interest in learning about the region and hearing what they had to say. For example, after stopping to warn me about crossing Route 6, Sachiko offered me a piece of practical advice: "Nobody walks around here. Without a car, you cannot get to anything in this rural region." She was right about that. Luckily, I had come to Minamisōma with a car, which my old friend in Aichi Prefecture had let me use for my fieldwork. It was a small silver Mitsubishi, which many locals referred to as the pillow car. The pillow car was a commercial vehicle for my friend's family-run pillow company and had promotional stickers stating "We produce everyone's 'Good Morning'!" on it. "You'd drive around Fukushima and do your research. That would help promote how my company is in support of the reconstruction of Fukushima," my friend said. I customized the car by putting a stand on the dashboard to affix a personal Geiger counter, since I believed that surveying airborne radiation in the region would guide my ethnography of fallout.

By 2013 it had become evident that the disaster recovery and reconstruction in Fukushima Prefecture lagged behind other hard-hit prefectures like Iwate and Miyagi. The TEPCO accident presented additional and unexpected challenges for the state's disaster-recovery and reconstruction (*fukkyū-fukkō*) protocols. Two years after 3.11, more than twenty-one thousand registered Minamisōma residents lived away from their original residences. No one knew how to decontaminate and remove radioisotopes from the environment, what level of radiation was safe, or how many more people, and who, would come back. Despite all those uncertainties, around forty-five thousand people still lived there in 2013.

Like my friend in Aichi, who lives about 580 kilometers away from Minamisōma, many people in Japan shared a generally supportive sentiment for the people in Fukushima. *Kizuna* (絆), or bonding, became the kanji of the year selected by the Japanese Kanji Proficiency Society in 2011 following 3.11 (Allison 2013; Bestor 2013; Gerster 2019; Sternsdorff-Cisterna 2015; Tokita 2015). At the same time, they did not hide their negative impressions and warned me about entering Fukushima. In the words of a middle-aged woman I met in 2011 in Sendai city, Miyagi Prefecture, about fifty-nine kilometers north, I should not go to Fukushima "if [I] ever want to have kids." Undeniably, there existed a collective imagination of Fukushima as the land of contamination. The abundance of technoscientific visualizations of the space and antinuclear discourses generated and reinforced this type of sentiment.[7]

In these representations, Fukushima Prefecture appeared as the source of radiation from which people should distance themselves. The international and national media also supported and actively contributed to this selective cartography, focusing on contamination in Fukushima while underreporting similar issues, such as the presence of contaminants in surrounding prefectures.[8] Physical distance shielded outsiders from exposure to local matters and made it difficult to imagine anything about Fukushima other than the visualized presence of radiation. But radiation is not the only invisible matter that needs to be visualized. The nuclear ghost emerged in the omitted lines of "people, places, and networks that create the most common space" (Kurgan 2013, 17).

Until I moved to Minamisōma in 2013, I, too, had a hard time imagining mundane life amidst the elevated risk of radiation exposure. Everywhere I looked, in the news and reports, it was clear that the TEPCO accident had released radioactive contaminants and people suffered as a result. Some cautioned about adverse effects in nonhuman species, while others criticized the dishonesty of the state and TEPCO. Yet general information about Minamisōma and coastal Fukushima seemed to be missing. Who lived there and why? Minamisōma remained a faraway place both physically and conceptually. Novelist Miri Yu (2020, 63), who commuted to Minamisōma from Kamakura, Kanagawa Prefecture, between 2011 and 2014 and then migrated to the city in 2015, wrote about the difficulty of traveling back and forth. It was hard enough to get to the city prior, she claimed, but "the nuclear accident cut its infrastructure to the south and the tsunami cut the infrastructure to the north, [which] worsened the already inconvenient access to coastal Fukushima and transformed it into literally 'the inaccessible corner' of Fukushima Prefecture."

Her description of coastal Fukushima's remoteness is not arbitrary, nor is it merely about physical distance. The imagination of distance is a recurring cultural theme of northeastern Japan (Tōhoku). Away from the former capital of Japan (Kyoto), one of the earliest, eighth-century documents, *Kojiki* (the Records of Ancient Matters), described Tōhoku in relation to the center as "Michinoku," or the depths of the way. Michinoku was where Indigenous Japanese (Emishi) lived and resisted the access by the settler-*yosomono*. Since then, central Japan conquered and colonized Emishi, but Tōhoku's imaginary as the "deep north" and its subordinate relationship to the center persisted in modern Japan (Akasaka 2007). According to intellectual historian of northeastern Japan Nathan Hopson (2017, 2), "For much of Japan's recorded history, Tōhoku had been synonymous with some combination of savagery and backwardness, a troublesome region either resisting or dragging down Japan." The TEPCO accident and the eerie presence of radiation in this cultural and historical margin added to the general sense of the unknown.

In 1878, Isabella Lucy Bird, a British explorer and writer, traveled to Tōhoku and Ezo/Hokkaidō to experience "the real" and "Old" Japan. In her book *Unbeaten Tracks in Japan,* Bird documented the deep north's rustic lifestyles and harsh living conditions, where "the houses were all poor, and the people dirty both in their clothing and persons" (Bird 1880, 206). Although she only passed through the western region of Fukushima (Aizu) on her way to Niigata, Fukushima was the entrance into what she named the "unbeaten tracks." Going through Fukushima, Bird wrote, "The infamous road was so slippery that my horse fell several times. . . . Good roads are really the most pressing need of Japan" (247). Since then, Tōhoku and Fukushima have changed, and their tracks have been paved, as Bird suggested they should be. Nonetheless, the imagination of distance has endured.

In recent history, scholars have mobilized Tōhoku as an idea and a place of reflection to critique post–World War II scientific modernization, expansionism (Hopson 2017), and economic inequality (Iwamoto 1994). Approaching Tōhoku as the perpetual colony of Tokyo (Akasaka 2014) and a place of constant suffering has enabled people to (re)discover hidden truths, just as locals imagined outsiders trying to locate the evils of the state and corporation in postfallout coastal Fukushima to advance their political ideology. However, there is an alternative way to approach coastal Fukushima, and this unpaved path is what I suggest we pursue.

Consider Miri Yu, for whom Tōhoku was neither a nostalgic place nor an underexplored idea in contrast with modern life in urban centers. Instead, Tōhoku was about a new thread, or *en*, that she wove with the residents while tracing her family history. Her grandfather, who escaped the Korean War, had illegally migrated to Japan and found himself in Minamisōma. Cultivating *en*—chance encounters that are believed to be designed by invisible threads—by living in Minamisōma, Yu hoped to attune herself to the experiences, feelings, and pains of residents and evacuees and to write something that "unshackles the wounded souls from imprisoning pains and sufferings" (2020, 37, 47). Unlike Yu, I had no personal link to the area, but *en* is not just about tracing preexisting connections. It is also about a "play of musement" (Sebeok 1981), a speculation. As Nozawa puts it, *en* is "the pleasure and pain of finding meaningful connections, without necessarily invoking a system of rights and obligations" (2015, 394). We make our own *en* by believing a new, old, and ongoing relationship to be meaningful and naming it as such. Naming my strange encounters with the nuclear ghost of Minamisōma as an *en*-inflected ethnography of fallout is my way of showing my commitment to the region and my relationships with its residents.

Japanese anthropologist and Tōhoku native Ichiro Numazaki claims that 3.11 caused "parachuting" research among *yosomono*, who often failed to consider the local cultures and histories (2012, 34). Although a few anthropologists in the West have tried to counter such a harsh claim by writing about the TEPCO accident from a distance or with visits to the region (see, e.g.,

Allison 2013; Bestor 2013; Gill 2013; Polleri 2019; Slater, Morioka, and Danzuka 2014; Sternsdorff-Cisterna 2020), much of coastal Fukushima remains "unbeaten" ethnographically.[9] Our task at hand is to explore and document the gray zone in order to provide views into how the residents have lived in the region and are currently living with radiation and under the weight of its various representations.

Caution: if you want to experience the nuclear ghost, then be prepared to suspend your preconceived ideas.

## A WINDOW INTO MINAMISŌMA

I often drove around the city and surveyed it. Each drive helped to render visible the multiple and diverse aftermaths of 3.11 that had fragmented the city, its people, and communities. From my car window, I gained my first perspective on how people confronted the same event differently. More than in any other municipality in coastal Fukushima, 3.11 and its aftermaths geographically divided Minamisōma and its residents.

In postfallout Minamisōma, to have meaningful and candid interactions you need the ability to quickly gather information to situate different individuals you meet in the city and surrounding regions. This ability proved to be far more crucial than knowing the scientific language of radiation or the proper use of a Geiger counter. Familiarity with the local geography both before and after 3.11 was something locals used to weed out many outsiders who often became extractive rather than supportive, who violated a tacit rule by walking across Route 6. As Sachiko advised, locals do not walk on Route 6 but drive on it. So, let me take you on a drive to lay out the city's postfallout geography.[10]

National Route 6, which locals refer to as Rokkoku, runs along the coastline and vertically cuts through the city in a north-south direction. It takes about thirty-five to forty minutes to drive it across Minamisōma from its northernmost to southernmost part, from Kashima to Odaka district, respectively (see map 2). Stretching over four hundred kilometers, the entire route connects Sendai city in the north and Tokyo in the south. Minamisōma is midway between them. Many locals would use the former Route 6 (Rikuzen Hamakaidō), located slightly inland and across the Jyoban train tracks, to avoid the countless construction trucks driving dutifully at the speed limit of forty kilometers per hour as they went back and forth

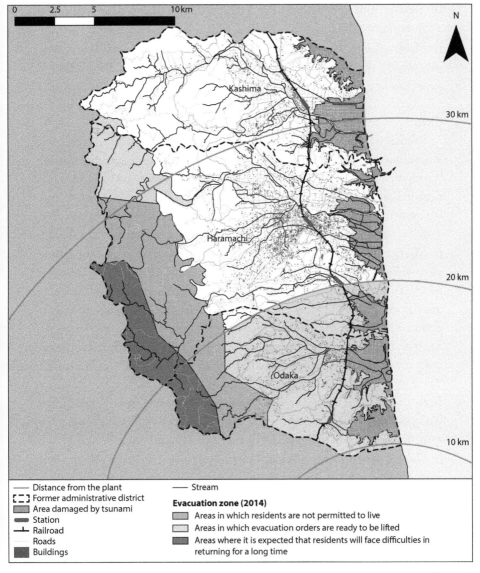

| Scale | | | |
|---|---|---|---|
| 0 | 2.5 | 5 | 10km |

N

30 km

20 km

10 km

Kashima

Haramachi

Odaka

—— Distance from the plant
⊏ ⊐ Former administrative district
▨ Area damaged by tsunami
▬ Station
⊥ Railroad
⋯ Roads
▓ Buildings

—— Stream

**Evacuation zone (2014)**
▨ Areas in which residents are not permitted to live
▢ Areas in which evacuation orders are ready to be lifted
▓ Areas where it is expected that residents will face difficulties in
   returning for a long time

MAP 2. Minamisōma in 2014. The evacuation order was determined by both the distance from the TEPCO Nuclear Power Plant and the level of estimated annual exposure. Taken from Zhang, Mao, and Zhang (2015, 6598).

to recover and reconstruct the tsunami-devastated coastal region of Minamisōma. A wide rut in Rokkoku, which made the drive rather uncomfortable, signaled the ongoing state-led disaster recovery efforts. In the summer of 2013, the traffic became even heavier on Rokkoku, and the number of convenience stores in the city increased once the national government, after a delay of a year and a half, started its decontamination project in the Odaka district.

Rokkoku is critical infrastructure for coastal Fukushima and its residents. According to locals, its presence contributed to Minamisōma residents being "culturally" closer to Miyagi Prefecture in the north than to Fukushima proper. For the residents, getting to the central part of Fukushima Prefecture meant driving about one and a half hours across the Abukuma Mountains, which made the trip more strenuous than cruising on smoother coastal roads to Miyagi Prefecture. After 3.11, due to the tsunami washing over several parts of Rokkoku,[11] many residents had no choice but to cross the Abukuma Mountains to reach Fukushima or further north, south, or west.[12] Running away from their homes and from the radioactive plume through the curvy and narrow mountain paths in the snowy March of 2011 is still lodged in many residents' memory. This cultural geography of the prefecture's center-coastal disconnect is also reflected in the structure of the local railroads. There is no train track connecting the coastal region to central Fukushima. In fact, since 3.11 it has been easier to get to Minamisōma from Tokyo by first going to Sendai (around eighty-one kilometers north of Fukushima city) by train and then transferring to the local line to Minamisōma.[13]

During the tsunami, Rokkoku functioned like a tsunami wall in some parts of the city, stopping waves and tsunami debris from reaching further inland.[14] In Kashima, one tsunami survivor found himself on Rokkoku sitting on his boat, which he had jumped into about two kilometers away as the tsunami overcame the coast around 3:35 p.m. on March 11, 2011 (Hayakawa and Matsuno 2012, 159–63). Many hamlets near the coastal region disappeared because of the tsunami. One such hamlet, located slightly outside the thirty-kilometer zone from the damaged nuclear power plant, was Minami-Migita in the Kashima district. Before 3.11, Minami-Migita's beach was a mecca for professional surfers.[15]

At Minami-Migita, the tsunami devastated all seventy households, killed fifty-four community members, and destroyed close to ten thousand pine trees by the beach. A group of survivors from the area took care of one thirty-meter-tall black pine tree that survived the tsunami.[16] Kazuo Goga, the leader of the preservation group for this "miracle pine tree," said the tree was "the proof that the hamlet existed there." He hoped the tree would help the surviving displaced community members to remember where they were from, despite the ongoing changes to the local landscape brought by the state's reconstruction projects. In spite of their preservation effort, the tree died in 2017 from salt damage, and the group used the tree's remains to make doorplates, which were distributed to the surviving community members.[17]

Situated mostly outside the thirty-kilometer zone, more than one-third of Kashima residents (around eleven thousand) stayed despite the city-organized mass evacuation between March 15 and 25 and the city's request that residents evacuate voluntarily on March 25. I met a few individuals in Kashima who had found themselves unable to evacuate with disabled or aged family members. Yamada was one such person who witnessed people disappearing from the city. "I wanted to evacuate," he said, "but I could not leave my ninety-year-old senile mother behind." He struggled to survive in the ghost town as food and other supplies stopped entering Minamisōma. "To be honest," Yamada shared, "I do not remember how we survived." Later, because of their decision not to evacuate, residents like Yamada were disqualified from receiving governmental compensation for the "evacuees." This, in turn, shaped their view of the somewhat arbitrary drawing of lines on the map and the voluntariness of the voluntary evacuation order, which served as the grounds for the state to treat residents in the same city differently. Being largely outside the thirty-kilometer zone, Kashima hosted many temporary housing units (*kasetsu*) for both tsunami survivors and TEPCO accident evacuees. The presence of nuclear refuges in particular bolstered Kashima residents' sense of inequality, especially with regard to the Haramachi district in central Minamisōma.

Passing Tōhoku Electric Power's thermal power plant, which is located between the Kashima and Haramachi districts and had been severely damaged by the tsunami, Rokkoku approaches the city's central area in the Haramachi district. Here many cars and construction trucks congested the roads around the city hospital, train station, library, city hall, hotels, and shopping areas. Immediately after the tsunami, many doctors and nurses had struggled to keep up with the flood of mud-covered survivors and bodies at the city hospital about twenty-three kilometers from the TEPCO Fukushima Dai-ichi Nuclear Power Plant (1F). As the conditions at 1F worsened and suspended fallout particles began exposing random black spots on X-ray-computed radiography images, the hospital experienced the second wave of people being transferred from the hospital in the south.[18] Keisuke Ota (2011), one of the doctors at the city hospital, describes the hospital during the 3.11 emergency as a place not for healing its patients but for safely transferring them outside the evacuation zone. The efforts of people like Ota and others at the hospital helped to sustain the region's population (Nabeshima 2020).

Haramachi is where over 60 percent of the city's population (around forty-five thousand residents) resided and remain today. In addition to *kasetsu* units, Haramachi also hosted multiple makeshift housing units for the reconstruction and decontamination workers. From a local perspective, those workers contributed to the need for the high density of convenience stores in the relatively small city. Even though the tsunami devastated the coastal area of the Haramachi district as much as it did in the two other districts and the Haramachi district is closer than the Kashima district to 1F, Haramachi felt like a typical rural Japanese city in 2013. Following the TEPCO accident, most of Haramachi fell in the thirty-kilometer stay-indoor order zone, but then the mayor, Katsunobu Sakurai, decided to remain in the city hall, located about twenty-five kilometers north of 1F, claiming that total evacuation would end the city. His decision made Minamisōma the frontline of postfallout coastal Fukushima, as it was the only municipality in coastal Fukushima with some of its parts falling inside the twenty-kilometer zone that did not enforce total evacuation.

In the summer of 2013, few cars would go further south of Haramachi on Rokkoku, passing by a busy 7-Eleven near the twenty-kilometer line, where police officers were stationed and dutifully watched every vehicle passing by. This line was where a Tokyo-based journalist and I were forced to turn around in the early spring of 2012. Uncultivated and weedy rice fields on the other side of the 7-Eleven signaled the agricultural activities that had been thwarted since March 2011. By April 2012 the state had modified its evacuation order and parts of the Odaka district had reopened, allowing residents daytime access to attend to their residences inside Odaka. One family, Mr. and Mrs. Kato, who owned a barbershop in central Odaka, reopened their business in 2012, bringing gallons of water from their temporary housing so that they could operate in the evacuation zone, which was without water and sewage (Watanabe 2020). It took four more years before the district fully opened.

When I entered the twenty-kilometer zone and went inside the Odaka district, most of the cars were marked as police from various prefectures or labeled with "City of Minamisōma," "Ministry of the Environment," or "Decontamination Vehicle," the latter surveying the ambient radiation levels or preparing an infrastructure for decontamination. Around the center of Odaka deactivated traffic lights signaled residents' absence, and many earthquake-damaged houses were left untouched, slowly deteriorating. In the center of the district there stood an empty train station. In front of it, a sign said, "Join the self-defense forces." One local woman, Kobayashi, an owner of a nearby Japanese-style inn, kept planting flowers around the drop-off rotary, near where overgrown vegetation had taken over the abandoned commuter bicycles. "I am preparing for people to come back," she said softly.

Sociologist Eiji Oguma (2013) wrote that "nobody dies in a ghost town." Nevertheless, I met many people who wanted to die nowhere else but in this ghost town, and because of this desire they waited patiently for Rokkoku and the district to reopen. Until September 2014, only cars with a proper permit could drive through Rokkoku past Okada and the 42.5 kilometers between Namie town and Naraha town in the south. About two years later, after Rokkoku's reopening in July 2016, the state lifted the evacuation order for the Odaka district, except for one highly contaminated area in the west, where there had been one household with two individuals before 3.11. As of 2022, eleven years after 3.11, around four thousand people, close to half of whom are sixty-five years of age or older, live in the Odaka district, and more than six thousand people are registered as residents. Although this is about one third of its predisaster size, the residents did come back, just as the woman with her train station flower garden anticipated. Now central Odaka has new restaurants, cafés, a sake brewery, pop-up stores, coworking spaces, and Yu Miri's café/bookstore/theater. Slowly but surely, residents are reconstructing the district.

The simple act of driving around the city exposes how even in the first few years after the TEPCO accident, contamination in Minamisōma was not a fixed state, as many imagined from afar. The residents actively negotiated their physical distance from contamination and radiation according to what was socially permissible. The patchy though determined presence in the evacuation zone of residents like the Katos and Kobayashi in 2013 asserted the possibility of being in proximity to the sources of radiation that the maps hypervisualized. At the same time, many other residents left the city and distanced themselves as much as they could to avoid exposure. Even within the same city, contamination and radiation exposure remained undetermined.

In the first few years after the accident, no one, including scientific experts, knew for certain whether low-dose radiation in the region would impact the residents' health and well-being. In fact, the experts disagreed with one another about the risk of living in the gray zone. While some groups suggested that the levels of radiation in the region were negligible, others countered that the radiation would have irreversible effects on locals' bodies.[19] Meanwhile, an overseas doctor, Ono, diagnosed the latter group as being haunted by a nuclear ghost. Caught between scientific, and often ideological, debates about radiation exposure, many remaining residents complained how their decision to remain in the city unwillingly enrolled them as experimental bodies.

The local geography along Route 6 reveals complex and ever-changing localities and distances of contamination, challenging the generalized representation of Fukushima as the homogeneous land of contamination. A stranger to the coastal region, I began immersing myself in this peculiar place from my car windows, where the city, infrastructure, and policies incessantly changed. At the same time, its residents remained committed to living in Minamisōma while many outsiders fixated on the ill effects of radiation exposure hidden or yet to come (cf. Auyero and Swistun 2008).

Ethnography—simultaneously a method and a product of written work (Gusterson 2008)—requires both the scientific rigor of systematic observations and the artistic openness to expressions. In my experience, ethnography sometimes requires solitary and aimless driving around to situate oneself in a field and immerse oneself in the place. I entered the field with a Geiger counter in my car, as if radiation and radioactive materials were the subject of my research and I was after their half-lives. As I drove around the city and constantly looked at my Geiger counter, I experienced firsthand how a radiation-centered perspective flattens the city into an irradiated space and homogenizes its residents as the exposed.

To articulate the necessity of suspending the radiation-centered approach and attend to residents' desires to live in this region, it is helpful to consider what anthropologist Johannes Fabian (1983) calls "a denial of coevalness." According to Fabian, anthropologists studying "the Other"—a group of people whose society and culture are thought to be radically different from the anthropologists' own—often deny the copresence of the researchers and their subjects in the same (conceptual) time and space.[20] This denial reproduces the exotic and allochronic image of the Other, a backward people from the past. The denial of coevalness, Fabian argues, rhetorically positions researchers and outsiders with power over their research subjects. Because of their backwardness, the logic goes, the cultural Other requires skilled researchers or time travelers, if you will, who can serve as a translator to speak on their behalf.

In the case of Fukushima, various researchers, experts, and outsiders spoke on behalf of the residents. One of these outsiders, a retired physicist and one of Japan's leading antinuclear figures, Hiroaki Koide, told me in the summer of 2012 at Kyoto University's research reactor facility in Wakayama prefecture that regardless of what the residents of coastal Fukushima think and feel, they should not be living there. When I asked him about the desires of some residents to remain in the region, he answered that it was not his concern as a scientist, who shares his knowledge on what he knows. Nonetheless, the opinions of the experts had determined the kind of life that was possible for the residents, contributing to the sentiment among locals that "the victims have been left out and forgotten" (Endo 2015, 106). Regardless of what constitutes expertise in postfallout coastal Fukushima, it was the residents and not the experts and outsiders who lived with contamination (Wynne 1996). Locals certainly did not need an anthropologist or anyone else to further cast them in a "suffering slot" (Robbins 2013).

I wanted to visualize the city not only as a contaminated space but also as a place where its residents felt compelled to stay despite the risks of exposure and other concerns, which meant I needed an alternative approach. For those who remained in and returned to the gray zone, their project was to get past the initial shock of 3.11 and to grapple with oscillating senses of hope and despair, recovery and deterioration, and continuity and renewal. By 2013, 3.11 was no longer an emergency, and what was needed was not what Slater (2013, 36) calls "urgent ethnography," which is grounded in "the mud and rubble of a disaster zone."

Despite all the uncertainties and precarities of the "post"-3.11 era, life was still happening to the residents in the present tense. They complained, cried, and were angry about the discrimination and unfairness of the system. At the same time, they also laughed, joked, and smiled to express their joy of living. As Hatsumi put it, "Like it or not, life goes on." Radiation certainly impacted their postfallout life, and the fear and worry—the chill—never ceased completely, but by 2013, radiation had retracted to the periphery of their lives, despite how it continued to be foregrounded on maps and in the imaginations of outsiders.

Shielded and distanced from the residents' lives as I observed them from my car windows, I realized that what I was immersing myself in was not a well-defined "contaminated community" (Edelstein 2004). Instead, in

Minamisōma the state, experts, media, locals, and outsiders constantly negotiated the fact and the degree of contamination and its potential effects in time and with time. Meanwhile, radioisotopes in the vicinity kept decaying at the known rates.[21]

Amid this technoscientific, ideological, and moral struggle to gauge postfallout human-radiation relations, the experiences of residents who decided to stay in the region remained a mystery to many outsiders, whose knowledge hinged on the radiation-centered half-life politics embedded in contamination maps. Those selective representations made residents who stayed in coastal Fukushima invisible, especially to distant observers. The representations also failed to capture how outsiders became a threat, like radiation, from which the residents tried to distance themselves, nor did the maps visualize the residents' hopes and desires that made them stay. Filling in such missing information requires a general knowledge of and interest in 3.11's broad impacts on Minamisōma and its residents.

## A BIRD'S-EYE VIEW OF MINAMISŌMA

The 395.58 square kilometers of Minamisōma city, about the size of Denver, Colorado, are located about 292 kilometers north of Tokyo. The tsunami devastated 40.8 square kilometers of its land and more than 4,000 households, causing 636 people to perish instantaneously. The number of deaths in Minamisōma is the highest of any city in Fukushima Prefecture, where the total number of tsunami-related deaths is 1,612. Moreover, at 520, Minamisōma has the highest number of *kanrenshi,* or disaster-related deaths, mostly due to evacuation, in the country. This number signals how many evacuees returned to or remained in Minamisōma to be close to their evacuated homes with the hope for their gradual return, like Kobayashi and Kato in Odaka. *Kanrenshi* has been increasing each year since 3.11; as of March 2022, the state confirmed 2,331 *kanrenshi* in Fukushima alone (the national total is 3,786).[22]

Minamisōma is distinct from other disaster-affected cities in northeastern Japan. It is the only municipality that fell under all three zones of evacuation (the twenty-kilometer zone, thirty-kilometer zone, and specially designated areas), but it also had territory (a large section of the Kashima district) outside the thirty-kilometer zone, which the state designated as a safe area.[23] Although more than half of Minamisōma's residential areas were subject to the mandatory and stay-indoor evacuation orders by March 15, 2011, the city did not enforce a total evacuation order.[24]

Prior to the TEPCO accident, Minamisōma was the sixth most populated city in Fukushima Prefecture, with a total population of 70,772 people (23,653 households). The city's economic structure is primarily based on the service sector, followed by manufacturing and agriculture. Even though agriculture is not the dominant mode of industry, around a quarter of the city's total land is used as farmland. Furthermore, approximately 42 percent of its land, mostly concentrated in the western part of the city, is composed of mountains and forests, which are more susceptible to the concentration of radioactive contaminants.

Historically, the northern part of coastal Fukushima (the Sōsō region) was a stable political entity. The domain lord Sōma and his family ruled most of the coastal region for around 740 years, since the fourteenth century (Iwamoto 2000). The former landscape of the Sōma domain extended as far north as what is now Sōma city and as far south as Ōkuma town, where the damaged TEPCO nuclear reactors sit. Iitate village, in the west of Minamisōma, which was heavily contaminated by the fallout even though it is physically further away from 1F than Minamisōma, was also a part of the old Sōma territory.

As a symbol of the unity and continuity of the spirit of the Sōma domain, Minamisōma cohosts the intermunicipal Nomaoi Festival in July every year (Abe 2014; Allison 2013). During the three-day festival, the residents, mostly those with samurai heritage, dress in samurai attire and march through coastal Fukushima with horses. Originating in the Heian period, more than a thousand years ago, the festival preserves this historical and cultural boundary by bringing together seven devastated municipalities in coastal Fukushima.

In the post-3.11 context, the festival has become even more meaningful to the participating municipalities since it allows differently situated people to come together. The fallout also impacted the festival's purpose; its final-day horse-offering ritual (*nomakake*) has been repurposed to wish for the resolution of the ongoing disaster.[25] Due to the TEPCO accident, the festival was suspended in 2011. It resumed partially in 2012, and in 2018 it was fully recovered. The residents' efforts to continue the festival have provided a sense of commitment to the local tradition across the contaminated communities in the face of the disaster in 2011 (Watanabe 2020). In 2020 and 2021, however, the global pandemic yet again caused the festival to be scaled down.

Minamisōma can be divided geographically into three sections: the coastal region in the east, the residential region in the center, and the forested region in the west. Rokkoku runs between the coastal and residential areas. The events of 3.11 introduced further divisions: the tsunami-stricken coastal part; the irradiated southern, western, and northwestern parts; the doubly affected southeastern part; and the least impacted and most populated central part. In this kaleidoscope of post-3.11 disaster cartography, Hatsumi's residence in central Minamisōma, about twenty-five kilometers north of 1F and about ten kilometers inland, was relatively less affected by the earthquake, tsunami, and the TEPCO accident, even though the area initially fell under the evacuation zone.

As I mentioned earlier, the Haramachi district was and is the most densely populated area of the city; most residents were neither direct victims of the tsunami on the coastal side nor state-recognized victims of the nuclear disaster residing within the twenty-kilometer zone. Nonetheless, Haramachi's initial designation as a stay-indoors evacuation zone and later as an evacuation cancellation preparation zone, where the state allowed residents to stay if they desired, impacted what compensation Haramachi residents could legally claim from the state and TEPCO, which others, like the Kashima residents, whose residences fell outside the zone, could not claim at all.

Immediately after 3.11, the city's population fell to fewer than ten thousand people. Thrown into the darkness of the unknown, one resident illuminated the vagueness of the situation in March 2011 with a poetic verse: "'No parent, no child, no house, what's there is the invisible radioactivity'" (Watanabe 2012, 178). As the situation at 1F remained unclear, the supply of goods stopped. Without any direction from the national government, city hall requested that Minamisōma's residents voluntarily evacuate on March 25, 2011, in order to address the city's acute material scarcity, and the residents living outside the official "stay home" evacuation order also evacuated. Later, many residents talked about which companies were not to be trusted as they had "escaped" to protect themselves at the time of Minamisōma's crisis.

Despite the uncertainty and confusion, the population quickly recovered to around thirty thousand by the end of April 2011 as the state reterritorialized evacuation zones and some schools in the Kashima district and Sōma city in the north reopened (Baba 2021).[26] Another spike of returning residents occurred in September 2011, as the state lifted the emergency evacuation order and many schools in the Haramachi district reopened. By the time I arrived, two years after the disaster, many aspects of residents' livelihood were returning to the predisaster state, except for those residents who were victimized by the tsunami, forced to evacuate their homes within the twenty-kilometer zone (over twelve thousand residents), or decided to remain evacuated voluntarily (*jishu hinan*) to avoid potential radiation exposure.

The events of 3.11 significantly shifted the demographics of the city. Before 3.11, more than half of its population (39,901) were of working age, between fifteen and sixty-five years old. According to the city, the loss of population (30 percent) after 3.11 was not evenly distributed across different age groups; there was more than a 44 percent decrease of young adults and a 54 percent loss of grade-school children. Age was a critical factor in residents' decisions to leave or stay in the city. While an aging society has been a general trend in Japan, where the national average rate of aging is 28.4 percent,[27] local politicians and national policymakers alike described Minamisōma as resembling the projected state of much of rural Japan by 2036.[28]

In sharp contrast to Minamisōma is Iwaki city. Iwaki is just outside the thirty-kilometer radius of the nuclear power plant in the south, where the wind directions in March 2011 spared the city from getting as much of the fallout plume. In Iwaki, the population (more than 320,000) increased across all ages after 3.11. Iwaki became the hub for an influx of evacuees from the power plant's neighboring towns (within the twenty-kilometer zone) and plant workers from all over the country. Many Minamisōma residents attributed the difference between the two cities to the differences in their position with respect to Tokyo: Iwaki city is physically closer to Tokyo, and it takes about three hours from Ueno station in Tokyo to Iwaki by train, compared to more than four hours from Tokyo to Minamisōma. Here we can see that the postfallout politics of distance are concerned not only with the physical distance from 1F but with other distances too. Minamisōma's provincial position relative to other municipalities within coastal Fukushima complicates the sweeping global representations of "Fukushima" as the land of contamination.

## "IT IS NOT A CONTAMINATED COMMUNITY BUT A DIVIDED COMMUNITY!"

The fact that the TEPCO accident made the homes of those living within the twenty-kilometer zone inaccessible, and thus alienated them from their previous livelihoods, had a more immediate effect on residents than any potential effects of low-dose radiation exposure on the body. Wanting to be closer to their abandoned houses, many elderly evacuees utilized vacant dwellings within the city as their *kariage* temporary housing.[29] Alternatively, some stayed in *kasetsu* units. In Minamisōma, the two types of disaster (natural and technological)[30] and different kinds of loss (from destruction and alienation) coexisted in the city's fragmented post-3.11 local geography.[31] This coexistence often posed social challenges for cooperation, which became especially palpable at the temporary housing complexes.

The makeshift establishment of *kasetsu* outside the twenty-kilometer zone (the Kashima and Haramachi districts) assisted evacuees in fulfilling their desire to remain close to their evacuated homes. As early as late September 2011, 2,004 units opened, enabling more than four thousand residents or evacuees from other municipalities in the south to reside in Minamisōma.[32] *Kasetsu* afforded a space where tsunami survivors from the Kashima district lived in the same complex as the nuclear refugees from the Odaka district or elsewhere. Nonetheless, treating *kasetsu* as a homogeneous solution for both kinds of disaster victims created numerous social issues.[33] One group complained to the other that they had lost their homes and their family from the tsunami, while the other claimed that although their houses remained intact, they were unable to return to them.

One resident who lost his house in Kashima from the tsunami told me that "they [evacuees from Odaka] complain to our family about little things, like we are noisy, and try to shun us from the *kasetsu* community socially. I want to tell them if I could how we have lost our house, but they still have theirs!" In contrast, an evacuee from Odaka living in the same *kasetsu* told me that "the Kashima people lost their houses from the tsunami, and I feel bad for them, but by losing everything they could move on easily. It is much harder not to be able to go back to your own house and not knowing if and when you could go back ever." Due to intersecting geographical, social, political, and economic divisions in Minamisōma, its residents negotiated "the damage" (*higai*) and the nature of loss from 3.11 differently, although the differences in their experience of the damage were omitted in visualizations of radiation and remained mostly invisible to many outsiders. The same evacuee said that "we were not in a contaminated community [*osen sareta comunitie*] but a divided community [*bundan sareta comunitie*]!"

This type of tension was not just the product of the postdisaster emergency efforts; it existed prior to 3.11. Minamisōma became the city it is now in 2006, resulting from the consolidation of three municipalities, the former

Kashima town, Haramachi city, and Odaka town. This state-initiated consolidation, referred to as the Merger Policy, was not necessary according to the towns' residents, and there are still lingering resentments and local conflicts about the forced consolidation. The residents of Minamisōma tend to have a stronger regional identity based on the hamlet or subhamlet level.

In his ethnographic study of the effects of the TEPCO accident, Tom Gill (2013) observes a corrosion of community in nearby Iitate village, which was initially included in the Merger Policy plan. Located farther away from 1F than Minamisōma, Iitate village, with a little over six thousand residents, was profoundly contaminated by the TEPCO accident. Despite the village-wide damage, the narratives of its residents emphasized the loss of their subsection within the village. Villagers identified their subsection as their home. Gill's observation suggests that people more strongly identify with each other within their districts than across the municipality. In Minamisōma, the preexisting district-based identity was further reinforced by the state-regimented evacuation zones, which roughly overlapped with the preexisting district boundary of Kashima (around thirty kilometers), Haramachi (around twenty kilometers), and Odaka (within twenty kilometers). This artificial zoning came to have real effects on the residents, as the zoning determined who could get compensated, for how much, and for how long. If the fallout led residents to ask themselves how far their residence was from 1F, then the compensation policy asked which district they belonged to.

This difference became more prominent a few years after 3.11, when the tsunami-affected residents started moving out of *kasetsu*. One resident who lost his house due to the tsunami explained the difference accordingly: "Seeing one's home completely swept away was hard, but it allowed to us to start from scratch. Unlike those people affected by the TEPCO accident in Odaka [in the south], I did not have to ask myself what to do with the house and the family grave, among other things. All I had to do is find real estate and buy a new house." Although on the radiation maps residents of Minamisōma appeared to be living at the front line of the state-designated fallout evacuation zones, in reality its residents were located in different zones and on less visible historical and cultural maps that predate 3.11. Kashima residents whose homes fell outside the evacuation zone complained about invisible walls that the state evacuation order created in the city while pointing out the irony that "walls cannot stop particles from traveling to Kashima."

Notwithstanding the fact that the zones were ineffective in protecting people from the boundary-defying airborne radiation, zoning produced palpable social effects on the residents. For example, different members of the same family— or, in some extreme cases, next-door neighbors—experienced the same event and its aftermaths differently by virtue of being in a different zone. In his seminal work on residential toxic exposure, environmental psychologist Michael Edelstein approaches toxicity at the level of community. According to Edelstein, a "contaminated community" is "any residential area located within or proximate to the identified boundaries for a known exposure to pollution" (2004, 9). However, the fragmented geography of Minamisōma defies any such neat categorization of its space and community as contaminated. Radiological contamination is one of many harms the residents experienced because of and since 3.11, but the real biophysical effects of this contamination are up for debate.

The grayness of Minamisōma manifests as the intersection of multiple harms where radiological contamination, its visualizations, scientific and political countermeasures, the media and academic discourses, and the public's moral judgment all impacted the lives and well-being of its residents. The ambiguity of harm, environmental sociologist William Freudenburg observes, leads to "corrosive communities," which threaten the sociocultural fabric (1997, 29–31). In Minamisōma, contaminants fell unevenly on top of preexisting social, cultural, and political boundaries, which residents continuously lived within and used to reorient themselves to the post-3.11 world. Given this new understanding, let us take another drive across Minamisōma, this time accompanied not by my Geiger counter but by Naoko, a nuclear evacuee.

———

# TWO

# *Spirited Away*

Today could be the day, for us to be spirited away
I thought I heard a child somewhere behind
I turned around, no spirit anywhere to be found
Something struck my spine, a chill, I felt odd
Alone in an open field, still, I stood.

JOTARO WAKAMATSU, *The Town Spirited Away*

## "ORDINARY" LIFE IN MINAMISŌMA

The former National Route 6 that connects central Japan to northeastern Japan also connected Naoko's temporary residence in Kashima and her evacuated home in Odaka, fifteen kilometers apart from each other, about a twenty-minute drive. For Naoko, a woman in her late seventies when I first met her in 2013, who farmed her family land, the route had become an essential path of her postfallout lifeworld. It was the route she would take to visit the local clinic and her relatives and friends, to transport household waste to a neighborhood garbage station, and to access her former residence in the Odaka district, which sat within the evacuation zone until July 2016.

My first encounter with Naoko was a rare happenstance, or what she later described as *en*. By attributing our chance encounter to *en,* she signaled the existence of some supernatural force that arranged our meeting. "One good thing about the TEPCO accident," she said earnestly, "is it brought you to my life. I mean, what would be the chance of meeting someone from western Japan who lives in the United States?" In the summer of 2013, I was accompanying the foreign film director we met in the introduction who had been making a feature-length documentary about the ordinary life of people after the TEPCO accident. Naoko was the grandmother of Toru, one of the main characters featured in the film project, and my first interaction with her was as a translator. It was the director's second visit, to follow up on his previous year of filming. The director wanted to feature Minamisōma because, according to him, the residents were living with radiation in their day-to-day lives. While I was with him, his film had the tentative title *New Normal.*

Naoko was very skeptical of her grandson's qualifications to be in the film because she thought he had nothing special to offer: "I do not know what the guy [the director] wants, but my grandson ain't nothing special! You tell him that." The director glanced at me, my cue to translate her words. I told him that she had said that her grandson was an "interesting kid," and he gave her a thumbs-up. As in this instance, I often decided not to translate everything I heard in Japanese, since the people I spoke too often had negative things to say about the director, who often demanded that they be visibly angry in front of the camera. He told me repeatedly that his goal was to capture the raw emotions of people, who should be angry at the state and TEPCO for the accident, yet Toru was only obliging him. Naoko once said of Toru, "He is very much like others in this region; he has a hard time saying no, and the director is very persistent."

As a translator trying to (dutifully) fulfill my responsibility, I often struggled to comprehend Naoko's words since she spoke using a thick local dialect. This communication issue, however, at times afforded me opportunities to step out of my role and speak to her as an individual, asking her to clarify the meaning of particular words. Every time I stopped her, she looked surprised. She found it hard to understand why I was confused. In fact, much later she told me that her first impression of me was that I spoke funny Japanese, with another dialect from western Japan. She thought I was a Southeast Asian who spoke Japanese fluently, yet she nonetheless felt I was very familiar to her and called me a *waganohito,* one of them.

After visiting the cast of the film, including Hatsumi, Toru, and a few others, in July 2013, the director and I disagreed about the residents' experiences. As I mentioned in the introduction, I eventually left the film project before its completion. After that, however, I stayed in touch with Naoko and her family. *En* materializes not in its discovery but in one's willingness to cultivate and sustain it. Sometimes people describe it as an invisible thread (*mienai ito*), and that was what I felt connected Naoko and me.

The closer I became with Naoko, the more often I would assist her with the tasks and activities that took her along the former Route 6. The TEPCO accident displaced her to an unfamiliar part of the city where she needed a car to get around, but she did not drive. She often complained how the taxi company would overcharge her, saying that "they take advantage of me, knowing that I am an evacuee and have no means to go around."

Driving with Naoko took my solitary drives with the Geiger counter to entirely new places. As Keith Basso (1996, 56), in his ethnographic monograph of the Western Apache, argues, "Relationships to places are lived most often in the company of other people, and it is on these communal occasions—when places are sensed together—that native views of the physical world become accessible to strangers." Naoko was one individual who offered me an experience of Minamisōma yet to be mapped. In the car, Naoko would tell me about the places we passed and share stories associated with them: a factory where her relative works, a small store that her old schoolmate runs, a school she went to seventy years ago, a train track where she had snacked on garden sorrel with her friend on their way back from elementary school, and a family-owned hair salon where she got her hair dyed before the evacuation. "When I was in elementary school," she said, recalling a memory from more than sixty-five years ago as we passed by a school in the Haramachi district, "shoes of kids from Kaibama [a coastal area in Haramachi] were black because in the olden days, there was a lot of iron sand in the coast." The iron-rich coast of Minamisōma benefited the old Japanese state between the eighth and tenth centuries as it attempted to colonize the deep north (Fujiki 2016; Iimura 2005).

Naoko had also discovered new places on the route that had become significant to her since the evacuation. When we passed by a crematory near her temporary residence and on the way to her evacuated house, she would often communicate, half-jokingly and half-earnestly, her plan for others to follow after her death. "I tell my son that if I die in the temporary house," she said, "just go to that crematory and drop my remains off at my family grave in Odaka before coming back to the temporary house." Some days she would report these instructions in a cheerful tone. Other days she sounded less hopeful.

The TEPCO accident displaced Naoko from her home for more than eight years. Given the length of time, her remark signified the symbolic and real distance between her temporary residence and evacuated home as well as the experiential rupture the TEPCO accident produced in the minds of evacuees. Over the course of my fieldwork I heard many versions of the same narrative (always with a hint of humor), especially from elderly residents— men and women alike—who expressed a desire to die at their evacuated homes. Many elderly residents desired to die well (*iishinikata*), and for their desire to be fulfilled, they felt it necessary to return to their evacuated homes regardless of the contamination.

I often drove Naoko to a medical clinic in central Minamisōma, between her temporary house and her evacuated house, for her routine monthly checkup. Naoko had a few health issues, such as high blood pressure, which is a common concern among people in the region. The first time I took her to the clinic, I observed that there were many other elderly people in the waiting room. Naoko talked with a couple of them, chatting about who was in which temporary housing complex, who had been evacuated to which city, whether they had plans to return to their "old" houses once the evacuation order was lifted, and so on. That day she also told the others that another friend had become sick and could not make it to the clinic.

The clinic had become a place for healthy people, geographically dispersed from the forced evacuation, to come together to socialize. The hospital had become a convenient place for many elderly people who evacuated within Minamisōma to gather and mingle, since one of the state compensations for the TEPCO accident was a medical exemption. Immediately after the accident, the national government granted free medical care to the evacuees and residents within evacuation zones to address health concerns related to the risk of radiation exposure. Because the exemption also applied to dental fees, a few residents told me that dentists had become busier than usual as many residents, especially the elderly, rushed to get new teeth—another ordinary event in postfallout Minamisōma.

That day in the waiting room, each of her friends asked if Naoko had come with her grandson, to which she responded that I was her friend and that I drove her places because she had no means of transportation. Her friends were all impressed, thanked me for doing a great deed, and lamented that young adults had become scarce in Minamisōma. Many local young adults, willingly or not, had abandoned their now-contaminated hometowns and left behind the elders, who were more physically and psychologically bound to their birthplace. One of Naoko's friends jokingly said it was a case of "spiriting away" (*kamikakushi*). Naming the disappearance of young people as "spiriting away"—a cultural imagination of the permeable boundary between the world of living and the dead—conveyed the remaining residents' experience of the borderline the TEPCO accident produced between this world (coastal Fukushima) and the other world. The generationally skewed impact of the TEPCO accident had exacerbated the general graying trend of

Japanese society. More immediately and tangibly than any scientifically discernible biological and teratogenic effects on the individual body, the TEPCO accident disintegrated the social organization (Gill 2013; Yagi 2021; Yamamoto 2017; Yamashita, Ichimura, and Sato. 2013; Tsujiuchi and Gill 2022; Yokemoto and Watanabe 2015). This disintegration occurred most palpably at the level of the household.

One woman waiting at the clinic remarked that her children would never come back because they must have realized the convenience of living in a bigger city. "Radiation is a good excuse for them not to come back to this boring countryside," she told us. "I am saying this because we all know that the radiation levels in many places in Fukushima and Koriyama city in the west [where many young people had evacuated to] are, in reality, higher than they are here, though we cannot tell them that." When I asked why her children had moved to a more contaminated area, she guessed that it was because of the physical distance from the nuclear plant. "We are closer to the plant here," she offered, "so they think it has to be more dangerous." Another chimed in, "If I had an option to leave my mother-in-law when I was young, I would have." Everyone chuckled, signaling the still-standard extended family structure in this rural region. The threat of radiation exposure since March 2011 had apparently become an excuse for some locals to resolve preexisting social and domestic issues and while circumventing moral judgments based on the region's traditions and norms.

Another of Naoko's friends shared that her son's family in Tokyo had told her never to send them anything, as she had done frequently in the past, because everything in Minamisōma was contaminated. What was upsetting to her was that her son, who had grown up in Minamisōma, did not defend her and instead told her that she did not know anything about radiation and did not care as much about its effects because her life was almost over. Disagreements over the perceived risk of radiation exposure had discouraged many locals from talking about radiation in private and in public. "We do not want to cause any trouble at home or outside by talking about it," Naoko lamented as she got up to wait in line to see a doctor.

Naoko herself had thoughts and experiences that she chose not to share with her family. On our way to the clinic that day, Naoko had unexpectedly confessed to me that she had been having a recurring dream for some time. She asked me not to tell her family, because she did not want them to think that she had gone crazy. I promised her not to do so and turned down the volume of the radio to suggest my willingness to listen to what she had to say. "You know," Naoko said, "my house is in Odaka," stating this fact known to both of us as if to establish the ground of her dream before she began. "In the dream, I am in my old house, but I am not inside my house. From where I stand, a big forest surrounds me, and it is very dark in there. From somewhere in the darkness, I start hearing voices of people, voices I recognize. I realize that they are dead people in my family, like my grandparents, mother-in-law, and close relatives. In the forest, I am feeling very sorry for them, guilty that I am in the forest and cannot come out from it."

When she finished reporting her dream, Naoko looked somewhat relieved. I did not offer any interpretation of her dream, although there were several obvious and not-so-obvious things I could have guessed. Nonetheless, I sensed that it was better to let it be and instead offered a general validating comment that she must have been tired from not sleeping well. Naoko nodded and said, "I will get some more sleeping pills from the doctor today, and that will help me to forget about it and sleep." We then both looked at the Geiger counter mounted next to the radio, which made clicking sounds to indicate the incessant disintegrations of radioisotopes in the vicinity, which read 0.12 μSv/h inside the car. Even though Minamisōma was supposed to be more contaminated than most places, the reading was no different from what one would experience in Tokyo or any other part of Japan.

It turned out that Naoko had been taking sleeping pills since not long after her sudden displacement in March 2011. When the enormous earthquake occurred on March 11, 2011, Naoko was tending to the family farm as usual. Without anything to hold onto in the open field, Naoko endured more than two minutes of the rattling earth. Her grandson, Toru, who was enjoying a day off from high school, came to rescue her in the field. Naoko remembered Toru yelling, "Grandma! Grandma!" Then she learned the news about the tsunami overcoming the coast, which she later discovered almost devoured her natal home, which was a few kilometers inland. Immediately following the second hydrogen explosion on March 13 at 1F, about twelve kilometers south of Naoko's residence, she fled with her family. Naoko did not bring much with her and left her cat because she believed that she could return in a few days. However, in less than a month Naoko and her family became some of the around 65,000 evacuees from Minamisōma, a number that swelled to more than 160,000 long-term evacuees throughout Fukushima Prefecture at its peak in 2012.

For Naoko the evacuation did not mean a transitory absence from her house in Odaka, as it did for some other Minamisōma residents. On average, evacuees from Fukushima relocated four times. After moving to several locations both within and outside Fukushima Prefecture, Naoko ended up in a house in the Kashima district. She made the decision with her family to go back to Minamisōma. For her, the sense of familiarity the city offered outweighed the potential risk of exposure. They all agreed that the place in the Kashima district could enable the family to tend to their evacuated home and till its land in the Odaka district. In April 2011, the state reterritorialization efforts designated the twenty-kilometer zone and most of Odaka district an exclusion zone. As the police barricaded people from crossing the twenty-kilometer radius line on Rokkoku, Naoko's son hiked up an unbarricaded hill to sneak into the family residence to rescue Naoko's homemade miso. "He said if we could not live in our house, we at least had to enjoy the familiar taste of Odaka," Naoko laughed. Within a year, as the salvaged miso stock was running out, Naoko discovered that the western part of Odaka where she had lived was more contaminated than the rest of the city.

Although she did not share her troubles with her family and friends, the experience of sudden displacement took a psychological toll on her. "I think I started taking the pill after I came back to Minamisōma from my temporary evacuation to various places," she said, "so it must have been since May of 2011." She told me that before the accident she had neither wanted nor needed to take sleeping pills. As a farmer who grew rice, cultivated vegetables and fruits in her garden, and continuously engaged in physical labor, she was always doing something outside and never had any problem sleeping, she said. Naoko implied that the limited amount of space available and the more urban setting of the temporary house contributed to her current state of inactivity and perhaps her sleeplessness.

The contamination posed a threat not only to people's lives but also to their livelihoods. What went unsaid by Naoko, however, was this sentiment shared by many locals: the sudden awareness of the radiological danger and hypervisualized contamination in the postfallout landscape across coastal Fukushima—the very reason she had to leave her home to begin with—made it undesirable for locals like Naoko to stay active and engage in farming. The general public's fear of the uncontrolled distribution of contaminants and governmental food-safety regulations discouraged those who had decided to stay in the region from the physical activity associated with tending to and tilling the land. The impacts of this significant lifestyle change manifested as extremely busy hospitals and clinics packed with elderly people with increased diagnoses of obesity, diabetes, hypertension, hyperuricemia, rheumatic diseases, and mental health issues (Hashimoto et al. 2020; Tsubokura 2018), which are often glossed as stress-related misalignments.

While waiting for Naoko that day, I noticed a few younger people in the waiting room. They were all wearing uniforms and were most likely disaster reconstruction and decontamination workers. Public health research in the region shows that decontamination workers, many of whom came from low socioeconomic backgrounds, often suffered from preexisting noncommunicable diseases such as hypertension, hyperlipidemia, and diabetes mellitus (Sawano et al. 2016, 2021). Participating in disaster remediation work afforded them access to health care that they might not have previously had. Unlike the elderly residents, these workers were usually alone at the clinic, attending to their smartphones. I wondered if they found any *en* in coastal Fukushima. Perhaps the reconstruction work was transitory labor and they preferred not to get entangled in the invisible local web of relations.

While I was waiting at the clinic for Naoko, who was seeing the doctor, I focused my attention on the waiting-room TV, which no one seemed to be watching. The TV was broadcasting news of Prime Minister Abe's presentation at the International Olympic Committee in Argentina on Tokyo's invitation to host the 2020 Olympic games. In the speech he proudly announced to the world in English that the situation in Fukushima was "under control."[1] I thought back to Naoko's dream and how she had used this same phrase to describe the *lack* of control she experienced in her dream. I knew that she had no control over her current situation—not only symbolically, as in her dream, but also in reality, since she had not been able to live in her own house, which was in the evacuation zone.

After I had waited a couple hours for her, Naoko came back to the waiting room. "I am sorry. There are just too many people inside! But the doctor told me I am fine, but I need to watch my high blood pressure," she said. "Stress might be causing many issues, the doctor told me, and we laughed how back in Odaka [where both of them had lived before the accident], stress [*sutoresu*] was never an issue!" Stress, she said, was not necessarily caused by the fear of radiation and its potential health effects; instead, it came mostly from the fact that she could not be at her house in Odaka.

When her name was called, she received her medical card and another card at the counter. Giving Naoko her cards back, a middle-aged receptionist at the counter asked her to confirm that she was an evacuee from the Odaka district. Having double-checked this fact, the lady said that Naoko was "qualified" for the medical exemption, so there was no charge for the visit, and they bowed farewell to each other. Earlier I had witnessed many similar exchanges. In most of them, the men and women were in the same category, as former residents of Odaka who qualified for the medical exemption. "Odaka and medical exemption": it sounded like a new and strange form of identity-profiling code in Minamisōma, which articulated potential difference in the treatment of the residents based on the distance of their prefallout residence from the damaged plant.

Naoko picked up her sleeping pills and medications for her high blood pressure from a nearby pharmacy, where a middle-aged male pharmacist explained to her that she was supposed to take only one sleeping pill a day if needed. While we were driving back to her temporary residence, Naoko noticed the concern on my face and assured me that only half a pill would put her in a dreamless sleep. She said that she was keeping her consumption of the pills under control. Listening to her, while also attending to the Geiger counter's clicking sound, I wondered for a moment about which one of Naoko's everyday life events—taking only half a pill to produce a dreamless sleep or having her recurring dream—was the condition of postfallout Minamisōma. Both the real and surreal seemed to coexist in the gray zone. But then I had a second thought. My solitary drives and my drives with Naoko illustrated that there was more than one Minamisōma, one technoscientifically recorded and represented and the other lived by the residents. Each had its own historicity, temporality, logic, texture, and mood. It made me think of a line from Murakami's *1Q84*, "Overhead, the two moons worked together to bathe the world in a strange light." What the residents, in their different versions of Minamisōma, had in common was that they simultaneously confronted both in their everyday lives.

### TEMPORALITIES OF THE ORDINARY

Prominent political scientist of Japan Richard Samuels (2013) observes that 3.11 did not cause much change to the Japanese political system at the national level. Locally, however, the TEPCO accident led to a radical change in the region's visual landscape. Radiation does not have a color, taste, or smell, yet the maps, technoscientific devices carried by individuals, and hundreds of radiation monitoring posts placed by the state made the invisible presence of radiation hypervisible in Minamisōma. Every day, the local TV news programs, radio shows, and newspapers would broadcast something about the

TEPCO accident in general and the damaged reactors in particular. There was "an excess of signs" (Parmentier 2012) that signaled the persistent presence of contaminants in the vicinity. The reversal of what was invisible and visible sharply contrasts with what Olga Kuchinskaya (2014) observed in post-Chornobyl Belarus, where the state made radiation "twice invisible" to its citizens by not communicating its presence and potential effects. In contrast, the invisible presence of radiation in Fukushima was made more visible than many other issues that the locals found more significant for their everyday lives.

In order to address the heightened concern about radiation, the city of Minamisōma began lending out personal dosimeters to its registered residents in June 2012. A dosimeter is a mobile device that allows its user to survey and record personal exposure to radiation in a vicinity. Immediately after the fallout in 2011, it became a crucial and familiar tool in coastal Fukushima for people to understand the blasted landscape and its potential danger to their bodies. I carried one everywhere I went in the beginning of my fieldwork.

In 2013, however, many people I met in coastal Fukushima did not carry their dosimeter, and if they kept it in their house it was left uncharged. For locals, the fact that I carried around a Geiger counter marked me as an outsider. My possession of the device was another reason, Sachiko in Kashima later revealed, that she had identified me as a *yosomono*. Similarly, a few residents joked that I lived in the time of antiquity, as if radiation detectors had been a popular fashion item between 2011 and 2012 but were out of style by 2013. At the same time, though, people expressed their vested interest in knowing current radiation levels and asked me to report how many sieverts (Sv) I recorded at which locations.

When I told them my readings, residents usually responded by saying either, "Was it still that much there?" or "Oh, it has gone down a lot," indicating their general knowledge about location-specific radiation levels. A few self-taught and more knowledgeable residents offered educated guesses to explain the variability in the readings. For example, they might point out what was happening at the power plant as reported in the news (e.g., the removal of debris around the highly contaminated sections) or describe the direction of winds and weather patterns of the day (e.g., if it were windy as opposed to raining) to estimate the possible movement of radioactive particles.

For locals, the Geiger counter's readings were significant only insofar as they indicated a difference from years ago, when the dosimeter was "in style" in the city. This was different from my own experience with the device; to me, each

FIGURE 1. A radiation monitoring post (MP) at Kita Hatohara Community Center in Odaka, Minamisōma, March 2014. MPs are placed by the prefecture and supervised by the Nuclear Regulation Authority. Photograph by the author.

value I saw seemed significant, indicating something that I could not see, smell, taste, or touch in the here and now. Nonetheless, my readings of the device lacked the dimension of historicity, which was critical to the way that residents understood each reading. This lack of awareness of historical information with which to interpret the present radiation levels also marked me as *yosomono*.

In general, using or possessing a personal radiation-monitoring device did not indicate or correspond to the degree of anxiety about the region's contamination. For the residents, the device was a symbolic item that indicated the extraordinariness of life after the TEPCO accident. "Even in case of an emergency, there is almost no need to use it," one Kashima resident in her early thirties, Midori, reflected, "because if something does happen, ultimately there are many monitoring posts around the city to look at."

A radiation monitoring post (MP) is a white pod about six feet tall with a solar panel on top powering its digital display. It measures and communicates the real-time ambient radiation levels of a specific location continuously to locals and *yosomono* alike (figure 1). Since the TEPCO accident, the local government installed over 3,000 MPs in the prefecture (over 250 in

Minamisōma), forming a formidable infrastructure for the radiation-warning system.[2] The ubiquitous presence of MPs meant that those in Fukushima Prefecture would encounter at least one MP almost every day. Each encounter with an MP would expose individuals to real-time readings of background airborne radiation levels. Unlike a personal monitoring device, which an individual could choose to use or not, MPs are out there, regardless of people's interest in using them.

Although various technoscientific devices placed throughout the region helped the residents to be informed and cautious, locals did not find much value in them, according to Midori, since "some part of me does not even want to know how much I am being irradiated. All I need to know is when I must run. After all, what the hell do those numbers *actually* mean anyway?" Similarly, a few middle school students mentioned that they never paid attention to MPs even though the schools and parks used by younger generations were heavily monitored. One of them observed that MPs were something "only anxious adults cared about."

Years after the accident, the quantification of radiation was not necessarily meaningful in and of itself for the residents. Personal radiation-monitoring devices, MPs, and locals' relationship to them were a mere reminder of the past threat. As Midori made clear in her comment, these devices cannot and do not remove the uncertainty generated by the continuous presence of the unknown in the present. The devices only made the invisible numerically visible and introduced people to unfamiliar and often confusing technoscientific terms, such as *sievert, becquerel, half-life, cesium,* and *internal* and *external exposure,* through which they were forced to communicate with the state, the experts, and sometimes outsiders.

In 2013, for many residents, MPs were nearly invisible, despite their pervasive presence throughout the city. Their presence had become so enmeshed with people's daily lives that they no longer noticed them. Not "seeing" MPs in their living environment indicated residents' increasing desire to avoid thinking about the "past" accident and to move on from it, if not forget it altogether. Hatsumi was one such resident who did not see monitoring posts around the city and expressed her desire not to let *yosomono* constantly bring her back to March 2011. If, as Avery Gordon (1997) conceptualizes, a ghost is something that constantly haunts people and draws them in affectively against their will, then for Hatsumi, the film director and I were the ghosts that demanded her to give repetitive attention to her recent past and to the MPs that had become invisible to her.

The first time I met Hatsumi, before she told me about the nuclear ghost of Minamisōma, I never imagined that I would become close to her. As I mentioned before, Hatsumi kicked the director and me out of her temple. Although for a while I avoided remembering anything about my initial failure, the more I become attuned to lives in Minamisōma, the more I realized how the director and I had violated Hatsumi's experience by projecting onto her our imagination of the TEPCO accident and its harm.

I remembered how when the director and I went to see her, Hatsumi, seeming somewhat annoyed, looked in my direction and ordered me to translate the following: "We just want to be left alone. I am sick of thinking about and talking about the TEPCO accident. In fact, I would like to forget what happened. It has been over two years now. Things have changed. I want to get back to my ordinary rhythm of life [*seikatsu no rizumu*]. I just told you about the change in my feeling. My answer is no."

Then we left. As we sat in the car parked outside Hatsumi's carefully maintained yard in complete silence, I listened carefully to the clicking sound of the director's personal Geiger counter that he had borrowed from a resident the previous day. The device clicked in concert with the noise of the cicadas outside and the broadcast of a local radio program, mechanically reporting data from the radiation monitoring posts distributed throughout the city: "Haramachi city hall 0.30 [μSv/h], city hospital 0.26, Obama public hall 0.15, Enei community center 0.29 . . ."

"I do not understand!" the director suddenly said, breaking the contemplative silence. He sounded very agitated as he took off the white N95 mask he wore whenever he exited the car to avoid inhaling the invisible particles in the air. "Why do they not want to talk about the disaster anymore? Stupid radiation is still everywhere!" He gave a glance at his Geiger counter as he yelled, "They were all cooperative last year. They wanted, it appeared to me, to talk about the disaster and share their fear of radiation. I am helping them to document the catastrophe they have been living with for their sake. They must know how truly significant this accident was and has been. It is not something they can or should easily try to let go by forgetting!"

Being a mere translator, I kept my mouth shut. Instead, I stared at a list of five people the director had handed to me earlier and wondered if the others on the list, which included Toru, would react in a similar way. Perhaps many Minamisōma residents would have avoided communicating their thoughts in this situation in order to avoid being categorized. As I later learned from Naoko and her friends at the local clinic, radiation was a sticky topic. The status as "the

exposed" is stuck to them like a permanent tattoo—one of the only valid cer-
tificates of their right for compensation—and it communicates the extent of
the disaster compensation they receive from the state and TEPCO. The
Minamisōma residents were not united nor in one common space, like the
various maps suggested. Hatsumi in Haramachi, Kobayashi and Naoko from
Odaka, and Yamada, Midori, and Sachiko in Kashima are all Minamisōma
residents who remained despite the TEPCO accident, but their postfallout
experience differed depending on which part of Minamisōma they lived in.

What the director seemed to have missed was that by 2013, a subtle but
definite shift was occurring in many parts of Minamisōma. In a city where
radiological contaminants had been present and hypervisualized for more
than two years, contamination had become more than a Geiger counter's
numerical output to inexperienced outsiders, for whom it confirmed the
validity of the continuity of the nuclear catastrophe. Although radiation was
everywhere, as the director observed, its levels were objectively lower in 2013
than during his first visit in 2012.

As silence resumed in the car, I meditated on the different rhythms of the
postfallout lives I had begun experiencing in Minamisōma: the spirited-away
city with the graying population; people's "ordinary" lives; the local radio's
mechanical, monotonous reporting of the airborne radiation; the random
clicking of the Geiger counter; and the onlookers' desire to freeze residents
in the past accident. They all ticked discordantly in my mind, giving a dishar-
mony of misplaced, though still precariously aligned, chords of bygone, real-
time, and latent hopes, desires, and suffering in Minamisōma.

The commonality across the different temporalities is that they all were,
are, and perhaps will remain uncertain; the losses, harms, and traumas of the
TEPCO accident had yet to be fully realized. The looming sense in
Minamisōma was that "the worst" was yet to come and was constantly being
deferred, engendering the sense of chill (samuke) that Baba would explain in
light of the global pandemic years later. To capture the cyclical tempo of
disaster, sociologist Norihiro Nihei (2012) suggests the use of the term saikan,
or "in-between disaster."[3] It was not that Hatsumi had forgotten about what
happened in 2011; it was that she might have been anticipating, and thus
preparing for, another catastrophe yet to come. She told me that one thing
she learned from 3.11 was to make sure her car's gas tank was always filled up
in case she might need to run immediately.

According to anthropologist Anne Allison (2013), for residents living in
this sort of entanglement of uncertainty—experiencing the precarity of life

in post-3.11 Japan—"things are still in the mud . . . [but] this is not altogether bad as long as people can stay with the uncertainty for a while and give both the lives that have been lost and the changes brewing in the current landscape . . . the time needed to prevent a mere repetition of the past" (203). Although seesawing between hope and despair may be the general rhythm of life, some people I encountered in Minamisōma deeply desired a return to the prefallout past. Sachiko policed outsiders to keep Kashima free of intruders in order to maintain the tight community where everyone knows one another, Naoko wanted to return to her "ordinary" life governed by the agricultural cycle, Kobayashi hoped for Odaka to be filled with people again along with her flowers, and the film director wanted to return to the footage of the "ordinary" Minamisōma he had documented through his camera a year previously. For each of them, the vicissitudes of time prevented the repetition of the past—the ordinary as they had remembered it. Every day sowed a seed for a new normal. For Hatsumi, on the other hand, the ordinary waited until the past could be put in its proper place through reorientation to the present.

Temporalities of the ordinary had been out of sync in Minamisōma, in addition to its ever-more-fragmented postfallout geography and demography (Morita et al. 2018). These spatial, temporal, and social disorientations made it, if not impossible, then dauntingly difficult and undesirable for residents to communicate not only their individual suffering and losses, but also their hopes and desires. Hatsumi's nuclear ghost wandered in this gray zone, where some were spirited away and others held on. Whatever the ghost was, it was not to be named.

UNDER CONTROL

In 2013, two years after the accident, a man in his late thirties named Oshima began working for the national government's decontamination project to remediate the irradiated environment of coastal Fukushima. Job opportunities burgeoned for young adults like Oshima, if they wished to remain in the region and were motivated to work. At a busy family restaurant franchise full of elderly customers in the Haramachi district, about five kilometers south from Oshima's residence in Kashima, recruitment signs for part-time work posted throughout the restaurant caught the eye of anyone who entered. The most noticeable message on these signs was the unusually high salary offered for jobs in a rural region; wages were printed in a large font: "1,200 yen per hour!!" (around nine dollars, or 1.5 times the average prefallout wage). One important qualification was printed in a much smaller font: "This job is for people younger than sixty-five." The presence of the multiple recruitment signs signaled that despite the exceptional wage, laboring bodies were missing, as if they had been "spirited away" by the TEPCO accident.

Noticing my gaze on one of the signs, Oshima commented, "There are many people in this region who do not need to work, or for whom labor has become antithetical to their new life based on compensation." As a resident of the Kashima district, located outside the state-mandated evacuation zone, Oshima had strong opinions about the nuclear accident compensation (*genshiryoku songai baishō*), which treated the residents in the same city and same district quite differently. "Look, an individual in Odaka would receive a compensation of 100,000 yen a month just for the psychological harm [*seishinteki higai*] until one year after the district reopens. For us [the Kashima residents], it was only for the first six months. Consider a household with five people; it would be 500,000 yen [$4,500] a month. Many of them make more money than they used to by doing nothing. It's terrible, isn't it?"

As the average individual annual income in the region was around $25,000 in 2011, who received how much money from the accident produced major social tensions within the already divided city (Yokemoto 2013). For example, even within the Odaka district, which Oshima described as the most highly compensated area in Minamisōma, there were multiple divisions. In April 2011 and then again in March 2012, the state reorganized its initial evacuation categories based on physical distance to more "objective" categories based on technoscientifically measurable contamination levels. This shift produced new divisions for those whose preaccident residences were inside the twenty-kilometer zone. Two individuals (1 household) in Odaka lived in the difficult-to-return or restricted zone, 510 residents (130 households) were in the restricted residence zone, and 12,238 residents (3,762 households) were in the evacuation cancellation preparation zone (Kansai Gakuin University, JCN, and SAFLAN 2015).[4] Moreover, the state also designated several hot spots outside the twenty-kilometer zone, including Haramachi and Kashima (152 households), as the special evacuation designation zone.

Each group qualified for different categories of compensation for different durations. Despite the divisions, psychological damage is one of the few common categories of compensable damage. Other common categories included compensation for evacuation (for the first year and two months), free medical treatment, tax relief, free rent, and free highway tolls.[5] In addition, highly individualized items highlighted the predisaster difference in class and wealth, such as compensation for lost or damaged property like land and forests, income, and other personal assets. Oshima's residence fell outside all the areas that were eligible for compensation, just like those of around 9,200 other residents in Kashima.

Lighting a cigarette, Oshima continued, "Getting money without working could make labor and life meaningless. It is a form of violence." He believed that the anticipation of long-term compensation had turned many evacuees into *takari* (beggars). He was sympathetic to the pain the evacuees had experienced from leaving everything behind, but he could not understand how compensation came to be distributed unequally. "How can we be the same Minamisōma residents when only some get compensated? For sure, the accident contaminated the entire city. But only some of us have been *yakuzuke* [drugged] with compensation."

Oshima shared that he had lost his previous job because the company had to shut down due to its location inside the thirty-kilometer zone. His residence is in Kashima, which meant he was not qualified for long-term compensation like some fellow Minamisōma residents, including Hatsumi in Haramachi and Naoko, who evacuated from Odaka and lived in Kashima like Oshima. "All of a sudden, I experienced an existential blackout, and I could not do anything for a while. Yet my mentor [*senpai*] convinced me that I had to work once again to take control over life for myself and my family." As a resident of Minamisōma but not an official "victim" according to the state compensation policy, Oshima experienced the TEPCO accident most acutely not as a transboundary environmental catastrophe but as a bounded social and political disaster that divided the city and its residents. If the postfallout mapping of the extent of the fallout divided the community geographically, then compensation came to cement such divisions.

While the radiation maps selectively visualized the threat of radiation exposure, other detrimental consequences, highly relevant and painfully real for the residents, were often left invisible. The TEPCO accident had produced many non-radiation-related burdens in the locality. Skyrocketing wages for part-time service-industry positions was just one of many signs of this. "I am precious here," Oshima commented. "Young working bodies have gone missing in this inaccessible corner of the prefecture [*riku no kotō*]." Although they had become precious, young, able bodies like Oshima's were particularly vulnerable to radiation exposure in the aftermaths of 3.11. Oshima did not hold a college degree and did not have special skills other than agricultural knowledge and experience from his previous job. For young residents like Oshima, who could not continue farming, decontamination was a viable opportunity. "What many outsiders might not understand is this irony. I am living outside the so-called danger zone, but here I am, hoping to make up for money that I could not receive from the accident by exposing myself!" Oshima raised his voice as he stubbed out his cigarette in an ashtray.

Oshima was critical of the entire situation. After complaining about TEPCO, other residents, and the problem with Minamisōma, he talked about the controversial September 2013 speech by Prime Minister Abe at the International Olympic Committee meeting. Since the speech, criticisms in the media and social media had exploded. I found Oshima's cynical commentary on it illustrative of the complex configuration of the everyday in Minamisōma.

"I think Abe is right about it," Oshima said. "I mean, I interpreted the speech as him being too honest, admitting to the world how we, the people in Fukushima, have been kept 'under control' by the government. Think about it. Have we done anything about the TEPCO accident? We have seen big antinuclear rallies in urban areas like Tokyo, but nothing in coastal Fukushima. Why?"

Without giving me a chance to reply, Oshima rushed to continue. "My grandma still refers to TEPCO with an honorific, Mr. Electronics [*Denryoku san*]. Meanwhile, the people in Tokyo judge us for being exposed from the plant they imposed on us. All we have done is to suppress our sense of uncertainty and the fear of the invisible with the bit of compensation we have received for our stupidly obedient obliviousness. The radioactive water the prime minister speaks of is a metaphor of us; we have been kept successfully at bay."

As sarcastic and bitter as his statements may sound, his interpretation of the prime minister's representation of Fukushima made sense to me. In the postfallout context, the residents appeared to distance themselves from the persistent public discourse on radiological contaminants in society and the indeterminate presence of radiation in their everyday lives. Even though, from a distance, the entire coastal Fukushima region appeared to be in radiological danger, what the radiation maps in Minamisōma illustrated was how the state and TEPCO sorted out who was to be compensated rather than what the residents feared or their anxiety about radiation exposure. More immediately than invisible low-dose radiation caused any biological effects, it caused geographical divisions in Minamisōma, producing multiple realities and tangible social disintegrations among the residents, who had different experiences in the same city. Writing about the unique effects of a technological disaster that damages the existing social fabric, Freudenburg notes, "Disruptions may be created not just by the accidents themselves, and not just by the experience of risk or threat, but also by subsequent actions that threaten the very system of agreed upon meanings that allow a complex social system to function" (1997, 34).

In an August 2011 press release by TEPCO, the party considered liable for compensating the victims of the accident, the company expressed their full commitment to adhere to the Act on Nuclear Damage Compensation (hereafter the Compensation Act), first established in 1961.[6] The Japanese state, on the other hand, claimed that their endorsement of the nuclear energy policy left them with only social responsibility for the accident, and stated that "the government will support TEPCO under the framework of the Compensation Act, basically aiming to minimize the burden to be placed on the public."[7] The state then provided TEPCO a list of eight categories of compensable damage for it to use when compensating the victims. According to TEPCO's "Overview of the Compensation Standards for the Major Categories of Damages (Provisional Translation)," those categories included: (1) damage due to evacuation instruction by the government, (2) damage due to the designation of the navigational hazard zone and the no-fly zone by the government, (3) damage due to the shipment restriction of primary industry products by the government, (4) damage due to other governmental instructions, (5) reputation damage, (6) indirect damage, (7) damage by radiation exposure, and (8) others.[8]

In their comparative studies on nuclear compensation in the United States, Japan, and China, Liu and Faure (2016) observe that although the Japanese Compensation Act does not specify the nature of compensable damage, the Dispute Reconciliation Committee for Nuclear Damage Compensation, which was established after the TEPCO accident, provided guidelines for a rather broad scope of compensation. According to the committee, this broadness was purposeful, as the nature of the accident itself remained unclear. Since the establishment of the primary guidelines, there have been four modifications to the extent of coverage, signaling the shapeshifting nature of damage from the accident.[9] After the tenth year of the accident, in June 2021, the state began negotiating with TEPCO for an additional revision to the guideline in order to account for potential economic and reputational damages from the release of contaminated water into the Pacific Ocean.[10]

For individual residents, the most relevant compensation category is the first one, "damage due to evacuation instruction by the government," within which there are ten subcategories, including evacuation expense, psychological harm, and property damage.[11] For each item, individuals or corporations needed to file a formal claim and provide supporting documents such as receipts, proof of residency, and, for medical claims, certification of treatment to TEPCO. Alternatively, a resident or a group of residents may utilize the state-established Alternative Dispute Resolution (ADR) to seek an out-of-court settlement on items not specified in the guidelines (Awaji, Yoshimura, and Yokemoto 2015; Jobin 2020; Kanbe 2021; Osaka 2020).

As of May 2021, TEPCO reported paying around $10 billion for a total of around 2,754,000 claims (1,008,000 individual claims, 1,296,000 claims for voluntary evacuation, and 450,000 claims by corporations and individual proprietors).[12] Since TEPCO itself is not capable of covering the full amount, the state established the Nuclear Damage Compensation Facilitation Corporation (Genshiryoky Songai Baishō Shien Kikō) to loan TEPCO funds.[13] Although the state has construed the compensable harms rather broadly, the guidelines are not legally binding. That is, TEPCO chiefly determines what is considered harm caused by the accident.

From the residents' perspective, understanding the different categories of compensable items, keeping up to date with changing categories, and preparing the supporting documents necessary to file claims with TEPCO took a lot of time and effort. Although Oshima, as an outsider to the compensation policy, imagined that applying for compensation was a lazy way to make a living, in reality it involved many steps, procedures, and emotional burdens. Often, TEPCO denied claims or asked for additional supporting documents. Hatsumi once mentioned that the act of constantly filing claims reminded her of her victim status and prevented her from moving on. "I had to constantly ask myself whether to charge 5,000 yen [$39] here and 2,000 yen [$15] there, which I spent to visit my daughters who were away for a while. Yes, I deserved to get that money back, but then, should I spend hours trying to file a claim for that? Am I that desperate?"

More than the act of reorienting oneself to life with radiation, living in postfallout Minamisōma meant navigating various harms (*higai*) and their associated legal, economic, political, and moral burdens and values. In the gray zone, where the nature and the target of harm constantly shifted, what exactly is the *higai* that the city and its residents experienced from the TEPCO accident and its aftermaths? Through what processes had postfallout compensation come to harm the coastal Fukushima residents and put them "under control"? To start answering such questions, let us dive into differently situated Minamisōma residents' kaleidoscopic experiences of *higai*.

# THREE

# Kaleidoscopic Harm

*No matter what we said, people would believe what they wanted
to believe. The more we struggled, the more vulnerable we'd be.*

HARUKI MURAKAMI, *Norwegian Wood*

## STUPID LINES

"What the nuclear accident taught me," Hatsumi commented in late 2013, "is
who my friends really are." As a resident in the Haramachi district who
received more compensation than most residents in Kashima but less than
those in Odaka, Hatsumi acutely experienced how the accident harmed her
preexisting interpersonal relations. According to her, the nuclear compensation changed people, and she discovered that "many of my old friends turned
out to not be my friends after all because they are from different districts and
were treated differently."

Hatsumi shared one such instance of a friendship ending. An old friend
from the Kashima district had stopped answering her calls in early 2012.
Worried because her friend lived with two teenagers, and Hatsumi had seen
news that bullying at schools related to contamination was on the rise,
Hatsumi kept calling her. "One day she answered," Hatsumi said heatedly
"and she was like, 'Sorry, unlike Haramachi people who get a lot of money
from TEPCO, we [Kashima residents] need to work hard to survive.'" As
Hatsumi recounted the exchange, she admitted that although she was in fact
qualified to receive more compensation than most Kashima residents, it was
not her choice. "After all, our area received more harm [*higai*] as a result of
the accident," she concluded.

Knowing that the radiation levels in the two districts were not so dispa-
rate, I asked Hatsumi what, precisely, was the nature of the harm she had
experienced, to which she responded, "We are closer to the power plant."
Hatsumi was not the only individual to talk about physical distance rather
than contamination levels when describing *higai* in Minamisōma. Despite

the state's declaration that Fukushima was under control, many remaining residents still feared that they would have to take flight again suddenly if something were to go wrong at 1F. Their focus on physical distance rather than on contamination levels signaled that regardless of the state's redesign of evacuation zones on April 22, 2011, to account for the actual contamination levels, the initial order based on physical distance stayed in their memory.

Here is a breakdown of how the state changed its evacuation orders:[1]

1. At 8:50 p.m. JST on March 11, 2011, the evacuation order was established for a two-kilometer radius from 1F.

2. At 9:23 p.m. on the same day, the evacuation order was extended to a three-kilometer radius, and a stay-indoors order was issued for a ten-kilometer radius.

3. At 5:44 a.m. on March 12, the evacuation order was extended to a ten-kilometer radius.

4. At 6:25 p.m. on the same day, the evacuation order was extended to a twenty-kilometer radius, impacting Odaka residents and some Haramachi residents.

5. At 11:00 a.m. on March 15, the stay-indoors order was issued for a thirty-kilometer radius, impacting most Haramachi residents and a few Kashima residents.

6. On April 22, 2011, the state reorganized the evacuation orders into three zones to account for the measured contamination levels: the evacuation-prepared area in case of emergency, the deliberate evacuation area, and the restricted area (see map 3).

7. In March 2012 the state further recategorized the existing three zones into the evacuation order cancellation preparation zone, restricted residence zone, and difficult-to-return zone.

Residents like Naoko and Hatsumi evacuated because their residences fell under the evacuation order determined by physical distance. Their experience of the accident differed significantly from that of people in Iitate village, which is located outside the thirty-kilometer zone and did not fall under the evacuation order until the state reorganized the evacuation zones to reflect actual contamination levels on April 22, 2011 (Kanno 2018; Owatari 2016; Ozawa 2012). Naoko, for example, did not discover that the area surrounding her residence was more contaminated and fell under the restricted residence zone until March 2012, while for Hatsumi, despite her initial evacuation, her

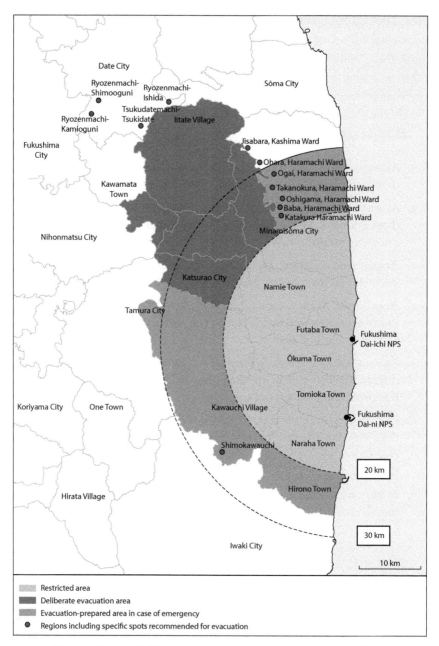

Date City

Ryozenmachi-
Shimooguni
Ryozenmachi-
Ishida

Sōma City

Ryozenmachi-
Kamioguni
Tsukudatemachi-
Tsukidate
Iitate Village

Fukushima
City

Jisabara, Kashima Ward

Ohara, Haramachi Ward
Ogai, Haramachi Ward

Kawamata
Town

Takanokura, Haramachi Ward
Oshigama, Haramachi Ward
Baba, Haramachi Ward
Katakura Haramachi Ward

Nihonmatsu City

Minamisōma City

Katsurao City

Namie Town

Tamura City

Futaba Town

Fukushima
Dai-ichi NPS

Ōkuma Town

Tomioka Town

Koriyama City
One Town
Kawauchi Village

Fukushima
Dai-ni NPS

Shimokawauchi
Naraha Town

20 km

Hirata Village

Hirono Town

30 km

Iwaki City

10 km

Restricted area
Deliberate evacuation area
Evacuation-prepared area in case of emergency
● Regions including specific spots recommended for evacuation

MAP 3. "Restricted Area, Deliberate Evacuation Area, Evacuation-Prepared Area in Case of Emergency and Regions Including Specific Spots Recommended for Evacuation," August 2011, taken from the Ministry of Economy, Trade, and Industry, https://web.archive.org /web/20211117055348/https://www.meti.go.jp/english/earthquake/nuclear/roadmap/pdf /evacuation_map_a.pdf. In March 2012, the state rearranged the areas into three zones. For detailed information regarding the evolution of evacuation orders, see https://web.archive.org /web/20220326043853/http://fukushimaontheglobe.com/the-earthquake-and-the-nuclear-accident/evacuation-orders-and-restricted-areas.

residence fell under the evacuation order cancellation preparation zone, for less contaminated spaces. In Minamisōma, where more than forty-five thousand residents lived in 2013, what mattered to most remaining residents was why they had to evacuate (their physical distance from 1F) rather than why some areas remained inaccessible (due to contamination levels).

In many locations within the city, I heard residents talking about their relationship to the accident by referring to physical distance. People would say, "I am thirty-one kilometers away, and . . ." or "This place is about twelve kilometers, so . . ." These narratives suggested that distance remained residents' principal reference point when for positioning themselves. They understood the TEPCO accident and experienced it differently according to the fixed physical distance from the damaged power plant. Importantly, while the state's April 2011 primary guidelines on the scope of nuclear damage extensively discuss the types of damage incurred as a result of the accident, it never defines "the victims," other than to suggest that the victims are those who qualified for compensation (Nuclear Energy Agency 2012, 89–101). Thus, within Minamisōma there were both victims and nonvictims based not on people's individual experiences of harm but on the compensation policy.

Hatsumi's friendship breakup indicated further ways that the compensation policy contributed to cementing distance as the accident's defining feature. Treating the accident not as an ecological issue but as geographically bounded one, the policy made real the projected divisions visualized in the various maps of the accident. According to this geographical framework, the harm done to people who lived thirty-one kilometers from 1F appeared to be qualitatively different from that done to people situated thirty kilometers away. Living inside one of the more highly compensated zones, Hatsumi was less critical of this somewhat arbitrary zoning, which someone like Oshima, who lived outside the zones, questioned.

Another Kashima resident offered a critique of the arbitrariness of the zones based on distance. "Before I decided to evacuate voluntarily," he said, "a barricade marking the thirty-kilometer radius from the nuclear plant was at the intersection over there." He pointed to the north side of his house. "When I came back about a month later, the checkpoint was moved to the bridge over there," he continued, pointing to the south side of his house. "Something happened during my short absence. Suddenly, my house was not included in the thirty-kilometer zone, and I was no longer qualified for what others could get."

According to a city official he talked to, he continued, the evacuation borderline moved because the national government changed the center of the accident from the outer boundary of the TEPCO nuclear power plant facility to the center of the four damaged reactors, between reactor units two and three. This small shift had significant effects on around 9,500 Kashima residents. Because his home was newly classified as being outside the thirty-kilometer zone, this resident lost about eleven additional months of compensation for psychological damage (100,000 yen per person) and his rights to other compensation that was provided to the "victims" inside the zone. The same individual continued angrily, "You see, I am harmed not only by the accident but also by those idiots who think drawing stupid lines on the map means nothing!"

Nonetheless, the state, TEPCO, media, and outsiders alike all cared about these "stupid lines" drawn on the maps and approached postfallout coastal Fukushima accordingly. For some Minamisōma residents, *higai* were not just the ecological distribution of radiological contaminants but also its representations, or how the maps and the compensation policy came to visualize the contours of the accident and contain radioactive spaces and residents (Morimoto 2015). How did someone like Naoko, whose preaccident residence fell within the twenty-kilometer zone, experience the drawing of the stupid lines?

## CHANGING HARM AND SHAPESHIFTING VICTIMHOOD

Naoko's family consisted of her grandson, Toru, and his parents. Naoko was the mother of Toru's father, Tengo. After moving in and out of the prefecture by staying at a local evacuation center, a relative's house, and a hotel, the family settled in temporary leased housing (*kariage jyutaku*) in Minamisōma, provided to them as a part of the nuclear damage compensation. Usually, evacuees preferred life in *kariage* since there was more privacy and structural security there than at the makeshift *kasetsu* units where most stayed. However, *kariage* residents were less visible, as disaster volunteers, relief organizations, various donations, and other types of support only went to *kasetsu* units, which for many outsiders were the iconic sign of the disaster and victimhood.

Legally speaking, the locations of evacuees' temporary residences were confidential, and thus not accessible to relief organizations. This legal barrier made it impossible to locate *kariage* evacuees. *Kasetsu* units, on the other hand, were visible from the outside like any housing project. Usually, each *kasetsu* had a community leader, and any outsider could talk to the leader to arrange an event like a giveaway, a research project, or sometimes a concert by a celebrity.[2] Naoko often felt that it was unfair that only those in temporary housing units received many items from relief organizations such as food, clothes, other commodities, and free trips outside the prefecture. Also, there was a transportation service that allowed *kasetsu* residents to travel to clinics, grocery stores, and other places where someone like Naoko could not easily access without paying for a taxi.

Naoko did not have a good impression of *kasetsu*. She told me in late 2013 that one time her friend, also an evacuee from Odaka who was living in one of the temporary housing units in Kashima, invited her to come to a giveaway event at a *kasetsu*. The experience was unpleasant for her; all Naoko saw was the *kasetsu* residents "being greedy." She said that they were trying to deceive the relief workers who were distributing goods by lying about the number of their family members in order to get more than they were allowed. "My friend told me that it was typical. She was like, 'We deserve it.' I did not think she was greedy like that, so I was disappointed."

Even though volunteer organizations did their best to distribute whatever items they had equally among the evacuees, they typically relied on the evacuees' self-reports. I experienced this when I helped one local aid organization distribute foodstuffs that they had received from an outside company in October 2013. As I arrived at a *kasetsu* community in Haramachi, many elderly residents came out to assess what we had brought. At least three different individuals told me to hand them two dozen eggs instead of one because their extended family members, staying at another unit, were away. At the same time, some others came back for a second time and pretended like they had not received something, claiming that the last time they had not been given, say, a bottle of mayonnaise, which I remembered handing to them. Surprisingly, many residents came out just to survey what we had brought from a distance and went back to their units unimpressed. Some even told me to bring something better, scolding me for my lack of imagination. "You think we are homeless?" one elderly man yelled at me.

As I struggled to be fair, the organizer, a Haramachi resident, told me to give them whatever they asked for because it was better than letting things spoil. "The problem is," he complained, "it is not like those people are poor. They get money from TEPCO. But people outside still send us things, like we just had the accident. People who truly need support cannot receive items because they are out working, or we do not know where they live," he sighed.[3] The compensation guidelines state that the harm itself is not definable. This meant that in the context of a prolonged disaster like TEPCO accident, who needed to be compensated and who deserved more than others—who the victims were—seemed to change over time. From the perspective of the victims, what support and items they needed and wanted also changed over time.

After seeing the "greedy" residents at one such giveaway event, Naoko felt discouraged from seeking free items. Naoko did not want to be like the people in *kasetsu,* who, according to her, had become used to receiving things for free. "The accident changed people," she commented. Expressing her discontent, Naoko said that she was an evacuee with some dignity, unlike some greedy people in *kasetsu.* Even though she could not access what the people at the *kasetsu* units could, she told me how lucky she was to stay at a house with her whole family, except for one of her grandsons. Naoko knew that when young people in a family evacuated from Minamisōma, the elders who were left behind had to stay in a *kasetsu* unit. The TEPCO accident forced families in the region apart, and, as both Hatsumi and Naoko observed, it changed people.

### DISCRIMINATION IN AND OUT

Even before the younger generation were spirited away after 3.11, they were already disappearing from Minamisōma; the TEPCO accident merely exacerbated this trend. Toru, Naoko's grandson, has an older brother who moved out of Minamisōma before 3.11, as many young adults in the area did after graduating from high school. One local described to me that young people tended to be *ryushutsusuru* (washed away) to outside places such as Fukushima, Koriyama, Sendai, Saitama, Ibaraki, and Tokyo. One reason for this was the lack of institutions of higher education, except for a technical college, in Minamisōma. Prior to the TEPCO accident, local youths wanted to work for TEPCO or its related companies (Kainuma 2011). Contributing to energy generation was what it meant to be successful in the region. After the accident, however, working for TEPCO lost its prestige. The accident harmed trust in TEPCO and the company's reputation in the region, causing the loss of an attractive industry that had prevented young laborers from disappearing from coastal Fukushima.

Toru remained in Minamisōma. By 2013 Toru had graduated high school and started working in Minamisōma at a convenience store. Like some of his friends, he considered working as a decontamination worker or on other reconstruction-related jobs, but his parents convinced him not to out of concern that his health might be jeopardized. In the winter of 2012 he had left the city for a short period, but he returned in less than a year, as he found living outside Minamisōma difficult, especially after 3.11. Luckily, Toru did not experience direct bullying and discrimination outside the city, like many other evacuees and residents did.

One teenager from Kashima, a tsunami survivor, shared with me his struggles at college in Hokkaidō in 2015. He recalled that his professors often teased him about coming from Fukushima and said that no one should be living there because of the contamination. He did not know what to say because he lived outside the evacuation zone, and the harm he had experienced was mainly the result of the tsunami, which no outsider seemed to remember or acknowledge. It was shocking and hurtful to him to learn how Minamisōma was seen by outsiders. In the postfallout period, evacuees who left Fukushima commonly felt compelled to hide or lie about where they were from, as Hatsumi did when she joined tours abroad. Discrimination was another harm, though less visible than the presence of radiation, that affected residents of Minamisōma when they left the area, but discrimination occurred within the city as well.[4]

Unlike what residents like Oshima imagined, all of Naoko's family members, except Naoko herself, worked during the daytime. This put the family in sharp contrast to the widespread stereotype of the region's nuclear evacuees. According to local rumors, the evacuees had become so used to living off state compensation and subsidies that many of them had stopped working. Indeed, some people did stop working and lived off compensation. However, some did so because the company they had worked for had gone bankrupt, they were not allowed to farm anymore because of the accident, or they had become too old and were not qualified to take any of the newly available jobs, like the part-time job at the restaurant that required applicants to be younger than sixty-five.

Nonetheless, in Kashima and further north in Sōma city, evacuees became marked as unproductive. Sachiko, whom we met in chapter 1, shared her observation that the evacuees from Odaka habitually went to pachinko parlors to gamble. When I asked her how she knew those pachinko players were evacuees, she responded confidently that there were more pachinko parlors along Route 6 since 3.11, and "they have all new expensive cars that they bought with their compensation money!" One evacuee from near 1F in his mid-thirties once complained that people went to pachinko parlors even before the accident since it was one of the few entertainments available in the region. These rumors spread, the same evacuee guessed, because of the economic disparities that the postfallout compensation policy had produced. Evacuees had to live in Minamisōma, where there were other residents, like Oshima and Sachiko, who did not receive compensation and could not afford things. Just as Sachiko questioned my presence, people were watching closely what unfamiliar faces in the district were doing and what new items they purchased. Toru once expressed his desire to move away from Minamisōma because "even though it is nice that you know many people, this place could get suffocating."

Toru's father, Tengo, once described his experience of living in Kashima like this: "I hate being looked at by people here [in Kashima] like we are criminals. I feel like the Kashima residents are watching us because we are *yosomono* [outsiders] and have been treated differently. The compensation policy is public knowledge, and they know we get more than they do. Sure, some people gamble all their money they get and are living a terrible life. But think about how meaningless one's life would become when suddenly what the person was always doing [i.e., farming] can no longer be done freely, let alone you cannot live in your own house."

In the earlier phase of evacuation, Tengo rejected any form of compensation. He kept ignoring the solicitations from the state and TEPCO to file a claim, and he also refused to go to consultation events that the local government hosted for the evacuees in the city. Tengo said he did this as his protest against the public imagination that Odaka residents depended on compensation for living. For Tengo, rejecting compensation maintained his dignity: "I work and I can live without any compensation!" Regardless of the state's categorization, both the "nonvictim," Oshima from Kashima, and the official victim, Tengo, problematized the postfallout possibility of life without labor.

Tengo preferred to talk about his postfallout experience only with his old neighbors and friends from Odaka. Some of his close friends left the city and decided not to come back, while others who stayed struggled, and a few suffered from alcoholism. The friends who left, whom Tengo sometimes envied for starting a new life, gradually lost touch with the city's changing social and political landscapes. The news outside the city, Tengo learned from his friends, focused more on contamination and less on the changes to the city.

Tengo told his friends more about the changing postfallout local landscape and less about contamination.

In his new residence in Kashima, little things bothered Tengo, such as how close he lived to others, unlike at his old home in a rural part of Odaka. He was used to having distance from the next-door neighbor at his farm-house. "I never had to worry about noises before," he said, "but now I feel like I have to. I do not want the neighbors to think Odaka people are trouble-some." Noise can be polluting. For Tengo, his work provided not only a way to maintain his sense of dignity but also an escape from his uncomfortable new life in Kashima. Before 3.11, Tengo worked for a company in Odaka just a few minutes away from his home. After the accident, the company relocated to Miyagi Prefecture. For many companies, it was important that their business avoid having Fukushima as their physical address. When his family returned to Minamisōma after a month of evacuation, Tengo went back to work for the same company. This meant that he had to spend more than two hours commuting each workday. Nonetheless, he wanted to continue work-ing, not depend on the compensation, and to be away from Kashima. He did not mind the drive, and he had other concerns. According to Tengo, "I worry about my mother, who always did farming and growing many vegetables and fruits in the garden back in Odaka. Now she cannot do much of anything and just stays inside the house day and night. It could make her go senile [*bokeru*]."

### A SMALL GARDEN IN KASHIMA

Inside the house was where Naoko always remained in 2013. She always had the TV on, as it made her feel like she had company. I often drove by her temporary residence by the former Route 6 to say hello. Naoko was usually there by herself and always welcomed company. Sometimes I saw her chatting with a mail delivery person. When I asked her what she would usually do during the day, she told me how she was busy with things like doing laundry, taking out the trash, preparing pickled vegetables, and making dinner. "I would wake up around five, eat around seven, and then at noon. After that, I also need to think about doing *omakane*." "What is *omakane*?" I asked, and she laughed. "You do not know *omakane*? That's when you cook for *bange*!" And I asked her again what *bange* (dinner) meant, and we both laughed. We would chat about mundane things like that.

As much as she enjoyed chatting with me, Naoko was very hesitant to talk about her family's situation. It was clear that she did have opinions, but she was reluctant to express them since she wanted to avoid causing tension in the house. "Look at all these documents with technical things I do not understand!" she said while pointing at a pile of documents from the national government, the city office, and TEPCO regarding compensation and other matters stacked messily in a corner of the living room. "I let young people take care and deal with things, and I do not say anything. I mean, look at those documents! I do not understand a thing they say. To be honest, I do not know how much longer I will live, so it does not matter. But I do not want to die here. Something is not right here. I miss how things were before the accident and doing what I used to do at my own house. Don't you think it is still possible?"

The precariousness of life was a topic that loomed over Minamisōma. Naoko's sense of precarity came from the uncertainty of the possibility or timing of her return to life as she had once lived it. Struggling to go up and down the stairs to the second floor of the *kariage* house (her previous residence was one story), she felt her body weakening each day. While keeping busy doing housework, she felt that she was turning the last corner of her life, and she did not want to her life to end in Kashima.

Naoko was not much concerned with the presence of radiation in her living environment; she was too old to worry about its potential risk, she thought. However, she could not use the same logic for young people like Toru. Unlike Naoko, who was approaching eighty in 2013, Toru had just turned twenty. Toru had much longer to live with the uncertain effects of low-dose radiation. Naoko was keenly aware that age mattered in the calculation of radiation risks. Similarly, the age difference between Naoko's elderly friend at the clinic and her son was the reason the friend had been accused by her son of being oblivious to the contamination in Minamisōma.

This generational difference created tension among many multigenerational households in coastal Fukushima. Even if Naoko desired to go back to Odaka as soon as possible, it did not mean that the family would want Toru to go back there as well until their property was properly decontaminated. Going back without Toru was not what Naoko wanted; she wanted to live with her family in Odaka and not by herself. The family had evacuated together and moved in to a *kariage* together. If moving back to their original residence required the family to separate, it made no sense to Naoko. Nevertheless, the state was not helpful in solving her personal struggle, as it did not provide any information about the date that Odaka district might reopen until late 2015.

Evacuees like Naoko thought that their evacuation in March 2011 would be a momentary escape from the potential threat caused by the nuclear explosions. However, it turned out to be the start of a multiyear displacement, which did not end until July 2016 for some and later for many others, including Naoko's family.[5] "As far as I am concerned, I would have felt so much better," Naoko moaned, "if we were told early on that we would not be able to go back. Then we [the family] could have thought about our future more realistically. Those people in Futaba district [south of Odaka] are much clearer about their future because they know that they will not be able to go back for a long time or never, so they can start a new life. I feel bad for poor young people being forced to decide."

Instead, in Naoko's case, the state had communicated that evacuees could go back once the radiation levels went down. To enable their return as soon as possible, the state planned to take necessary measures (decontamination) by remediating the contaminated environment. In doing so, the state emphasized that it was on the side of the residents, honoring evacuees' willingness to return as well as their health and well-being. However, the state and the experts failed to communicate to the residents what it meant for once-delimited areas to be "livable" again or for whom a specific location was too "radioactive." It was left to each resident to make this judgment individually, and the only information the state gave to residents was a technoscientific one. The state would do their best to keep radiation levels below the 1 mSv per year standard recommended by the International Commission on Radiological Protection (ICRP).[6] Only one thing was clear: the opening of the district meant the end of compensation but not of contamination. The source of the contaminants, 1F, will not be fully contained for a long time, probably not until Toru reaches Naoko's age.

Naoko wanted to believe the state and TEPCO, where people she knew worked. She wanted to believe them for her own sake. For her own well-being, she desperately needed to return to her house, to farm the family's food, and to make miso as she always had, yet what she saw on TV made her doubt both the authorities and her own decision. Watching TV at home alone, Naoko often saw many so-called experts and commentators criticize the 1 mSv standard as inhumane and claim that Fukushima was unrecoverable. TV also exposed Naoko to a view held by outsiders that it was a citizen's right to live without any radiological contamination at all.[7] Despite the confusion, Naoko could not consult anyone, including her family, because everyone had different opinions about radiation. For Naoko, radiation mattered not because she

FIGURE 2. Tengo preparing a garden at his temporary residence in Kashima in 2015. Photograph by the author.

could be exposed to it but because it could break up her family unit, which she could not risk. Alone at her *kariage* residence, all she could do was purse her lips tight, remain silent, and wait for the state's decision in her favor.

In 2015 I helped Naoko and Tengo transform a small segment of their *kariage* housing's front yard into a garden plot (figure 2). Using a machine that he brought from storage at the Odaka residence, Tengo dug into the hard, infertile ground filled with small rocks. Naoko and I carefully removed the rocks before adding new soil and fertilizer. As we gathered three buckets full of rocks, Tengo used another machine to mix the soil. We put more potash fertilizer in the mix than usual. We had learned from other farmers in the region that potassium-rich soil could prevent vegetables from absorbing cesium and becoming contaminated (Nemoto 2017). The small garden, which was probably one-thousandth the size of the one Naoko had worked previously, became a new place for her to grow vegetables like snap peas, cherry tomatoes, Japanese eggplant, perilla, red chile peppers, and cucumbers. While the tiny garden enabled Naoko to be more active at her *kariage* housing, it also signaled the family's acceptance that their displacement would continue.

### IS A HARMFUL RUMOR AN ACTUAL HARM?

In Shinchi town, Sōma district, about fifty kilometers north of 1F and thirty-eight kilometers north of his original residence, Hiroshi Miura, an evacuee from the Odaka district, began growing rice with his son in 2012. Downscaling from his previous farming operation, which consisted of 6,000 square meters of rice fields, 500 square meters of vegetable gardens, and three hundred chickens, Miura secured around 3,500 square meters in Shinchi town for his son. Although his son had initially evacuated to an area near Tokyo, he was training to be a farmer, and he expressed an interest in farming only in Fukushima. According to state-sanctioned zoning and technoscientific monitoring, fifty kilometers was far enough away from 1F to safely farm. However, farming anywhere in the region was highly challenging for two reasons. First, evacuated farmers had to find new locations to farm and start almost from scratch. Second, Fukushima had acquired a negative reputation because of the TEPCO accident, and consumers and traders avoided foodstuffs produced there. Miura moved as far away as Shinchi town to avoid being associated with the name "Sōma," since, he said, "the name 'Sōma' became well known, especially after the mayor's SOS YouTube video became famous." The market was cruel.

Since 2012, to improve the transparency of contamination in food, all the rice grown in Fukushima has had to be tested for radioactivity. For rice and other general foodstuffs such as vegetables, fruit, and meat, the state set a

TABLE 1 Comparison of international regulations for radioactive materials in foods

| Radionuclide | Japan | Codex Alimentarius Commission | European Union | United States |
|---|---|---|---|---|
| Radioactive cesium (Bq/kg) | Milk 50 Infant foods 50 General foods 100 | Infant foods 1,000 General foods 1,000 | Milk 1,000 Infant foods 400 General foods 1,250 | All foods 1,200 |
| Upper limits for additional doses | 1mSv | 1mSv | 1mSv | 5mSv |
| Assumed percentage of foods containing radioactive materials | 50% | 10% | 10% | 30% |

SOURCE: Modified "Food and Radiation Q&A" published by Consumer Affairs Agency.

NOTES: The Codex Alimentarius Commission is an intergovernmental body created in 1963 by the Food and Agriculture Organization (FAO) of the United Nations and the World Health Organization (WHO) for the purpose of protecting consumers' health and ensuring fair-trade practices in the food trade. The commission establishes international standards for foods. Standard limits incorporate effects of the amount of food intake and assumed percentages of foods containing radioactive materials. Therefore, the values are not suitable for inter-comparison. Indicated standard limits for drinking water are the WHO guidance levels of radioactive materials, which are referred to in respective countries, and standard limits for radioactive materials vary by country due to differences in adopted preconditions. Therefore, the values are not suitable for inter-comparison.

limit of 100 becquerels (Bq) per kilogram for cesium-134 and cesium-137 combined. Unlike after the Chornobyl accident, the Japanese state responded quickly by stopping the circulation of potentially contaminated foodstuffs and by lowering the allowable levels of contamination in food in the market to 500 Bq per kilogram until March 2012 and then to 100 Bq/kg (see table 1).[8] Those newly set limits were much stricter than the internationally recognized Codex Alimentarius Commission guidelines for contaminated food, which is set at 1,000 Bq/kg for cesium, and the U.S. standard for all foods, which is 1,200 Bq/kg for cesium.[9]

*Becquerel* (Bq), like *sievert* (Sv) and *cesium*, is a technoscientific term that the TEPCO accident introduced to the lay public. Unlike the sievert, which measures the health effect of ionizing radiation on the human body, the becquerel is a unit of radioactivity (1 Bq is equivalent to one nucleus decaying per second). The higher the quantity of becquerels, the more active a given contaminant is.[10] In addition to the presence of radiation in their everyday lives, residents had to deal with this type of technical language, or half-life politics, with which the state and experts explained the degrees of contamination and constructed postfallout policies. Thus, not unlike survivors of the Bhopal disaster whom anthropologist Veena Das has observed (2000), the state forced coastal Fukushima residents to translate their individual experiences of suffering into the quantified register of the physical and life sciences.

This scientification of the accident and the hegemonic radiation-centered approach alienated people like Naoko. She often complained about how the unfamiliar terms, like becquerel and sievert, did not mean anything to her. A vegetable with 150 Bq/kg did not have any reality for her. She avoided farming before 2015 not out of concern about radioactivity and scientific measurements but because people around her told her not to farm. While contamination was a chemical and objective phenomenon, Naoko experienced it as a psychosocial problem. Her relationship to the land, food, and other people determined the desirability or the lack thereof of a particular food.

Vernacularly, locals used different categories than becquerels and sieverts to make sense of their relationship to their region. They called unsafe places or edibles "*takai*" (high radiation level) and safer ones "*hikui*" (low radiation level). For certain hot spots around the residential areas (e.g., gutters, ditches, and water holes) and near the mountainside, they used the word "thick," or *koi*. *Koi* was a term that expressed the locals' understanding of some predictable patterns in the distribution of contaminants in their lived environment.

Calling a general area or foodstuff *koi* indicated that radioactive materials tend to concentrate in specific areas or foodstuffs but not others. For example, a resident would say that mushrooms were *koi* because they knew mushrooms tend to grow in highly radioactive areas and are more likely to absorb contaminants from their surroundings than rice. In its use, *koi* was like the term *hot spot,* but unlike hot spot, it carried affective weight. Put differently, the technoscientific language of becquerels did not adequately capture how each resident felt about particular areas or foodstuffs. Nor did it mean that the same measurements of different items carried the same affective weights, since they felt that some contaminants were "thicker," and therefore creepier and more chilling, than others.

It was not just the consumers; local farmers themselves wanted to know if it was safe to consume foodstuffs they had produced prior to the accident and if they could continue farming after it. Miura, one of the first to host a testing site for the public, described the food monitoring practice in 2013, saying, "The testing is mostly for local farmers because no one outside the area would really buy their rice. That's how badly Fukushima is imagined to be contaminated by the rest of the world." Yet non-Fukushima residents were not the only ones to think this way. A restaurant owner in a nearby area confessed that even though she used the tested Fukushima products for the restaurant at times to cut costs (the accident significantly lowered the value of Fukushima products), she bought non-Fukushima products for her home. She told me that she was afraid of consuming potentially radioactive foods and the numbers were not enough to make her feel secure. "I am okay, but for my kids I want to be careful because I probably only have a few more decades ahead of me, but my kids will live much longer. What if they get exposed to *koi* contaminants?" she said.

This concern for self and family spoke to an elevated sensitivity among the residents about food and radiation in general. The consumers' concern underlined that rice, vegetables, fruit, or anything grown in Fukushima would not sell outside Fukushima or even *inside* it. The fear and hesitancy were bolstered by news reports on some test results—though extremely rare even in 2013—that surpassed the newly set limits. The question became, could a more robust testing regime and stricter standards enable Fukushima farmers to overcome the negative reputation? Miura's answer was no.

At Miura's agricultural cooperative in Shinchi town, which I first visited in 2013, every thirty-kilogram bag of rice went through a machine that measured the number of becquerels per kilogram. After one bag passed through the machine, the computer connected to the machine read "9 Bq/kg," which was much lower than the limit of 100 Bq/kg. "Twenty Bq/kg is the highest amount we allow for a thirty-kilogram bag of rice," Miura's son, who was operating the machine, told me while attaching a sticker that guaranteed the safety of the rice in the bag. A barcode was printed on each of these stickers, promising the transparency of the testing process. This practice, started only in 2012, allowed anyone to trace the source of an individual bag of rice down to the location of production and testing by scanning the barcode.[11]

Setting a significantly more stringent standard than the country required was Miura's attempt at countering Fukushima's negative reputation in the media and consumers' minds. Despite his efforts, however, most of Fukushima foodstuffs did not, in fact, sell well after 2011, and even if they were sold, they sold for a much lower price than before (Igarashi 2018; Kikuchi 2013).[12] "In reality," said Miura's son as he measured bags of rice, "Fukushima products are the most secure [*anzen*] of all from the perspective of monitoring and testing. No other place in Japan has been testing as rigorously as we do here." Pointing out the disadvantage that the TEPCO accident imposed on Fukushima farmers, he continued, "they [non-Fukushima farmers] might be afraid to test and find out that their produce is more contaminated than Fukushima's, but they are not required to test. What I have learned from testing countless bags of rice is that the name of Fukushima has been contaminated, perhaps more so than the land itself."

"The name of Fukushima has been contaminated." Such is the extent of the damage done by harmful rumors (*fūhyō higai*). Here is how the Secondary Guidelines on Determination of the Scope of Nuclear Damage, by the Dispute Reconciliation Committee for Nuclear Damage Compensation, define a "so-called *fūhyō higai*":

> Although there is no established definition of so-called "rumour-related" damage, in these Guidelines "rumour-related" damage refers to concern about the risk of contamination with radioactive material in relation to products or services, due to facts that are widely known through media reports, leading consumers or trading partners to refrain from purchasing the product or service, or stop trading in the service or product, resulting in damage. "Rumour-related" damage is eligible for compensation if there is a sufficient causal relationship to the accident. The general criterion for this is as follows: when a consumer or trading partner is concerned about the risk of contamination with radioactive material resulting from the accident in relation to a product or service, and their psychological state of wanting to avoid the product or service is reasonable from the perspective of an average, ordinary person. (Nuclear Energy Agency 2012, 110–11)[13]

According to sociologist Naoya Sekiya (2011, 2016), unlike other types of harm, *fūhyō higai* is considered indirect, reputational damage that results in quantifiable economic losses. The first incidence of *fūhyō* damage is traceable to the 1954 Lucky Dragon Incident, when U.S. testing of a hydrogen bomb in the Marshall Islands exposed a Japanese fishing boat and contaminated tuna, which later circulated in the market. The event caused a national panic among consumers and was featured prominently in the media. Consumers' avoidance extended beyond tuna to other types of fish, causing significant economic losses. This incident led to the development of the legal concept of indirect damage (*kansetsu songai*) in 1956 and was later applied to a few other incidents of environmental contamination.

The state considers damage concerns to be *fūhyō* when consumers and traders alike continue to believe that something is contaminated even though the extent of the contamination is scientifically proven otherwise. While this technoscientific framing of *fūhyō higai* acknowledges the reasonableness of people's avoidance of foodstuffs near the site of a nuclear accident or other environmental pollution, it fails to acknowledge the following two interrelated social factors. First, the state defines what is reasonable only narrowly, in terms of scientific evidence. Thus, more risk-avoidant people, such as mothers of young children, are vulnerable to being perceived as irrational or neurotic (Kimura 2016). According to the state definition, until 2012 people's avoidance of Fukushima rice was based on a rational choice, since no data was available. Here, *fūhyō higai* was based on the fear of the unknown, stemming from the lack of data. However, in 2013 Fukushima Prefecture tested 11,006,534 bags of rice, and of those, only 28 bags (0.0003 percent) did not pass the newly set 100 Bq/kg standard. Given the data, the state's definition of *fūhyō higai* reframed the consumer-producer relationship. With the presence of data, *fūhyō higai* results from consumers' irrational avoidance of Fukushima rice based on "rumors."

Second, technoscientifically quantifiable magnitudes of safety (*anzen*) do not equate to consumers' subjective feeling of safety (*anshin*). In his ethnography of a grassroots food-monitoring organization in Tokyo, Sternsdorff-Cisterna (2015, 2020) observes that the food contamination issue in the aftermaths of the TEPCO accident was a symptom of a lack of trust in the state and the experts. Ordinary citizens wanted to test foodstuffs themselves to confirm the security of food promised by the state, but, more importantly, they wanted to regain their lost sense of trust in the food's safety. Observing this widespread trend in postfallout Japan, Sternsdorff-Cisterna writes,

"Safety is more than laboratory tests; safety is also a social relationship. It can only exist insofar as people trust that the products they are eating are indeed safe" (2015, 463). Participation in grassroots food testing enabled concerned citizens to share their anxiety and also led to increased scientific literacy.

While data transparency is critical to combatting the so-called *fūhyō higai* and helped to alleviate the general avoidance of Fukushima food (Kunii et al. 2018; Sekiya 2016), the lasting effects of the TEPCO accident revealed consumers' diminishing interest in Fukushima. When their interest level is low, media coverage decisively impacts consumers' and traders' impressions of Fukushima. For example, Yasumoto and Sekiya (2018) found that non-Fukushima people tended to interpret any news stories about Fukushima evacuees in their region as an indication of the lack of progress in Fukushima. Media representations of the presence of displaced people sustained the impression that Fukushima was still dangerous. What further complicated Fukushima's reputation was that it was not difficult for consumers to find the same items produced elsewhere in the globalized, neoliberal market. Miura's son commented with an ironic tone that "maybe people had no previous image of Fukushima. There is no major reputation like how many people think of Niigata and rice. Radiation came to fill the empty reputation of Fukushima, and there has not been any alternative to overcome it."

Observing the messy politics surrounding food safety, Miura expressed his concerns in Izumi Katayama's 2015 book *Fukushima Rice Is Safe [Anzen], but I Am Fine If You Do Not Want to Eat It*. The title of the book captures Miura's stance toward the social disintegration produced by the TEPCO accident. Fighting at the frontline of *fūhyō higai,* Miura warned that the conflict over what is safe or unsafe only served the state and TEPCO, since it helped them to displace the issue from their lack of accountability to the tension between consumers and producers. While Miura acknowledged that the more testing farmers did, the more frustrated they became about consumers' avoidance, he claimed that the real issue lay elsewhere. Neither consumers nor Fukushima farmers were the problem; the TEPCO accident was.

Decoupling consumers' right to choose the foods they ingest from the economic losses among Fukushima framers, Miura asked producers and consumers alike to remember how the TEPCO accident had disintegrated the social fabric of trust, without which the food industry is not possible. Miura argues that *fūhyō higai* is actual harm (*jitsugai*) for which the state and TEPCO need to compensate farmers, as much as the public needs to educate themselves about the science of radiation and the current state of Fukushima.

Nonetheless, many residents, whether they ate or avoided local foods, experienced psychological challenges related to radiological issues in their everyday lives.


## A NEW SPECIES OF HARM

In March 2011, the sudden awareness of radiation's ubiquitous presence in Fukushima made everything seem strange to the residents, but it was not the end of the story. Saeki, from Kashima, lived in a supposedly safe area in Minamisōma. A mother of a teenager, she told me in 2013 that she had made a point to avoid reading anything that might remind her of the potential adverse effects of chronic low-dose radiation exposure. Mulling it over, she said, would not help with the postfallout conditions of her life and would only produce more anxiety.

"It is not that I do not care," she explained to me, "but I know that I must live with radiation, [and so] I want to put it in the background of my life. Radiological contamination is scary and can change everything, like how I experience the familiar land, food, and even myself. It is just everywhere once you become aware of it. When I see a flower and think it is beautiful, I do not want to believe that it might be contaminated and dangerous. By trying not to think about what I cannot sense and trying to trust what I see and feel as they are, I can at least mentally distance myself from the contamination and not be reminded of its presence all the time. Left alone, I worry about what could happen to my kids ten to twenty years from now. But it is not like I get compensated for my anxiety like others."

In her comment, Saeki hinted at the emerging inequality in postfallout coastal Fukushima. For example, in some parts of the Odaka district—especially its coastal region, where her relatives lived—the average radiation level was no different from that at her residence in Kashima. This fact confused Saeki. She often asked herself about the nature of the TEPCO accident. If who can get compensated had little to do with the actual radiation levels they were exposed to, she asked, "who and what is the state compensating for?" She initially thought that maybe the state considered those who evacuated to be the victims of the TEPCO accident, yet over time this seemed unlikely, since Saeki and her family had also evacuated for a month, like many others, but she did not get compensated like the others. Even though the city had asked Kashima residents to evacuate, the state considered her family's evacuation to be voluntary.

Scholars such as Beck (1987), Lifton and Falk (1982), and Masco (2006) discuss how a radiological danger leads people to experience a sudden disorientation of their senses and their relationship to the world. Kashima residents like Saeki, however, experienced the TEPCO accident not just as a threat to her senses. More than anything, the accident gradually manifested as a logical and moral flaw in the state policy. After finding out that her family was not qualified for the same compensation as other residents in the city, she remembered feeling a chill; she realized that, for the state, she was disposable. Saeki said that the feeling of disposability was much more dreadful than the fear of the invisible radiation. Poet and Minamisōma resident Jotaro Wakamatsu (2012) has described the coastal Fukushima residents as *kimin* (abandoned people).

While defining the nature of the TEPCO accident's harm ambiguously as "anything determined to be reasonably correlated" (*sōtou inga kankei no aru songai*) to the accident, the state and TEPCO's radiation-centered half-life politics had underplayed the ongoing nature of the TEPCO accident and the residents' changing relationship to it. After observing the Three Mile Island nuclear accident in 1979, Erikson (1991, 34) discussed the plotless nature of a nuclear accident and its lingering danger. He writes, "Radiological emergencies violate all the rules of plot—some, but not all, have clearly defined beginnings, but to the victims, they never end; the 'all clear' never sounds." In a context in which there was no clearly defined victimhood, Fukushima residents began reasserting their "ordinary," despite the ongoing Nuclear Emergency Situation first declared on March 11, 2011. For them, the TEPCO accident was not just a never-ending event; it was an ongoing catastrophe whose harm was ever changing.

As we have witnessed thus far, by 2013 the TEPCO accident was no longer merely a radioactive danger to individual bodies. Its damage had become increasingly psychosocial, including a routinized sense of uncertainty of radiation exposure; increasing mistrust of the state authorities, the experts, TEPCO, and one's own senses and judgment; social disintegrations; and secondary health issues stemming from long-term evacuation (Tsubokura 2018).

The state's intention to use nuclear compensation to "remedy" the victims and TEPCO's responsibility to do so failed to acknowledge experiences that were not quantifiable. As Minamisōma residents' experiences attest, the harm of the TEPCO accident went beyond radiation and its imagined, known, and represented threats to individuals and the environment. Here, the logic of correlation is a limited framework for approaching a nuclear accident, since such reasoning implies that with the passing of time, a link between the accident to a phenomenon becomes looser, less evident, or confounded with other variables. If, as Saeki worried, her children became sick in ten to twenty years, it would be hard to prove the connection between a sickness in 2031 and the accident in 2011.

Furthermore, the TEPCO accident was not an event but a process. At different stages of postfallout, different types of harm have emerged. Individual residents might not have been exposed to radiation beyond the ICRP recommended limits (Tsubokura 2016), but their communities were disintegrated, families were split up, and many future possibilities were curtailed.[14] It is clear from the residents' experiences that nuclear compensation itself came to be harmful to the community. As the residents experienced, the TEPCO disaster in the first decade was about a series of (sur)real social discriminations.

To capture the boundless, lasting, and messy impacts of a nuclear accident, Erikson (1994) describes a nuclear accident as "a new species of trouble" that is qualitatively different from a natural disaster. Although a nuclear accident is far from being "new" in the twenty-first century, the undefined victimhood and unnamed postfallout harm percolating in Minamisōma provided a fertile ground for a new species of local politics and economy of contamination to emerge: the compensation game.

_____

# The Compensation Game

Our responsibility begins with imagination.

HARUKI MURAKAMI, *Kafka on the Shore*

## *KATARINIKUSA* AND *IKINIKUSA*

Imagine you are a resident of Minamisōma. Depending on which part of the city you are situated in, your experience of the TEPCO accident, low-dose radiation, and ecological contamination would be very different. This is not merely because, empirically speaking, radioactive debris traveled and fell haphazardly due to the weather patterns and variations in the local terrestrial and built landscape or the fact that different individuals have different tolerances to radiation, both biophysiologically and psychologically.[1]

Following the accident, you may decide to remain in the city since you imagine yourself situated outside the state-sanctioned evacuation zones like Saeki, Oshima, Yamada, and Sachiko and you try your best not to acknowledge the presence of radiation in your vicinity, or you might move out of the city as a voluntary evacuee to maintain physical distance because you have a child and are concerned about potential radiological exposure to yourself and your family. If so, you might feel alienated in an unfamiliar place and struggle with the lack of support from the national and local governments. At the same time, unimaginative people might accuse you of being too sensitive, paranoid, and neurotic, as they did the risk-avoiding mothers who were labeled as "irrational" (Kimura 2016; Morioka 2014).

You may see yourself like Hatsumi, who lived in the residential part of the Haramachi district, where the average level of airborne radiation was relatively low but where you would still receive some compensation. Living there, you might wonder about what the residents within the twenty-kilometer zone might be qualified to receive or feel guilty that you might receive something other residents in the city could not. But remember that you would be close to the still-unstable 1F, and every time there is an earthquake or a typhoon, you might wonder if you will need to evacuate yet again. This newly acquired risk awareness may make you fill up your car's gas tank rather obsessively before the gas gauge reaches half empty. You might not want to repeat the trauma of the gas shortage in March 2011, which made it difficult for many residents to evacuate promptly. Nonetheless, you might feel that various aspects of your life in Haramachi appear normal: schools are open; people are out shopping; children, although few, are playing at a park with or without masks; and almost no one carries a Geiger counter or dosimeter.

Alternatively, you might feel closer to evacuees like Naoko and Miura and imagine yourself having lost access to your residence in Odaka. If so, you might imagine yourself either remaining within Minamisōma or moving to a nearby city to a *kasetsu* unit, where you would be uncomfortable with your new neighbors' gaze, or you might imagine leaving the city altogether to live elsewhere and hide your identity as an evacuee from Fukushima. If you were far away, it would be harder for you to make sure that your evacuated residence was not overcome by wildlife or vegetation. Moreover, you might not receive much information about the city, although you would still need to constantly engage with the state and TEPCO to work out your compensation. The local news would not be local to you if you left the area. You would see no reporting of real-time radiation readings nor any news about the city's recovery. All you would see and hear would be negative representations of Fukushima. People around you might speak differently than the familiar dialect spoken at home by your elders. You might want to change your license plate immediately so that people wouldn't identify you as an evacuee.

Prominent scholar of the TEPCO accident Hiroshi Kainuma (2015) describes postfallout Fukushima as "difficult to tell" (*katarinikui*). He argues that there is no one truth, one reality, one narrative that can capture everything that has happened since March 2011. Some people left Fukushima and never came back; around 278 children and young adults have been diagnosed with thyroid cancer as of June 2022; and others died from the tsunami or the long-term evacuation. Simultaneously, many have struggled with discrimination, alcoholism, mental health issues, and secondary health problems. By 2021, 118 people in Fukushima had died by suicide because of loss of hope or financial and other struggles.[2] Meanwhile, others profited from the accident.

Despite all the harm and struggle, residents like Hatsumi, Naoko, Saeki, Sachiko, Yamada, Oshima, and Miura remained in Minamisōma and confronted the social disintegration that various evacuation orders had produced, which were later cemented by the nuclear compensation. Depending on whose perspective you take, favor, or ignore, different images of Fukushima emerge. If you are sympathetic to risk-avoidant individuals, you might position the TEPCO accident as the beacon for the rise of anti–nuclear-energy movements.[3] Meanwhile, others might side with the residents who remained and instead problematize the compensation scheme, while others might approach the accident scientifically and believe that the public's lack of scientific literacy has exacerbated the accident's aftermaths. There is no right or wrong position, and variously situated individuals inside and outside of Fukushima subscribe to different positions. As a result, there are tensions and contentions. Remember Naoko who took the option of being silent to survive.

Postfallout Fukushima is not just *katarinikui*. For residents, it has also been difficult to live (*ikinikui*). The difficulty has stemmed not only from how they have lived with radiation but also from how this invisible phenomenon came to be rendered selectively (in)visible in their daily lives. As we approach how people lived in postfallout Fukushima, what the residents desire from us is not "intolerance, theories cut off from reality, empty terminology, usurped ideals, inflexible systems" (Murakami 2005, 168), which they feel that the state and the public alike have imposed on them. Instead, they desire our ability to imagine and take the chance to experience something we do not already know or cannot search on the internet. An act of imagining the invisible, which I am calling *en,* allows exploration of different lines— whether it is a line connecting points or rays of radiation cutting across a space, place, or body—perspectives, choices, and livelihoods of individuals yet to be visualized. Journalist Satoru Ishido (2021) calls such an act "walking on the invisible lines" of 3.11. What is (in)visible in coastal Fukushima is up to your imagination. One thing I know is that the nuclear ghost has a message, and it requires a willing receiver of its message like you.

Now, let's imagine being in postfallout coastal Fukushima. As one of Minamisōma's residents, regardless of what you think about radiation exposure, what you see and hear in the media and social media, or what you learn about how the state and TEPCO have been handling the situation, one thing would become clear to you: the logic behind the compensation policy is the political "acknowledgment" of where you lived on the map and how different lines were drawn and redrawn. No lines would be added; they would only be removed. Each removal of a line signals a few things. It could mean one more step toward normalization, or the ending of compensation. In this sense, the accident might appear to you as an event that occurred in the past, and contamination might also seem like an issue of the past. You might look back at the visualizations of Fukushima on the map in chapter 3 (map 3) and think they represented both its contamination and its containment.

You would notice that compensation does not account for the sustained presence or ecological movements of contaminants within the region, nor for individual anxiety and changed life circumstances. Radiation still looms in the areas where the state lifted its orders as well as outside those zones. If anything, the accident led to the realization that you are always already exposed. As one Kashima resident stated, there is no invisible wall preventing particles from traveling. In a city with more than forty-three thousand residents in 2013, any worries about radiation you might still have might be considered excessive, whether in public or even within your own household. You would no longer know or trust where your friends, coworkers, neighbors, and family members stand regarding radiation. You might choose to remain silent like Naoko.

With your deliberate silence and that of others, radiation slowly but gradually disappears socially in Minamisōma. Remember Oshima and his interpretation of Prime Minister Abe's announcement to the world in September 2013 that the situation in Fukushima was "under control." Implied in the official statement was the state's understanding that after March 2011 no new contaminants would be added to the surrounding environment. The series of maps of coastal Fukushima with decreasing color intensity made this fact visible. Unofficially, however, it could mean, as Oshima imagined, the pathetic state of affairs in Minamisōma, where the residents chose to mute themselves. Meanwhile, evacuees and outsiders from afar might talk only about radiation. Further away from the radiation's sources, that talk only becomes louder.

As the ambient radiation levels across coastal Fukushima began to decrease after March 2013, the socially and politically permissible dose of radiation exposure changed. The bad news was that such changes only allowed for more exposure of residents, not less. As a resident, before the accident you might have heard local politicians, experts, and electric companies claim that Chornobyl was different from Fukushima and that there would not be any accidents like that in Japan. The same group of people might have told you that Fukushima is no Chornobyl. Indeed, in one sense it is not; scientifically, it is a different experimental field entirely, one in which the international scientific community has been actively learning how much chronic low-dose radiation exposure from accident-driven radioactive materials is acceptable. It has been an extremely rare and valuable opportunity for them. But for you, it is different: You did not enroll yourself in the research. You just happened to be there.

Regardless of your feelings, the state made a political decision to gradually lift an evacuation order and science followed. Initially, one of the reopening criteria that the state, following the ICRP recommendation, set for the region was to achieve the 1 mSv additional-dose standard. In other words, no individual should be receiving more than 1mSv per year of radiation exposure from the accident *in addition to* the known average dose one would receive from naturally occurring sources like cosmic rays and radon in the air. According to the Ministry of the Environment, the average annual exposure for the Japanese people is around 2.1mSv per year, 0.3mSv lower than the world average.

In November 2013, the state declared that the ICRP recommendation was aspirational and, though it should remain the long-term goal, it should not be understood as the primary criterion for reopening a zone.[4] Following this interpretive shift, the state targeted for reopening any region where it seemed possible to achieve the condition of less than 20 mSv/y additional exposure. You might feel that, without choosing any of it, you have gone through multiple statuses in a few years: natural disaster victim, nuclear accident victim, exposed person (*hibakusha*), and experimental subject (*hikensha*).

Hearing the news about shifting exposure thresholds, you may realize that the decision to open or not open a region turned out to be not just about the presence or absence of radiation in a specific locale. It was a political and community decision. Reopening had its benefits, but they were not necessarily beneficial to you. The state wanted to reduce the amount of money they loaned TEPCO to compensate victims for their evacuation. The length of evacuation has affected the rest of Japan, since the state has drawn compensation from tax money. In this way, the state policy implicated its citizens and the national interest in favoring the early reopening of the zones.

The state also wanted to show the world that the country was steadily overcoming these "unprecedented" disasters. The state was not the only actor here. Concerned about the sustainability of their communities, histories, and cultures, some municipalities also wanted to resume normal life and reconstruct it. We know that Naoko privately wanted the same thing. Even if you disagreed with people like Naoko who desired to return as soon as possible, each month and year of waiting made it harder for a community to recover. However, not everyone was interested in reopening their communities. Some residents already had given up, started a new life elsewhere, and had no intention of returning. Conflicts emerged among the residents and between communities. Whether or not to reopen communities generated substantial political tension among the residents and sometimes within individual households, since opening a zone meant cutting compensation for a group of people.[5] Again, where you were situated mattered greatly.

Let me tell you this: each decision you made—each struggle, conflict, and moment of stress, anger, fear, confusion, guilt, shame, and despair you might have experienced—none of it was your fault. You did what you could under the extraordinary circumstances of the TEPCO accident. There was no right answer, and you were not alone in thinking and feeling like you did. The state, experts, and public alike made you believe there was a correct answer, but they never gave you many options to choose from. You either had to be quiet or leave, but to where?

In social science, what you are vicariously experiencing can be called *structural violence* (Farmer 1996; Galtung 1969). It is a kind of violence that Oshima saw in Abe's statement about the situation being under control. The state, experts, and TEPCO made residents feel like they had no option but to follow their decisions. In the process, residents were made to feel that everything was already decided without consideration of their experiences and that there was no room for them to participate other than to follow the stupid lines drawn on the maps. As Farmer (1996, 12) puts it, in cases of structural violence, "historically given (and often economically driven) processes and forces" conspire to constrain individual agency. This structuring of the violence in coastal Fukushima is the half-life politics through which you have been made to believe everything has to do with radiation and its half-life but has no regard your or your family's lives and livelihoods. The most alienating experience for the residents was that they could not tell who made what decisions and whom those decisions served. No one claimed accountability for the objectified residual presence of radiation. Not only were the residents' complex experiences nameless, but namelessness was also the cause of those experiences. Hatsumi called this "the nuclear ghost of Minamisōma."

Despite what some might have thought and continue to envision, the TEPCO accident was not and has not been merely a radiological disaster. As your imagined experience as a resident suggests, it also disintegrated social and domestic fabrics, making your life hard to tell (*katarinikui*) and hard to live (*ikinikui*). All your losses, nameless experiences, and lost voice cannot ever be fully compensated, nor could they easily be heard when your interlocutors would rather trust their technoscientific instruments like Geiger counters and monitoring posts. Still, it was your rather involuntary participation in the compensation policy—the state and TEPCO's attempts to make up for your losses—that concretized social, economic, and moral differences and radically changed the postfallout local landscape.

## THE COMPENSATION ECONOMY

Now imagine yourself an evacuee from the Haramachi district, inside the thirty-kilometer zone. Large parts of Haramachi fell under the stay-indoors order on March 15, 2011, and then, on April 22, the state placed the area in the emergency evacuation preparation zone. This meant that you had the option to stay near the unstable power plant, as around nine thousand others did, or to stay away from it. Then the situation changed on September 30 of the same year, as the state lifted the order. After the reopening, you had to decide whether to return to Haramachi. Your decision would be affected by multiple factors, including, but not limited to, the condition and type of your evacuated residence; your age, the number of people in your family and their ages, your job prospects after your return, the location of any other family members who lived outside your household, and so on.

You might want to consult with your friends and neighbors, but you would need to think carefully about who to talk to. There could be no guarantee that they would think similarly about radiation and its potential effects. Sharing could risk exposing your or others' "bias." Some might make you feel guilty for staying away from the city longer than others. They might ask you how much longer you plan to be away. They would not mean to hurt you, but small things like this might make you doubt their intentions, or make you feel as if they were blaming you for being too sensitive about the risk of exposure. Here is a warning from Saeki from Kashima: "Minamisōma is a small city. People are very caring, but they can be too nosy. If I complain about radiation in public, then someone would point at me to accuse me of being still stuck in the past or say that I complain because I am envious of others who receive compensation. Similarly, if I'd say we should move on from the past and not talk about radiation, then someone would tell me that I do not care about people's sufferings or let the government and TEPCO have their way. In Minamisōma, the smart thing to do is to be quiet and keep things to yourself, not to stir anything up, to avoid being picked on." Minamisōma could offer you a kind of intimacy you might have missed while living in an unfamiliar city during your evacuation, but, as the young resident Toru put it, Minamisōma can be "suffocating."

We must talk about money. After all, the accident and its aftermaths caused grave inconveniences, and you deserved to be compensated. Regardless of your decisions, one of the most time-bound and general forms of compensation available to you would be compensation for the "psychological harm concerning evacuation" (*hinanseikatsu touni yoru seishinteki songai baishō*). As a resident in Haramachi, in the former evacuation preparation zone, you would be automatically qualified for this until August 2012, one year after the lifting of the order in September 2011. That is, you would receive 100,000 yen for seventeen months, which equals 1,700,000 yen, or around $14,000. If you had a reason to continue your evacuation after October 2011 (for example, due to work, schooling, or avoiding the separation of family members), you might be qualified to receive compensation for longer. Moreover, if you or someone in your family was pregnant or required special care (due to health or other issues) during this time, you might be qualified for more money. For this, you would need to file an additional claim to TEPCO.

Another important fact is that this type of compensation was given to each member of your family, so the more members, the more money the family would receive. This information might make you wonder about what your neighbor with a larger family would get. As a reference, Toru's family of four (his brother is not counted, since he lived outside Minamisōma) would be qualified to receive 400,000 yen per month. Since they were evacuees from the twenty-kilometer zone, they were qualified to receive compensation for psychological harm until March 2018—that is, 400,000 yen for eighty-four months, which equals 33,600,000 yen, or roughly $250,000.

Notice that the reopening of a zone came to indicate two facts. First, reopening signaled that the level of radiation had fallen low enough that the locale was considered technoscientifically livable or, in the case of the evacuation preparation zone, that the experts had determined that the difficulties at 1F were now under control. Toru's family had to wait much longer than others to return to their former home since they had to wait for the state to finish cleaning up contaminants from the built environment, which took a long time. Second, reopening meant the eventual termination of compensation, which usually happened one year later. In this way, the presence or absence of radiation became linked to residents' compensation-based economy. The earlier the possibility of return, the faster the disappearance of radiation in the area *as well as* the disappearance of benefits. Inversely, the later the possibility of return, the slower the disappearance of radiation in the area *as well as* the greater social and financial differences produced by the compensation received among differently located residents within Minamisōma. As a resident of Haramachi, what you could receive was around $60,000 less than what Naoko was entitled to, but you would have been able to go back to your residence at least seven years before Naoko could. Would $60,000 be enough to make up for the psychological suffering of seven more years of displacement? Only you would know.

Notably, even within the same district, this logic of compensation produced an ideological division between those who desired to go back as soon as possible and those who decided never to go back. While the former group was willing to accept the loss of compensation in exchange for returning to their original homes earlier, for the latter group it was financially beneficial to prolong the length of their compensation. Imagine you are struggling at the place to which you evacuated, having a hard time securing a stable job or watching your child being bullied at a new school. Imagine further that returning to Minamisōma was not an option for you because you and your family could not be near the damaged plant. In this situation, you would hope that the state delay reopening your district, since its reopening means you would eventually lose your stable monthly payment. The amount, 100,000 yen per person per month, might not be much, especially if you chose to live in an urban area, but it does help. You may, as a result, disagree with the state's decision to open your district. Notice that the compensation economy could drive your relationship to radiation.

Learning that reopening a zone requires both the reduction of radiation levels and an agreement by the local government and community members, you could try a few things. You may feel compelled to criticize the judgment of the state and experts, demand reassessment, point out the unreadiness of various functions in the city, or convince others in the district to do the same. In the process, you might come to develop a more extreme position on radiation and its danger. Moreover, you might come into conflict with your fellow community members who desired to return early. They might tell you that you are only interested in money, not the future of the community. In turn, you might wonder why people had become blind to the risk of radiation exposure.

For the compensation-based economy, it was no longer the presence or absence of radiation that articulated what was at stake in the city. Instead, the central issue became the presence or absence of compensation associated with the technoscientifically detectable level of contamination. Such a mode of living with radiation in Minamisōma illustrates how the selective distribution of compensation itself came to be an additional harm, a mechanism to selectively visualize radiation in vicinities. For residents, whether they remained in the city or lived elsewhere, their forced participation in the compensation economy had come to stand in for their enrollment in the half-life politics.

According to this politics, compensation had to do with the objective presence of contaminants, on the one hand, and, on the other, their objective absence. In this sense, the compensation policy framed the TEPCO accident as a problem purely about radiation and not about the social and domestic struggles you have experienced because of the elevated levels of radiation. As such, the end of compensation marked the end of the TEPCO accident. According to this logic, the accident appeared to be guilty only of causing radiological danger.[6] Here, compensation not only justified the "stupid lines" drawn on the map that produced your lived experience of geographical and social divisions, but it also came to contain the social, political, and moral consequences of radiation.

### THE COMPENSATION GAME

This relationship between compensation and radiation made the individual calculation of risk and (un)willingness the focal point of negotiation, or what I call the *compensation game,* which has at least four distinct positions: *nuclear economy, nuclear rights, nuclear disavowal,* and *nuclear kimin.* Each position represents a specific stance toward radiation. The compensation game is a heuristic to help us further imagine the half-life politics enacted by the compensation policy on Minamisōma residents and the ways that the postfallout policy contributed to further dividing the community.

Imagine yourself approaching nuclear compensation purely economically. The best strategy for maximizing economic gains from nuclear compensation is to physically distance oneself from radiation in order to reduce the potential risk of exposure while simultaneously emphasizing or even exaggerating its presence from afar. That is, one would subscribe to the perspective that Minamisōma is a contaminated and unlivable place. This position, *nuclear economy,* allows both the avoidance of the potential risk of exposure and the maintenance of the right to receive financial compensation. In contrast, when thinking from the perspective of one's psychological and spiritual attachment to the land and one's home, like Hatsumi and Naoko, the best strategy is to selectively ignore the presence of radiation and hope for the earliest return, even though you might risk exposing yourself more by staying near the edges of evacuation zones. This oppositional position to nuclear economy, *nuclear rights,* suggests that Minamisōma is not so contaminated and is livable.

What complicates this negotiation process is the existence of two additional positions. There are people who do not care or who tend to ignore the entire situation. People in this position are likely to avoid participating in any social and political discourse related to the accident and its associated recovery and reconstruction policy. People in this category tend to be young people who stayed but are planning to leave or who evacuated far away and do not have much attachment to the land. Of the people we have met, Toru may be closest to this position. Although Toru's life was impacted by the accident, he quickly moved on and adapted to his new life in *kariage.* Toru toyed with the idea of leaving the city to start a new life elsewhere. Moreover, he did not have to deal with the paperwork necessary to receive compensation, since his father took care of everything on his behalf. Similarly, Toru's brother, who lived far away, may also fit into this category. I call this position *nuclear disavowal.* Someone in this category would say that whether or not this place is contaminated is not my problem. Finally, some people, including Saeki and Tengo, either selectively ignored the presence of radiation or resisted the right to receive compensation. Residents in this position had come to mistrust the state and TEPCO's ability to change the situation. That is, the postfallout conditions showed that the state had abandoned people. I call this group *nuclear kimin,* following Jotaro Wakamatsu (2012).

As a structuring mechanism, the compensation game had at least three essential conditions.[7] First, only those who were qualified based on the state-regimented evacuation zoning participated in the game. Second, the distribution of compensation was an open secret. While the amount and the length of a few compensable items—such as psychological harm, medical service, and highway tolls—are publicly known, the total amount of compensation one received was subject to individual negotiations and was usually kept private. This open secret produced speculation, suspicion, and rumors, especially among the economic outsiders.[8] For example, Sachiko from Kashima read someone's newer-looking car as a sign that the individual was from Odaka and was thus receiving a lot of money from TEPCO. Third,

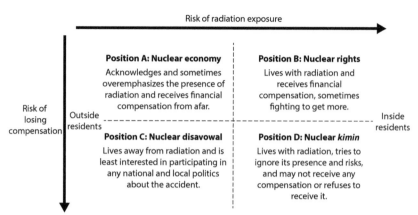

FIGURE 3. A conceptual diagram of the compensation game in Minamisōma. Description of the game: 1) An individual either actively holds one position or is categorized into one. Therefore, even within the family, different members may take different positions. 2) Each position is fluid and in a continuum, and the rule and extent of the compensation policy and the measurable presence of radiation may change over time. However, crossing the vertical line is more challenging than crossing the horizontal line. It requires physical relocation and/ or change in the compensation policy. The diagonal categories are the most opposed to each other. 3) Each individual position influences the overall compensation policy and the timing of the reopening of the city. Diagram by the author.

members of the same family might hold different stances toward the situation or might live in different areas of the city and hence would qualify for different amounts. For example, as a resident of Kashima Saeki was not qualified for much, but her parents lived in Odaka, so she had knowledge of the extent of compensation available for the insiders (figure 3).

The compensation policy placed the residents in a complex local negotiation dilemma in which they were enrolled by virtue of their predisaster residency within the city. I call this dilemma a *game* since participants could potentially manipulate their position vis-à-vis others to gain more—for example, by hiring a lawyer or by using the state's Alternative Dispute Resolution (ADR) service.[9] For instance, many so-called voluntary evacuees launched civil actions against TEPCO to get additional compensation (De Togni 2022; Jobin 2020). More importantly, it is a game with social and moral consequences. Imagine yourself in Naoko's position. She could not voice her opinion because she wanted to avoid conflict with her son or with Toru. Naoko's stress, which her doctor diagnosed as causing her health issues, resulted in part from being caught in this domestic struggle. Anthropologist Victor Turner uses the concept of *social drama* to explain a multiparty "sequence of social interactions of a conflictive, competitive, or agonistic type" (1988, 33). Whereas Turner claimed that such dramas begin when a member of a community breaks a rule and others take sides, there is no rule breaker for the compensation game. It *is* the rule breaker.

Although unqualified individuals were excluded from participation, they nonetheless influenced the internal dynamics of the game by having opinions about each position. They were also influenced by the dynamics themselves. People inside the dilemma could be their friends, relatives, significant others or family of significant others, or parents and/or children living away. Moreover, it is important to note that from the outsiders' point of view, all participants in the game, regardless of individual qualification, belonged to the contaminated community. This is a common misunderstanding of outsiders, who tend to homogenize all the residents as "victims." It is assumed that people like Saeki, Oshima, and Sachiko had received money from TEPCO. Although they did to some extent, it was very little compared to that received by others like Naoko. In this dilemma, no one won. One of the outcomes of this game was that it produced quantifiable differences within the community and thus created divisions and fissions. Based on his observation of social harm across various technological disasters, Freudenburg argues that "it can be the very 'overlooking' of supposedly 'local' concerns that may actually create the greatest risks to the social fabric" (1997, 34).

By positioning the residents in different categories, I am not suggesting that they fit neatly into one of these four positions. In reality, these positions existed along a continuum, and residents cycled through different positions. Hatsumi, for example, admitted that she felt abandoned by the state and TEPCO once she was no longer qualified to receive compensation in late 2012. She also revealed that around the same time, she tried not to care about radiation and its potential adverse health effects, just like Saeki from Kashima. This was why Hatsumi, who supported the film director in 2012, no longer welcomed him in 2013. Hatsumi's experience is one example of how the presence or absence of compensation influenced the residents' experience of a life with radiation and its (in)visibility in the city.

The compensation game individuated *higai,* produced different positions toward radiation, and prevented the residents' voices and concerns from coalescing against the state and TEPCO. It restructured the existing social fabric by overemphasizing residents' differences, thus shifting the responsibility away from the state and TEPCO. The corrosive effect was most damaging in traditionally tight-knit communities like Minamisōma. This type of postfallout social division adds nuance to the observation by Daniel Aldrich (2019), a political scientist of Japan, that one's social capital and social networks helped to increase an individual's chance of surviving and recovering from a natural disaster. In the case of the TEPCO accident, existing social capital (especially stronger bonds [*kizuna*] like biological kinship and long-term friendship) contributed to heightened tension among the residents. One elderly resident in Kashima, who expressed her distaste for the collective national sentiment of fostering *kizuna* after 3.11, reminded me that the Chinese character for *kizuna* (絆) has another meaning in addition to bonds: a shackle (*hodashi*).

Miura, an evacuee from Odaka whom we met in chapter 3, warned about falling into this half-life politics. He said that approaching food safety as a tension between consumers and producers produces further divisions and only benefits the state and TEPCO. As the state and TEPCO translated the invisible presence of the radiological danger through the presence of compensation, this postfallout policy fragmented the community and its ability to articulate the name of the harm done to their collective body. Meanwhile radiological contaminants remained in the atmosphere, water, and species and across terrestrial environments. It seems like, in its aftermaths, the invisible radiation kept changing shape. It was not radiation that changed, however. It was the social, political, and economic drama surrounding radiation—the compensation game—that made radiation visible for some and invisible for others.

If the invisibility of radiation is what caused the struggle of interpretations and ensuing social divisions and fragmentations in coastal Fukushima, does seeing radiation help us imagine and understand different lives in Minamisōma better? This was precisely the question one Sōma resident, Tanno, posed to me. It turns out that it is easier to "see" radiation itself, technoscientifically speaking, than to imagine the surreal social life of radiation in coastal Fukushima.

### THE CLOUD CHAMBER AND THE (OMNI)PRESENCE OF ATOMS

Living in Sōma city, outside the state-mandated evacuation zone, Tanno was spared from the severe disintegration of the community. As a fisherman, he lost his boat from the tsunami and the ocean from the TEPCO accident. When we talked in the fall of 2013, he told me how he missed going out in the ocean and fishing. For Tanno, the radiation issue was real, and, like the ocean, it was unbounded. Because radiological contamination is a transboundary phenomenon that had an impact beyond the state-sanctioned zones, he reasoned, it should bring people together instead of breaking them apart. For Tanno, the postfallout reconfiguration of Minamisōma was surreal, especially the way that residents seemed to be divided about radiation. When I shared with him an early sketch of my compensation game diagram (figure 3), he thought none of the categories applied to him as someone who received compensation only for tangible losses like his job and boat.

"If we let the state make us believe radiation is like a ghost [*obake*]," he said, "then there is no way we could win. We go nowhere by just fighting over its presence/absence. I know this because people believe our fish is contaminated regardless of rigorously testing it for safety. We first agree that it exists, and it impacts our lives, and only then can we have a real dialogue on how it impacts us." Throwing rocks at the Pacific Ocean, Tanno half-jokingly offered a solution for the social disintegrations in Minamisōma: "I wish we could make radiation visible by coloring it depending on how strong it is, like the high level is red, the medium level is yellow, etc. Oh, thick (*koi*) areas should be purple! But if I could see radiation in colors in my everyday life, I would not be living here! Somewhere there," he continued, pointing west to the mountainside, "would look very red. Isn't that scary?"

He laughed. "But if we could see it, there is no lying or fighting over its presence or absence, or who gets how much money for what cannot be seen." As I was witnessing Minamisōma residents' everyday struggles with the atomizing effect of the compensation game, his words did not make sense to me at the time. I thought he was just trying to be funny and provoke a reaction in—or even scare off—a curious outsider. But in the last month of 2014, I finally saw radiation, not in coastal Fukushima but about 150 kilometers south of Minamisōma, in the place where nuclear energy in Japan began.

In December 2014 I visited the Japan Atomic Energy Agency (JAEA) in Tōkai Village, Ibaraki Prefecture, where one of the technoscientific milestones of Japan's atomic age is located.[10] JRR-1 is Japan's first research nuclear reactor,[11] installed in 1956.[12] The tiny reactor with a generating capacity of fifty kilowatts was imported from North American Aviation Inc. for $25,800. The reactor used 20 percent enriched uranium as its fuel, which the U.S. company Mallinckrodt Chemical Works produced. The same company purified uranium oxide for the Manhattan Project, through which the United States developed the two atomic bombs later dropped in Hiroshima and

Nagasaki in 1945. Japan paid approximately $35,000 per ten kilograms for the enriched uranium. With this all-U.S. technology, Japan initiated its great leap forward, entering its nuclear renaissance after a period of atomic censorship under the U.S.-Allied occupation between 1945 and 1952.

On August 27, 1957, the front page of Japan's major national paper, *Asahi Shimbun*, reported on this symbolic day on which Japan lit the first "atomic-powered fire" (*genshino hi*): "The Japanese nuclear research institute's effort to catch up [with the world] by lighting up 'the second fire' in Tōkai village after the atomic bombs in Hiroshima and Nagasaki 12 years ago has enormous significance" (quoted in Asahi Shimbun Investigation Team 2014, 11). In Tōkai village, Japan, like many other countries (Jasanoff and Kim 2009; Osseo-Asare 2019), initiated its postwar efforts to control and sublimate the technology that had devastated the country.

JAEA's heavily guarded entrance gate contrasted markedly with the friendlier-looking Ibaraki Science Museum of Atomic Energy across the street. Out of curiosity, I walked inside the museum, where a middle-aged woman greeted me: "Welcome, where did you come from? You must be *yoso no hito* [a person from outside the village], right?" Taken aback by her quick profiling, I asked her what made her think so. Her answer situated Tōkai village as a contested space within the technopolitical map of "nuclearity" (Hecht 2012a) that had been drawn around the TEPCO accident.

"Well, no one from the village would come here alone," she replied. "This is a PR center for nuclear energy, exhibiting positive aspects of nuclear science. We usually have a group of visitors from schools. Since the nuclear accident in Fukushima, many people have come to visit and complained, asking how I could continue propagating the safety myth [*anzen shinwa*]. People don't understand that it is just my job. What's here tells how we conceived of nuclear energy, and the museum tells the facts of atomic science. It does not reflect my personal beliefs."

After I told her that I was a student from the United States visiting JAEA, she apologized, explaining, "I thought you were one of those newspaper people, so I got defensive." But then she said that many scientists from the United States had visited the museum since the TEPCO accident. Some of these visitors, she had been told, were famous, but she did not know who they were. I asked her what kind of people had been complaining to her, and she told me that they were usually journalists, both domestic and international.

"They are very rude," she said. "They simply want to hear that I am worried about living near the nuclear facility and if I say no, they tell me I am a believer of the safety myth like I am a fool. The fact is, those journalists themselves did not even care about nuclear anything before the disaster. They had participated in the safety myth they suddenly speak so arrogantly of! I have been living here for decades, and from my experience, it is not terrible to be in Tōkai village. To me, it would be harder to imagine this village without nuclear facilities. Living here is probably like living near a big factory anywhere in Japan. There is always some risk; 3.11 has proven to us there is no risk-free place anywhere in this country."

Many journalists have visited and continue to visit this small village of around thirty-five thousand people and consisting of about thirty-nine square kilometers of land, in large part because Tōkai village is home to the largest nuclear research complex in Japan. Unlike the city of Minamisōma, where three separate towns had to merge to maintain the city's population size and financial and administrative power in 2006, Tōkai village remained a village, since the nuclear facility enabled the village to remain financially secure and independent. A merger with the surrounding towns would benefit only the other towns, not Tōkai village itself. Since the appearance of JRR-1 in 1956, 13.5 percent of the village's land has been occupied by fifteen nuclear-related facilities. In Tōkai village itself, there are three nuclear reactors, including JAEA's fast-breeder reactor (FBR), a U.S. boiling water reactor (BWR), and Japan's first British-made commercial reactor (a gas-cooled reactor), owned by the Japan Atomic Power Company (JAPC).[13]

The high traffic of journalists has also had to do with the history of nuclear accidents and radiation exposure. Tōkai village was the site of a severe nuclear accident in 1990, resulting in the death of two workers and the severe injury of another from acute radiation syndrome stemming from direct exposure to a high-dose neutron beam.[14] One of them, a thirty-five-year-old male, received a radiation dose of more than 16 to 20 Sv and died eighty-four days after the accident. In 2013, most of the residential areas in Minamisōma on average exposed people to 0.20 μSv per hour. Twenty Sv is one hundred million times more doses of radiation than 0.20 μSv. Another worker, a forty-year-old male, received more than 6 to 10 Sv and died 211 days after the accident. One surviving fifty-four-year-old worker was exposed to approximately 1 to 4 Sv.[15] This so-called JOC criticality accident caused *fūhyō higai* among the local fishing and agricultural communities and the food-processing industry. Their negotiations for compensation for the economic loss caused by consumers' and traders' concerns helped shape the nuclear damage compensation policy applied after the TEPCO accident (Nakajima 2013; Sekiya 2003, 2011).

Like everything else in the small village, the museum looked elegant and expensive. A cartoon Albert Einstein wearing a T-shirt with the equation $E = mc^2$ is the museum's mascot and accentuates the museum's scientific and educational ambience. "Would you like to see some exhibits?" the woman asked. Intrigued, I asked her to show me the most exciting thing in the museum. She brought me to a section called the Atomic LABO, right next to the area named Einstein Square, where she guided me to look at a large black container with illuminated transparent glass on the top (figure 4). The sign by this strange instrument read "Cloud Chamber."

"If you press the button, you can see radiation [*houshasen*]," she told me proudly and added, "It is one of the biggest in the world." And indeed, the chamber magically produced intricate, arabesque-like lines of water vapor, projecting the crisscrossing paths of various waves of radiation, which, to my eye, cut arbitrarily across the contained space inside the chamber. The Cloud Chamber, or Kiribako, consists of a sealed structure with two plates, a warmer one on top and a cooler one on the bottom. At the top, liquid alcohol such as methanol or isopropanol vaporizes and saturates the chamber. As the vapor cools, it condenses on the cooler bottom plate. When this happens, various rays of radiation can be visualized as they cut through the supersaturated environment. Because a cloud chamber uses either dry ice or liquid nitrogen to keep its lower plate cool, as I approached the chamber, I felt a chill.

Mesmerized by the patterns, I noticed that the sign next to the chamber read, "The goal of this exhibit is to show how radiation is omnipresent and, therefore, not scary." The unfamiliar image of vapor-rendered radiation in the Cloud Chamber brought to mind Tanno's comments about coloring radiation. As I stood in front of the Cloud Chamber I saw naturally occurring radiation, although it would otherwise be undetectable in everyday life. "Seeing" radiation in the box as the vaporous mutilation of space, I found myself agreeing with Tanno. In theory, if we could visualize radiation, it would become less creepy, since people often attribute their fear of radiation to its ghost-like invisibility.

But it might even be scarier when something that always remains below the threshold of our senses is suddenly brought to our senses—either through vision or physical damage—as happened to the workers who were exposed to high doses of radiation inside the Tōkai nuclear complex. At the edges of the evacuation zones in coastal Fukushima, the radiation levels were not high enough for residents to be able to "see" and "sense" its effect or to "experience" its physical damages immediately. Instead of experiencing radiation

FIGURE 4. Cloud Chamber at the Ibaraki Science Museum of Atomic Energy in Tōkai village. It is described as one of the biggest cloud chambers in the world. The text on the front of the chamber body reads, "Cloud Chamber: You can see radiation around you!!" Photograph by the author.

directly in their lived environment, various radiation maps, evacuation zones, and the unequal distribution of compensation have come to color some—but not all—radiation in their vicinities. In a way, Tanno was wrong to assume that radiation had no color. In Minamisōma, where various mediations of the invisible radiation coexisted, its residents experience its physical, social, political, and moral effects as grayness.

Let's once again imagine living in postfallout Minamisōma, where the TEPCO accident suddenly forced its residents to confront not only the invisible radiological contaminants—their omnipresence—but also their various representations in their everyday lives. In that situation, residents struggled to determine the meaning of radiation and their relationship to it, as its presence and absence were selectively visualized and monetized. As many of its residents gradually returned from their short displacement or chose to stay away, Minamisōma had become an *ikinikui* place, where its residents could not be in dialogue with each other and share their experiences. Yet in this place, this city, many people remained, and people like Hatsumi, Naoko, Kobayashi, Kato, and many others desired to live and die in Minamisōma. Their continuing livelihoods in Minamisōma demanded an inquiry not into the science of radiation and its half-lives that had produced the compensation economy but into a more holistic understanding of what it means to live well. If radiation is omnipresent, as the cloud chamber illustrates, but its presence has been rendered selectively visible, how could it impact people's well-being? What if, despite popular imagination, radiation exposure was not the central problem? These are the questions a few locally embedded experts started exploring.

# FIVE

# Radioactive Mosquitos and the Science of Half-Lives

Existence is communication, and communication, existence.

HARUKI MURAKAMI, *A Wild Sheep Chase*

## RADIOACTIVE MOSQUITOS OF MINAMISŌMA

On April 16, 2011, a group of Minamisōma residents started a local radio station, Hibari FM. Initially a hub for providing postdisaster recovery information to the region's residents, the station evolved into a communication platform for addressing local postfallout issues, such as radiation monitoring results, information for coping with low-dose radiation in postfallout life, interviews of residents and evacuees, discussions by postfallout migrants, and conversations among young residents. Until its termination on March 25, 2018, Hibari FM grappled with the lasting and shapeshifting effects of the TEPCO accident on the region's remaining residents.

In particular, the radio station worked toward fulfilling the need for open dialogue among the divided communities. This resident-initiated effort reflected the slogan the city established to state the central goal of its revitalization plan: "All hearts united for the rebirth of Minamisōma."[1] First released in December 2011, the slogan still lives amid the global pandemic in 2022. In the city's monthly magazine commemorating the tenth anniversary of 3.11, the current mayor, Kazuo Monma, wrote that the pandemic reminded him of the fear and uncertainty residents experienced from the invisible threats they faced after the TEPCO accident. The lessons learned and the wisdom gathered by the surviving residents of 3.11, Monma argued, would help them overcome the pandemic.

One of the fifteen-minute programs Hibari FM broadcast between 2012 and 2013 was *I, Tsubokura, Will Answer Your Concerns*. The premise of the program was the following: Dr. Masaharu Tsubokura, a specialist in hematology who researched radiation health at Minamisōma's city hospital, communicated empirically grounded information about radiation exposure to concerned residents.[2] As his show was a platform for expert-resident dialogue, Dr. Tsubokura received questions from anonymous local listeners. Some questions he received over the show's twenty-five episodes related to the potential effects of internal exposure, the biological half-lives of different types of cesium compared to their physical half-lives,[3] radiation exposure through food intake, the difference between naturally occurring and artificial radioisotopes, and differential contamination levels in local produce. Listeners also sought advice on how much longer to wear a mask, what to do after accidently ingesting dirt, and who should be most concerned about exposure. This radio program captured the types of questions and concerns the residents had on their minds between 2012 and 2013, which they otherwise struggled to communicate publicly due to the various fissions and divisions initially produced by zoning and later cemented by the unequal distribution of compensation (Yagi 2021; Yoshida 2018).

In the sixth episode, which aired in the summer of 2012, one question seemed to take Dr. Tsubokura off guard. The radio host, Konno, read the question, as he always did, on behalf of the listeners. Before Konno began, he first announced it as an intriguing question, and then he read, "Is it the case that we take radioisotopes into the body from mosquito bites?" The question seemed to indicate the listener's pragmatic concern about radiation and its potential pathways into the human body. Konno continued, "Mosquitos tend to hatch eggs in the bottom of ponds or puddles where radioisotopes are known to be highly concentrated. That's the kind of environment where mosquitos grow, and in the process, they probably absorb lots of radioisotopes inside their bodies. Therefore, mosquitos must be highly radioactive. I wonder if radioisotopes would get transferred intravenously and we would get contaminated with a poison they leave in the body when those radioactive mosquitos bite. Would mosquitos this year be safe?"

Taken aback by this question, Dr. Tsubokura laughed nervously, but he kept his cool. Dr. Tsubokura's reaction hinted that this was something he had not thought of as a medical expert. In contrast, residents imagined many likely crossroads between radioisotopes and their everyday lives. A mosquito bite was one of them. What background knowledge about radiation exposure allowed this local listener to imagine a mosquito dilemma when medical science had not? Understanding the significance of this specific communicative instance between an expert and a layperson requires a description of why internal exposure had become a matter of great concern for the residents in Minamisōma.

From the very beginning of his radio program, Dr. Tsubokura was particularly keen to distinguish between external and internal exposure. In Japan, there is a tendency to generalize all radiation exposure as one cultural-historical category of *Hibaku*, a violent and dehumanizing form of suffering that is exemplified most acutely by the atomic bomb victims of Hiroshima and Nagasaki. External exposure (*gaibu hibaku*) occurs when radioactive materials in the vicinity emit different types of ionizing radiation (alpha, beta, gamma, and neutrons) and individuals are exposed to them.[4] Exposures resulting from the U.S. bombings of Hiroshima and Nagasaki are extreme cases of external exposure. Environmentally, humans are constantly being irradiated from naturally occurring radiation sources, such as radon in the air and cosmic rays. The body is also subject to external exposure in an artificial environment, such as during a CT scan.[5] Depending on which parts of the body the scanner is applied to, one scan could result in a few mSv up to 20 mSv of external exposure.[6] In contrast, internal exposure (*naibu hibaku*) occurs when individuals inhale or ingest radioactive particles inside their bodies and those particles emit radiation inside the body. Some examples include inhaling radon in air, ingesting radioactive materials from contaminated foods, and ingesting polonium and lead from smoking cigarettes.

The heart of the mosquito riddle touched on the gray zone between what is external (a mosquito) and internal (the human body) and the permeability and transferability of radioisotopes. By mid-2012, when this episode was aired for the first time, local listeners were not only concerned with the mere presence of radioisotopes in the environment, which technoscientific instruments like dosimeters and Geiger counters were able to detect and which various monitoring posts in the region constantly visualized. They were also concerned about the possible paths by which the haphazardly spread fallout residue in their environment could enter into their bodies. This shift in people's concern about internal radiation coincided with the more rigorous monitoring of contamination in foods, a development that was closely dealt with by farmers like Miura, whom we met in chapter 3.

During the radio show, Dr. Tsubokura struggled to answer the mosquito puzzle and ended up offering a mundane answer, emphasizing how small a mosquito is and how each bite could inject only a trivial amount of (radioactive) poison into the body. He concluded that there was no reason to worry about mosquito bites causing significant internal radiation exposure. Instead, he argued that being bitten might increase other health-related risks.

Dr. Tsubokura's consistent message was that radiation is only one of many risks we face.

According to the global medical community, an exposure event of 700 mSv or more is known to cause acute radiation sickness, as in Hiroshima, Nagasaki, and the JOC criticality accident in Tōkai village. In these instances, individuals were irradiated with a high dose of radiation and experienced severe symptoms, such as loss of hair, skin problems, gastrointestinal issues, cataracts, and extreme nausea (Hall and Giaccia 2019).[7] In comparison, an exposure of around 10 to 20 mSv in a discrete event like a CT scan is not immediately detrimental to one's health, although it could result in long-term issues. Nonetheless, the delayed onset of disease makes it difficult to empirically determine a causal link between a past exposure event and present condition. The longer the time between the exposure and the onset of disease, the more likely it is that something else, such as other toxicants or unhealthy lifestyle choices, could have contributed to the development of the disease.

If Toru develops cancer in 2031, science will most likely deny there is a direct link between his exposure in 2011 and his cancer twenty years later. Since Toru chose not to be tested constantly, it would be harder for him to prove otherwise. He reasoned, however, that "even if I get tested constantly and diagnosed with a cancer later in my life, they'd use the data to say that I did not get exposed enough." He was right about that. According to exposure estimations based on field survey research, right after the accident Minamisōma residents were exposed internally to an average of 0.1–0.2 mSv and externally to less than 2 mSv of cesium-134 and -137, two of the most prominent radioisotopes in the region (Tsubokura et al. 2015; see also Harada et al. 2014). Medically, their exposure is considered negligible.

The issue of the increased incidence of thyroid cancer among youth in Fukushima illustrates the difficulty of determining a causal relationship between radiation exposure and a disease.[8] Medical experts and the lay public have been divided on how to interpret about 287 cases of thyroid cancer among Fukushima's youth,[9] reflecting the politics surrounding radiation exposure.[10] Both national and international medical communities have pointed out the possibility of "screening bias" rather than the TEPCO accident itself as the primary factor accounting for the increased incidence of thyroid cancer in Fukushima. Dr. Tsubokura supports this claim and lists two well-known rationales. First, new and improved screening technology enables the detection of smaller cancerous tissues that were not previously detectable. Second, thyroid cancers in Fukushima were being discovered much earlier than they had previously had been after Chornobyl. The data from the Chornobyl accident suggest that it takes a longer period, usually ten or more years, for an iodine-exposed thyroid to grow cancerous tissue.[11]

From the residents' perspective, however, the TEPCO accident led to the discovery of a cancer that did not exist previously. Therefore, experientially, the experts' scientific reasoning is hard to swallow. While the victims are seeking accountability, the experts are concerned about empiricism. The thyroid debate exemplifies not only the politics surrounding the interpretation of scientific data but also the difficulty of communicating the scientific knowledge of radiation, which at times invalidates individual subjective experiences in the interest of the objective discovery of facts.

Anthropologist Gregory Button (2010) discusses the role science plays in producing uncertainty in the wake of industrial environmental contamination. "Like all other disciplines," he writes, "science is more than politics, but it cannot evade the ideological currents in our culture any more than other sectors of society" (179). Similarly, Saeki from Kashima once expressed her growing mistrust of what was presented to her as science this way: "If scientific uncertainty led to social anxiety, wouldn't scientists be accountable for causing a social disaster? People talked about how the TEPCO disaster was a human-made disaster [*jinsai*]. I would include science as a part of this *jinsai*." The technological accident, they felt, justified scientific hegemony. The state and experts forced residents to accept science's cryptic terminology as the descriptors of their postfallout livelihoods. In this context, the question of the radioactive mosquitos of Minamisōma was an attempt by a resident to translate science back into more mundane experience.

Although many of the health impacts of radiation exposure remain gray in coastal Fukushima, one international significance of the TEPCO accident is that it offers some baseline quantitative data for understanding future incidents of chronic low-dose radiation exposure. The TEPCO accident has produced a rare opportunity for the scientific community to monitor and track the exposure and health outcomes of a large cohort of people. In other words, the TEPCO accident transformed Fukushima into an experimental site for scientists to amass previously missing data on the potential effects of chronic low-dose radiation exposure on human bodies.[12]

Prior to the accident, the gold standard for data related to external radiation exposure came from the longitudinal studies of the victims and survivors of Hiroshima and Nagasaki (see, e.g., Hall and Giaccia 2019; Lindee

1994, 2016). Hatsumi once commented on the irony of this historical continuity in the following manner: "It is unfortunate that Japan has come to offer, yet again, the data for *hibaku*. It is the capital of radiation exposure research!" Naoko similarly felt uneasy about this and decided not to enroll herself in exposure research, even though she knew that testing for exposure (both internal and external) would be free to her. "I am not a state's experimental body," she said. Many residents expressed to me their feeling that radiation and its concerns had forced them to live a life in contradiction: in the public eye, they had to live as the exposed victim (*hibakusha*); for the medical community, they had to live as the experimental body to disprove their publicly perceived victimhood.

ALL WE NEED IS ALARA

In the summer of 2019, I spent forty hours over a period of two weeks completing a Radiation Safety Officer (RSO) training course at the Mailman School of Public Health at Columbia University, taught by various Columbia faculty members. Designed to meet the requirements for the United States Nuclear Regulatory Commission Class C Radiation Safety Officer, the training covered topics such as the fundamentals of radiation, health physics, dosimetry, handling of radioactive waste, radiation protection, and risk communication.[13] As the only anthropologist and nonlab affiliate in a class full of chemists, oncologists, lab managers, and healthcare personnel, I found the lecture on radiation safety particularly fascinating, unlike the rest of the participants. From their bored looks and multiple yawns, it was clear the topic was familiar to most in the room. Nonetheless, it had a very different meaning to me, especially since I had spent a long time in coastal Fukushima, where low-dose radiation exposure had become the central theme of everyday life regardless of occupation. The training gave me a new appreciation for the work that experts like Dr. Tsubokura have done in coastal Fukushima, where he consistently communicated to the residents state-of-art knowledge about radiation safety using locally specific data.

The technoscientific lessons of radiation safety were straightforward and consistent across social, cultural, and national landscapes.[14] I had heard the same explanations over and over in Japan since the TEPCO accident. Radiation is everywhere in the environment—from a piece of banana, a fragment of rock, the air, and even medical procedures. It is a part of our everyday lives, and we are always already exposed (and we are radioactive beings!). Radiation exposure is a mundane experience, and no one is exempt from it.[15] Nonetheless, too much exposure to ionizing radiation, though rare, can cause irreversible molecular damage. One example of ionizing radiation is ultraviolet (UV) rays. The sunlight releases a weak form of UV rays, primarily UVA rays. Long-term and "too much" exposure to the sun can cause skin damage such as sunburn, wrinkles, and, in the worst case, skin cancer.

The dilemma is that there is no consensus about what constitutes "too much" radiation. Traditionally, an average white male has been used as the "Reference Man" to calculate when exposure is above or below the thresholds.[16] In one lecture, an instructor discussed the equivalent risks of medical radiation exposure. For example, receiving one abdominal CT scan would expose an individual externally to around 10 mSv of ionizing radiation. Its equivalent risks include driving six thousand kilometers in a car, flying forty thousand kilometers in a jet, or who knows how many mosquito bites in Minamisōma. In this scientific understanding, low-dose radiation exposure causes probabilistic effects. That is, a risk of radiation exposure is encapsulated in a number that can be compared to other risks. Despite the fact that the risk from low-dose radiation exposure is small, it is a risk nonetheless. Thus, what is scientifically sound is to for people to avoid "unnecessary" exposure, or, in other words, to limit exposure to what is "as low as reasonably achievable" (ALARA). To do so, one must consider the following three safety measures: time, distance, and shielding. Minimize your time near a radioactive source, keep away from it, and put something between yourself and the source. Simple, just like that.

The underlying assumption of the ALARA principle is the specific human-radiation relation. Although radiation is a natural part of our lives, our exposure to it can be and should be controlled by manipulating the environment. A safety officer's job is to determine, using technoscientific instruments, various sources of exposure to minimize human-radiation encounters both spatially and temporally.[17] The expertise required is the ability to identify the sources and know where they are. The message of ALARA is the following: radiation is essentially a manageable risk, and a lack of knowledge can lead to "unnecessary" exposure. In this technoscientific regime, "information equals reality, and thus reality can be *created and transformed in the shaping of information and information policy*" (Beck 1987, 156; emphasis in original).

Before the establishment of radiation safety, many prominent scientists such as Emil Herman Grubbe, Wilhelm Conrad Röntgen, and Marie Curie worked intimately with radioactive materials and exposed themselves and others (Jorgenson 2016). Just as the residents of coastal Fukushima became experimental bodies used to study the effects of chronic low-dose radiation exposure, these scientists' exposure led to safety knowledge as well. More detailed data on radiation safety came from collective bodies, often exposed involuntarily, such as the victims and survivors of Hiroshima and Nagasaki, nuclear workers throughout the world, victims of Chornobyl, medical patients, and many underprivileged and marginalized communities. The job of the radiation safety expert is to use the body of epidemiological data to manage known risks and educate those who do not know the "correct information" about the science of radiation. "All we need to know is ALARA," was the mantra that the Columbia instructors repeated.

In the context of the training, the TEPCO accident was an example not of a serious nuclear accident but of a failure of risk communication. By 2019 the scientific community had learned about the extent of the accident, and the Fukushima case appeared minor compared to the Chornobyl accident in terms of the total volume of contaminants released. As far as radiation experts were concerned, the TEPCO accident offered examples of both good and bad risk communication, one based on science and the other on pseudoscience. When referring to the TEPCO accident, the head instructor repeatedly discussed misinformation that circulated online, in newspapers, and on social media and pointed out the existence of "radiophobia," which causes people to react "unscientifically." Masahiro Ono, the epidemiologist we met in chapter 1, called radiophobia and people's unscientific reactions to radiation a "nuclear ghost." To combat people's phobia-driven reactions, a safety officer needs to know and provide scientific facts and educate others about ALARA. Following Ono's reasoning, then, a safety officer is a ghostbuster.

This "scientific" approach to the TEPCO accident radically differs from writings by scholars like Aya Kimura (2016), whose research challenges the scientific paternalism in risk communication. In her study of the citizen-based radiation-monitoring centers that have burgeoned in Japan since the TEPCO accident, Kimura describes the struggles among her research subjects, mostly mothers. Many concerned mothers participated in grassroots efforts to measure the radiation levels of foods as a result of their growing mistrust of food-safety authorities. Although their efforts were aimed at keeping their and other young people's potential exposure "as low as possible," they met with adverse social and political consequences. Some of these mothers were thought to have *housha nō*, or "radioactive brain," because they were not being "reasonable" in believing in Ono's version of the nuclear ghost and ignoring the existing scientific data.[18]

Kimura (2015, 4) claims that one lesson from the TEPCO accident is the ineffectiveness of the top-down communication style that is prevalent in risk communication. Risk communication for issues such as radiological contamination is structured around a hierarchical system in which scientists deliver their expert knowledge to the lay public, and "the goal is thus that local residents will ultimately think in line with the experts." In this model, going against expert judgment—failing to comply with recommendations—signals ignorance, irrationality, and illiteracy. This model displaces the issue of trust in authorities and experts to the lay public's lack of knowledge, thereby devaluing individual experiences as irrelevant. Indeed, this was the shared sentiment in postfallout Minamisōma. The TEPCO accident not only exposed the residents internally and externally to fallout debris; it also made them feel that they were exposable members of society.

At Columbia, this pattern of risk communication was being reproduced. The two-hour lecture on risk communication started with a quote from a blog post on the *Scientific American* website by author and risk communication consultant David Ropeik.[19] Written on March 21, 2011, nine days after the first hydrogen explosion of Reactor Unit No. 1, the blog post quickly diagnosed the situation in Fukushima accordingly: "The peril is not just the radiation. It's people's fears of radiation. Whether those fears are consistent with the evidence of the actual physical risk (they aren't) doesn't matter." Effective risk communication, the instructor continued, is based on facts. He then went over general points, including the data that researchers like Dr. Tsubokura carefully collected over time in Fukushima. Those data led to a scientific position that the levels of radiation in most areas of coastal Fukushima were negligible and, therefore, the fear was unscientific.

The lecture culminated in an exercise to craft a letter demystifying misinformation. One example was from a *San Francisco Chronicle* article that discussed the discovery of traces of radioactive cesium from the TEPCO accident in wine produced in Napa Valley, California. The instructor shared the following sample letter he had written with his students and sent to the newspaper, which ended with a note on the responsibility of different stakeholders: "The ability of scientists to detect ionizing radiation at incredibly tiny levels is both astounding and genuinely frightening to the general public. This capacity is far greater than our ability to detect pesticides, household chemicals, or industrial toxins in our food, drink, and air. As a result, the media, state officials, and scientists share a special responsibility to educate the public about health risks from radiation as compared to other everyday activities and actions." When I read the sample letter, it made me think of Fukushima residents who received thyroid cancer diagnoses—their agony, frustration, and sense of abandonment by the state and experts. For them, the science and its facts must have felt surreal.

Thanks to researchers like Dr. Tsubokura, in 2019, when I took the training, many aspects of the TEPCO accident had become well-known facts. What the factualized knowledge shared among the scientific community did not cover, however, was the strenuous negotiation processes through which the experts and the lay public confronted the TEPCO accident and the politics surrounding radiation exposure. This lack was evident when I posed the riddle of radioactive mosquitos in Minamisōma. The lecturers and participants alike took my question as a joke and an invitation to elaborate culturally specific fantasies of radiological contamination in the West, like Spider-Man, the Hulk, the Teenage Mutant Ninja Turtles, and SpongeBob SquarePants. Their response reinforced how science can be used in dismissive and oppressive ways, just as the Minamisōma residents experienced.

A specialist in radiological-risk media communication, Tanja Perko (2016, 685) has observed a trend in the European media of referring to the memory of the Chornobyl accident when communicating about the situation in Fukushima. This trend suggests to Perko that "Fukushima will become a reference in risk communication related to nuclear incidents and accidents" for future nuclear emergencies. If Fukushima is the reference point, what were some lessons people could learn from it for better risk communication?[20]

## EVERY MESSAGE HURTS SOMEONE

Dr. Tsubokura arrived in coastal Fukushima one month after the accident.[21] Born in Osaka city and a graduate of Tokyo University Medical School, he previously had no *en* to the region. In April 2011, when he was attending a conference in France, a phone call from his mentor changed his life path. Only after the phone call did Minamisōma become an actual place on his mental map.

After years of commuting to coastal Fukushima from Tokyo, in 2020 he took a professorship at Fukushima University, where he now leads the Department of Radiation Health Management. I first met Dr. Tsubokura in 2013, when he was giving a talk to a *kasetsu* community in the Kashima district. I was a frequent listener of his radio program while driving around coastal Fukushima and wanted to meet him to learn more about his experience. When he spoke, his Kansai dialect, which was more familiar to my ear than the Sōma dialect, was audible. Among the elderly residents at the *kasetsu,* he looked very young. Despite his outsider status and age, his commitment to the health and well-being of residents and sustained face-to-face engagements with them gained their trust, unlike some other experts who crunched numbers from afar. He once explained his dedication to coastal Fukushima as his small gesture of bereavement for the suffering experienced by countless people due to the TEPCO accident.

His interactions with the locals started with emergency volunteer work. In early 2011, as Dr. Tsubokura attended to the evacuees in temporary shelters and *kasetsu* units, he encountered many concerned residents who wanted to know about radiation and its dangers. They knew why they had to evacuate but did not know precisely what the threat meant to them and their families. Information and opinions about radiation mushroomed in the news and online spaces. Bombarded by the excess of information that followed the earthquake, tsunami, and nuclear meltdowns, some residents described 3.11 as a combination of natural, human-made, and "information disaster" (Takahashi 2011). While providing emergency medical care, he offered his knowledge about radiation and its probable risks to whoever asked. This simple act seemed to relieve confused residents. At the same time, his visit to the evacuation center in Iitate village in May 2011 exposed him to the broader impacts of the TEPCO accident, which not only released radioactive debris but also displaced countless locals.[22] From his early exposure to the community in crisis, Dr. Tsubokura knew that medical care in coastal Fukushima required more than a knowledge of the science of radiation; it also required being able to communicate the risks of radiation.

Nonetheless, his attempts at communicating the potential risks of radiation exposure met with diverse reactions. While some residents were put at ease when they learned that radiation is omnipresent and the amount detected in the region was below medically significant levels, others doubted that information, accusing him of normalizing the situation and spreading misinformation. Since the accident, different experts had weighed in on the potential dangers of radiation exposure. Even though they based their judgment on science, the experts themselves disagreed and criticized each other. Producing more confusion than clarity, the experts gradually lost their credibility with the people of coastal Fukushima.

The more residents and evacuees Dr. Tsubokura interacted with, the clearer it became that locally specific data was missing. Residents wanted to know what the TEPCO accident could mean to their bodies and not to a hypothetical body.[23] Tengo, an evacuee from Odaka, once articulated his frustration with what he heard from the state, TEPCO, and the experts in the following way: "They all repeated that there was no immediate danger, saying that this was nothing like Hiroshima and Nagasaki, or they would say how Chornobyl was worse than Fukushima. But my question was what could happen to my family and me in Minamisōma!" For residents, Fukushima was not Hiroshima and Nagasaki, nor was it Chornobyl. It is a real place and not a conceptual space of contamination.

There was a dire need for more specific data on radiation exposure to back up Dr. Tsubokura's general medical knowledge. The residents needed to know—they had the right to know—the risks from the elevated levels of radiation where they lived. In the sheer chaos of the combined disasters of the earthquake, tsunami, and nuclear meltdowns, nobody seemed to be collecting data, and this was the void Dr. Tsubokura decided to fill. Within a few months after the phone call from his mentor, he started working part-time at Minamisōma Municipal General Hospital, where he quickly set up an internal-exposure monitoring regimen in July 2011. Because the instrument was not readily available, the first whole body counter (WBC) in Minamisōma came from Tottori Prefecture, more than nine hundred kilometers away. Dr. Tsubokura was a specialist of hematology, not radiology, but he quickly taught himself to operate the machine. Many residents rushed to enroll for testing, filling up appointments for the next six months, and the team at the hospital tested the internal exposure of around twenty people a day.

By 2012 Dr. Tsubokura had collected over ten thousand WBC data, and in 2014 he published the results with a local nonprofit organization in a booklet for local mothers (figure 5). Since he was collaborating with a local women's group, it was essential for Dr. Tsubokura to think carefully about the language he employed to communicate the data. From his experience, technical terminology alienated residents and the ghostbusting strategy made them less receptive to the message at hand. If the booklet reproduced the state and the nuclear industry's risk communication tactic, he thought, the booklet was not necessary. Instead, he communicated that the risk was never zero, despite the scientifically negligible risk stemming from both internal and external exposure in the region (Akiyama et al. 2015; Tsubokura et al. 2017; Tsubokura et al. 2015; Yamamoto et al. 2019).

FIGURE 5. The front page of the booklet "First Held in Minamisoma Radiation and Health Seminar by Dr. Tsubokura." By the Veteran Mothers' Society (2014).

Dr. Tsubokura's message was different from the state's, which emphasized using objective knowledge from the science of radiation to correct people's "excessive" concern about radiation. For example, the residents repeatedly heard popular health rhetoric similar to what I heard at the Radiation Officer Training at Columbia University, which went something like this: in everyday life, humans are internally irradiated from the consumption of potassium (for example, from bananas) and from the intake of airborne radon in the environment. While this communicates the known fact of internal exposure, it also conveys that people's concerns about contamination are unscientific, if not pathological.

Against this top-down ghostbusting communication strategy, Dr. Tsubokura took a more action-oriented approach. He suggested that the residents who lived close to 1F do what they could to reduce their total amount of exposure from both internal and external exposure. For example, even if individuals avoided going outside and received less external exposure on average, if they did not pay attention to the food they ate they could still experience a high level of internal exposure. The radiological threat had an intimate relationship to lifestyle: where one lived and went and what one did and ate each day mattered. Even though it would take a long time for the radioactive traces of the accident to disappear, everyone could act to reduce the total amount of exposure individually and collectively. Dr. Tsubokura's WBC data on the residents of Minamisōma, Sōma, and elsewhere backed up this claim.

With his risk communication strategy, Dr. Tsubokura never intended to suggest that the TEPCO accident had not harmed the environment and people both inside and outside of Fukushima. Similarly, the residents did not simply ignore the potential effects of living with chronic low-dose radiation exposure. They tried to shield themselves from the public discourses on it by resisting the scientific equation of their bodies with the bodies of Hiroshima, Nagasaki, and Chornobyl. Instead, they collectively showed that exposure is locally specific.

The accident did haphazardly release artificial radioisotopes across international boundaries. In fact, the TEPCO accident released a significant amount[24] of various radioactive isotopes such as iodine-131, xenon-133, cesium-134 and -137, strontium-90, tellurium-132, and others into the air, adding to contaminants from the atmospheric nuclear-weapons testing in the 1950s and 1960s and the Chornobyl accident in 1986.[25] However, the actual medical concern for local communities was the residents' total amount of internal and external exposure, not exposure from different types of radioisotopes, especially in a chronic low-dose exposure situation.[26]

Dr. Tsubokura's medical-scientific position is that his and others' longitudinal studies of internal and external exposure indicate no empirically significant increase in people's exposure to cesium.[27] The residents' exposure levels also dropped a year after the accident. The amount of exposure, both internally and externally, turned out to be less than what people received from nuclear weapons testing and the levels in many European countries after the Chornobyl accident.[28] Most of the machine-detectable exposure cases that his team identified had been preventable. The cases were among either those who had consumed certain contaminated foodstuffs (e.g., boar meat, mushrooms, edible wild plants, citrus fruits, and demersal fish), or those from outside the region, who were less careful about exposure. More importantly, they tended to be socioeconomically disadvantaged (Sawano et al. 2019).

Using the locally specific data he and others gathered over time, his advice to the residents became more specific and nuanced than the initial, more general advice he had offered to the concerned residents in early 2011. Dr. Tsubokura's empirically grounded message was that the residents could effectively avoid unnecessary radiation exposure individually and collectively. More importantly, the data revealed that excessive avoidance of radiation by limiting outside activities, avoiding local produce, and living away from one's home could open individuals to other health risks (Tsubokura 2018). Local data enabled Dr. Tsubokura to communicate to the residents that reducing their exposure to "as low as reasonably achievable" (ALARA) was not the only goal and to teach them to think about what "reasonable" exposure could mean to differently situated individuals in postfallout coastal Fukushima. Different individuals perceived the risk differently, and their beliefs had to be honored.

Dr. Tsubokura's research with the locals betrayed both public impressions about the TEPCO accident and the related media and antinuclear discourses. Evacuation posed far more significant risks to the health and well-being of evacuees and residents around the evacuation zones than radiation.[29] Most recent studies of Odaka residents from 2021 by Dr. Tsubokura's team (Kobashi et al. 2021) suggest that elderly people who returned to a former evacuation zone required less long-term care than those who remained evacuated. The implications of this finding were colossal considering that the accident produced more than one hundred sixty thousand evacuees in Fukushima Prefecture alone. Although Dr. Tsubokura's data revealed a fundamental flaw of the state's emergency evacuation protocols, its postfallout countermeasures were not a complete failure. As described in chapter 3, unlike during the Chornobyl accident, the Japanese state responded quickly by stopping the circulation of potentially contaminated foodstuffs and by lowering the allowable levels of contamination in food to 500 Bq/kg until March 2012, and then lowering them to 100 Bq/kg.

It is not often discussed that the success in lowering contamination levels in food is owed to the many evacuees and residents who sacrificed their cows, produce, and fertile land.[30] For example, Matsumoto, a dairy farmer in Katsurao village, southwest of Minamisōma, was one of many farmers who had to kill his cows to prevent the circulation of milk and meat from Fukushima following the TEPCO accident. Witnessing his oblivious cows being euthanized en masse, all Matsumoto could do was cry and apologize to the cows, whose carcasses were mercilessly thrown into a pit.[31] In Ōtomi hamlet in Odaka, there stands a memorial tablet (figure 6) commemorating the cows that suffered from starvation, experimentation, and extermination.[32] It says, "Cows that cried continuously for food without knowing the existence of radiological contamination and starved to death. Cows that were forced to be experimented on and killed for humans. Cows that died due to our selfishness. Please forgive us."

Although his empirical data on radiation exposure provided good news for concerned residents of Minamisōma and elsewhere, Dr. Tsubokura revealed his frustration to me in June 2015 at the city hospital. While showing me the special "babyscan" whole body counter (WBC) customized for measuring the internal exposure of young children, he shared that his research data had unintentionally come to tell a narrative that denied the residents' and evacuees' subjective experiences of the anxiety and fear of living with radiation—a story that normalized the TEPCO accident, like the one I later heard in 2019 at Columbia University.[33] Speaking strictly in terms of his professional, academic perspective as someone trained in medicine, he lamented, "Many people inside and outside often accuse me of being a pronuclear doctor working for TEPCO and the state because, as a scientist, what I tell people based on the data is this: there is almost no significant radiological risk from the accident for people around the evacuation zones. I have been consistently communicating this to many people inside and outside of Fukushima, hoping to combat the stigma many residents are facing and might face. But some people think I am [full of] hocus-pocus [detarame] or trying to please TEPCO and the state by promoting the safety myth [anzen shinwa]."[34] Reflecting on his comment for a second, he continued, "Perhaps

原発事故放射能の存在すら
知ることもなく空腹に鳴き続け
息絶えた牛たちよ
人間のため試験に供され
命を絶たれた牛たちよ
全て人間の身勝手により絶命した
牛たちよ許して下さい
合掌

FIGURE 6. Commemorative
tablet for cows in Ōtomi,
Odaka district, Minamisōma.
Photograph by the author.

what the people want and need after the emergency phase is not the medically grounded knowledge someone like me could offer, but more psychological support to help reduce the stress and the anxiety of living with radiation and its associated negative images."

As a medical doctor and researcher working closely with the residents, Dr. Tsubokura learned that every risk communication risks hurting someone.[35] He confronted this moral dilemma most acutely while working to put together the radiation safety booklet with the local women's group. As requested, the booklet was prepared for mothers who remained in the region. It was not for those who had evacuated. Supporting the decision of women who stayed, however, unintentionally implied that those who stayed away were being irrational. "Whatever the message is," he said in a 2017 interview by a online magazine,[36] "it always hurts someone."

Dr. Tsubokura's struggle with communicating the invisible to the lay public reveals a dilemma of the knowledge production of nuclear things. This dilemma is what Ui ([1971] 2006) identified over fifty years ago as the surreal dimension of environmental disaster, the social discrimination it instigates. The medical knowledge of an expert like Dr. Tsubokura was much valued when the residents felt uncertain and scared about the invisible presence of radiation. Once his studies gradually visualized the medically negligible intersectionality between the half-life of radioisotopes and residents' lives, his expert knowledge became less valuable, even to the point that some people began to devalue it as ideologically skewed. Dr. Tsubokura said to me, "I am struggling with my position here recently because, as a doctor, I feel as if I have been doing a disfavor to some residents, for whom the effects of radiation need to be kept opaque to get what they need [compensation]."

His experience offers a view into the limitation of science's role in risk communication. In a time of crisis and uncertainty, scientific data are critical. The data produce facts and help inform the public. However, the data do not necessarily validate or speak to individuals' subjective experiences, and, at times, the data might deny them. Should the experts' job be calling out people and correcting their "irrational" thinking if the lay public's concern is about radioactive mosquitos or honoring their curiosity and interests in the science of radiation? That's not a scientific question, but it is a question with social and moral implications.[37]

One lesson Dr. Tsubokura learned is the following: the data that scientists and authorities use should be consistent when communicating radiological risks. In contrast, the Japanese state and the experts kept shifting their stances, changing what was "permissible," and ultimately deferring the risk assessment to individual residents. Dr. Tsubokura believes that the lack of trust in experts emerged from this lack of consistency. Instead, experts must communicate based only on data, on the known facts, and they are responsible for explaining if and when new data challenge previously known facts. More importantly, they must take responsibility for those facts. On the other hand, risk communication should be done by other experts, like social workers, counselors, community leaders, and other care specialists, who can better attend to people's situated experiences and specific needs and concerns.

Dr. Sae Ochi (2021, 193), a medical doctor at Jikei University and a part-time physician in Sōma city, also observed the broader effects of the accident among her patients. "The residents may not think of radiation as science in the first place," she writes, and when they talk about their concerns, they are talking about "life instead of science." What is the role of the radiological protection experts and scientific facts in developing a heuristic for the residents and evacuees to make "an informed decision" (Schneider et al. 2019, 269)?

As the state increasingly relied on the "objective" science of radiation and engaged in half-life politics to form its postfallout policy, residents felt that any decisions they made could be wrong, such as the decision to stay in or leave one's hometown, to return or abandon one's house, to eat or avoid local produce, to drink tap water or bottled water, and to raise children at home or elsewhere.[38] This was especially the case when "there is no explicit guidance about what should or should not be done in daily life" (Leppold, Tanimoto, and Tsubokura 2016, 60). Inherent in each decision was the risk of potential "unreasonable" radiation exposure. How could residents and evacuees attend to their broader livelihood when radiation risks had become socially and politically inflated?

## FROM THE SCIENCE OF HALF-LIVES TO LIFE ITSELF

Although Dr. Tsubokura is an iconic figure of radiation exposure research, since late 2012 most of his studies have addressed lifestyle and secondary health issues overshadowed by the selective attention given to radiation exposure. Those issues include diabetes, hyperlipidemia, hypertension, decreased motor function, and mental health issues.[39] As on-site researchers moved beyond general radiation-exposure concerns, the TEPCO accident emerged as more than mere radiological danger; instead, it impacted the surrounding communities' health and well-being, broadly construed.

This is not to suggest that, as communication scholar Olga Kuchinskaya (2014) observed in postfallout Belarus, the state and experts kept the affected population's awareness of radiation risks low. On the contrary, the mosquito inquiry on the radio in Minamisōma and various citizen-centered radiation-monitoring projects throughout the country (Brown 2021; Kimura 2016; Murillo and Bonner 2021; Kuchinskaya 2019; Polleri 2019) indicate the exact opposite of a lack of awareness in Japan. Moreover, Dr. Tsubokura spent a significant amount of time giving seminars at temporary shelters, city events, and local schools (see, e.g., Kuroda et al. 2020; Tsubokura, Kitamura, and Yoshida 2018).

His public outreach work was one way to address indirect radiation issues, such as the loss of trust in local produce despite evidence that it was under the regulated contamination level. When schoolchildren were routinely tested for internal radiation, Dr. Tsubokura's team asked their parents to fill out a questionnaire. One of their answers caused the team concern. Even though the WBC data suggested that there was no internal contamination among local adults and children, more than 60 percent of parents answered that they avoided local vegetables, rice, and tap water because of fears of contamination. This result led Dr. Tsubokura to focus on educating youth. He felt that it was essential for them to have accurate knowledge of radiation, not necessarily so that they could think scientifically but, more importantly, so that they could regain local pride and combat the potential discrimination and stigma they might face as they traveled outside the prefecture. For residents, life mattered as much as—if not more than—the science of half-lives.

Perhaps the difference between postfallout Belarus and Minamisōma lies precisely in the consistent presence of figures like Dr. Tsubokura, who enabled the residents to "articulate the signs of radiation danger" (Kuchinskaya 2014, 37). In a way, Dr. Tsubokura's WBC and public outreach work contributed to establishing an infrastructure of radiation knowledge. According to Kuchinskaya, such knowledge provides "available instrumental resources (tools and spaces for articulation) as well as interactive resources (opportunities for interaction with other perspectives)" (37).

The booklet Dr. Tsubokura coproduced with the local women is another example of this point. The booklet speaks to the struggle of experts and laypeople to communicate with each other. In this collaborative process, the residents had to negotiate the significance of the potential risk imposed by the TEPCO accident and to regain their confidence in the livability of the city. To do so, they not only had to learn the science of radiation, but they also needed to trust in their own experiences and the place they lived. Sometimes this meant asking a radio show a question about radioactive mosquitos.

A group of researchers, including Dr. Tsubokura, have discussed their collaborative process of preparing an "information booklet for returnees" with Iitate village locals: "The nuclear power plant accident has changed the lives of many villagers. . . . There is no magic solution. But no matter how trivial the problems appear to be, it is important to work on those small things together with villagers in order to help them 'make sense' of life after the accident" (Kuroda et al. 2020, 314). Dr. Tsubokura's long-term engagement with coastal Fukushima residents suggests that any effective risk communication strategy requires, first and foremost, an understanding of life as defined and lived by the community in crisis. Radiation is omnipresent and radioactive materials are everywhere, but there are locally specific pathways through which they travel and affect lives.

Equally important, however, is that establishing an infrastructure of radiological knowledge led to adverse reactions both inside and outside the region. As Dr. Tsubokura hinted, for some people the TEPCO accident had to remain dangerous so that it could continue to support antinuclear or anti-governmental political positions. For others, the contamination from the TEPCO accident had to be deemed unsafe in order to maintain the local compensation-based economic structure. For others like Naoko, however, the persistent presence of contaminants and the associated compensation were signs of an extended period of alienation that needed to end for her to be able to die well at home. What different individuals considered to be risky varied significantly. The perceived risk of radiation was more closely related to a particular livelihood in a specific place than to the contamination and various half-lives visualized in bodies and the environment.

As I discussed in the previous chapters, the technoscientifically visualized presence or absence of radiation had more than a scientific life. In the entanglement of science, politics, and economy in a country with a traumatic history of radiation exposure in Hiroshima, Nagasaki, and elsewhere, Dr. Tsubokura's research data emerged as a potential threat; his locally specific findings appeared to support the state as well as TEPCO's efforts to "underplay" the degree and extent of the accident. Such an interpretation is far from accurate, since his conclusion was not that there is no risk from the TEPCO accident, but rather that radiation is not the primary health risk among the residents living around the evacuation zones, for which the state and TEPCO had failed to be accountable.

Dr. Tsubokura's struggle to tame the selective attention to radiation exposure with local data and redirect the public's attention to secondary health issues illustrates the limits of science as the dominant approach to confronting an invisible hazard in society. Ochi, who similarly grappled with risk communication issues in coastal Fukushima, discusses the violence of scientific vindication that "sometimes hurts the victims of a disaster even if they [the experts] do not intend to be paternalistic" (2021, 193). Ochi points out that the primary concerns of disaster-affected populations show that "a context-specific rationality among residents" had "a more potent influence on decision-making than epidemiology and statistics" (193). All they need, Ochi argues, is "life communication."

The radioactive mosquitos in Minamisōma that Dr. Tsubokura encountered in the summer of 2012 captured the tension residents experienced between the science of half-lives and life. A bite might not expose an individual internally or externally, and thus it bears no medical or epidemiological significance. Nonetheless, a bite might expose the individual to other risks. Anxiety and fear arising from the lack of an adequate answer about the bite's meaning might lead the same individual to seek information on the internet, in the media, or on social media, where scientific (mis)information and (mis)translations live and propagate (Hasegawa et al. 2020; Pascale 2017; Tsubokura et al. 2018; Valaskivi et al. 2019).

The TEPCO accident led the state, experts, media, national and international activists, and lay public alike to attend to the invisible by making its presence and its relation to humans visible. While this was a necessary response to the unprecedented crisis, the selectively visualized half-lives in the region gradually became obstacles that got in the way of residents and evacuees envisioning their lives after the accident. This discursive and interpretative shift in society from "life itself" to a radiation-centered "half-life" alienated the residents from their lived experience. It rendered the technoscientifically measurable presence or absence of nuclear things as the central social, political, and moral project.

Dr. Tsubokura and others' struggles with the surreal or extrascientific dimensions of the TEPCO accident exemplify how the selective attention to radiation fissured the atoms of the community. While Beck (1992) argues that the individualization of risk is the defining characteristic of modern society, what is individualized in postfallout is not the risk but its perception. Experts like Dr. Tsubokura, who tried to expose the locally specific interrelationship between the residents and radioactive materials through the establishment of radiation knowledge infrastructure, became additional victims of the discriminating process of radiation-centered, half-life politics.

Half-life politics renders radiation as a single, purely objective technoscientific reality independent of any social, cultural, political, historical, and economic contexts of those living with radiation. For the residents, the technoscience of radiation generalized their differences and experiences in its knowledge production. It also invalidated their situated sufferings of personal, social, and political kinds that are not empirically falsifiable. In postfallout coastal Fukushima, the invisible particles and their rays became more visible, and their various half-lives appeared to matter more than the residents' lives and well-being. How, then, can we read against this archive of what has been rendered visible? Hatsumi experienced her postfallout life blown off course by the storm of the science of half-lives and gave these disappearing lives in Minamisōma a label, the "nuclear ghost." In order to attune to the nuclear ghost of Minamisōma and suspend the alienating effects of half-life politics, I suggest we drop our Geiger counters and hear how the local winds sing.

# SIX

# *Between* Fūhyō *and* Fūka

You can hide memories, but you can't erase the history that
produced them.

HARUKI MURAKAMI,
*Colorless Tsukuru Tazaki and His Years of Pilgrimage*

## NAVIGATING THE TWO WINDS

Back on Route 6 again. On May 29, 2015, more than four years after 3.11, I
was on the way to the Fukushima Dai-ichi Nuclear Power Plant. In January
2013, TEPCO had established its Fukushima reconstruction headquarters in
Naraha town, about twenty kilometers south of 1F, to signal their commitment to assisting in the recovery and reconstruction of coastal Fukushima's
local communities. By the end of 2014, TEPCO's public relations department had loosened the visitor policy of 1F, which had previously opened its
doors only to scientists, engineers, politicians, regulatory bodies, and selected
members of the media. This signaled at least two subtle but significant
changes in TEPCO's public relations strategy. First was TEPCO's confidence in the increasing stability of 1F; working conditions inside the power
station had become more manageable and safer than before. Second, TEPCO
had changed whom they considered to be the eligible stakeholders (*kankeisha*) of the accident.

As containment began at 1F, TEPCO realized there was an incommensurable gap between the experts and the lay public in terms of the perception of
radiological risk. Although the experts used technoscientific tools to monitor
the risk and gather data to communicate the progress at 1F, the public persistently questioned TEPCO's lack of transparency (*toumeisei*). For the public,
radiation was an alien species and the perpetrator of dehumanizing violence
inflicted upon the irradiated victims of Hiroshima and Nagasaki.

To address this gap, TEPCO opened its iron gate and transformed 1F into
a place for the public to experience. By 2015 1F was no longer a forbidden site,

to be discussed and critiqued from afar.[1] Just as a group of scholars and activists in 2013 had proposed should happen (Azuma 2013), the transformation of 1F and coastal Fukushima into a tourism site was in progress.[2] For TEPCO, this meant a return to the 1970s, when the company used the tour of the facility to promote the safety of nuclear energy and 1F had received about six thousand visitors a month, for a total of seven hundred thousand visitors between 1970 and 1979 (Tahara 1979).

In recent years TEPCO has exposed many interested individuals, including residents, researchers, and students, to the inside of 1F: 10,000 people in 2017, 12,500 people in 2018, 18,900 people in 2019, and 4,322 in 2020.[3] TEPCO encouraged the visitors to share their experiences via social media and other means by providing visitors with photographs of their tour. The presence of visitors, in addition to plant workers, made 1F one of the most crowded places in the region. According to TEPCO, since 2012 there has been an average of five thousand laborers, about half of whom are locals, working inside the plant daily. In the first year of the accident, about twenty thousand workers exposed themselves in order to stabilize the damaged reactors and improve the working conditions within the facility.

More than thirty kilometers south of 1F, an Iwaki resident, Riken Komatsu (2018, 2021), works for a fish paste (*kamaboko*) company. He is also a local activist who is one of the founding members of Umilabo, a citizen-run organization that monitors radiological contamination in fish around the 1F property. To experience the "unavoidable" local heritage of coastal Fukushima, Komatsu joined the 1F tour in 2015. As he wrote in his 2018 book *Shin fukkō ron* (New reconstruction theory) and again in its updated edition in 2021, while he was experiencing 1F from inside the tour bus, a thought came to him: "Even if humanity ends, the radiological contaminants from the 2011 accident will not disappear. What an irony. It is not us, but radioisotopes—that which shuttered our hearts—that will keep on telling the story of the accident. We are powerless against nuclear power" (2021, 249).

Living through the accident and its messy aftermaths as a local stakeholder whose present and future livelihood depends on the safety and progress at 1F, Komatsu diagnoses Fukushima's postfallout conundrum by identifying the "two winds," or *kaze*/風, blowing simultaneously. One wind is "harmful rumor," or *fūhyō*/風評, spelled in Japanese as wind and reputation, and the other is "forgetting," or *fūka*/風化, spelled in Japanese as wind and shapeshifting. We have seen through residents like Hatsumi and Naoko that as time passed and the initial fear of the TEPCO accident gradually receded, many locals and nonlocals alike stopped talking about Fukushima. The sheer complexity of the postfallout issues (*katarinikusa*), ideological divisions, technical jargon, and rapid changes in policies together discouraged the public from updating their knowledge about Fukushima, and their early impressions of Fukushima as a dangerous place to be avoided lingered. Although Dr. Tsubokura tried to update people through his radio program, it was aired locally and had a limited reach. Moreover, many of his research findings have been published in English and mostly for the professional community.

Hiroshi Kainuma (2015) contends that the general avoidance of the complicated topic of Fukushima led people to subscribe to stereotypes, such as "Fukushima is an unsafe place" and "everything and everyone in Fukushima is contaminated." This, in turn, led to a confirmation bias where the public sought information that confirms those impressions. His 2015 book *Hajimeteno Fukushima-gaku* (The first time studying Fukushima) tries to counter this selective remembrance of Fukushima and update people's understanding by offering up-to-date facts about Fukushima and the TEPCO disaster in a Q&A format. The book was somewhat controversial. While people who are already interested in and somewhat familiar with Fukushima and its postfallout used it to update themselves, Kainuma was also accused of being pronuclear, supporting the state and TEPCO in underplaying the magnitude of the accident, and especially its impacts on human biology.

In his discussion on "slow violence," environmental humanist Rob Nixon (2011) argues that a slow onset and attritional disaster, like chemical and radiological exposure, tends to escape the modern, fast-paced, technology-oriented society, in which people are increasingly trained to refresh or slide screens to retrieve more up-to-date content. He writes, "In the long arc between the emergence of slow violence and its delayed effects, both the causes and the memory of catastrophe readily fade from view as the casualties incurred typically pass untallied and unremembered" (8–9). Between the spectacular image of radiation exposure and the slow, shapeshifting consequences of the accident, *fūhyō* and *fūka* are intertwined. If Nixon is right, people are more likely to pay attention to and remember spectacular things like irradiated bodies than residents' invisibilized everyday struggles, such as Naoko's sleeping issue or the bullying that many evacuees experienced in and out of Fukushima, even if the latter caused an equal level of violence to the residents, if not more. In coastal Fukushima, the conundrum of the two winds is *what* and *how* to remember about the ever-changing TEPCO accident and its more-than-radiological effects. Radioisotopes do not remember, nor do they tell any story. They just decay.

To successfully navigate the two winds in Fukushima, Komatsu (2021) argues, the public needs to learn about the current state of the region. For this to happen, each individual needs to cultivate broader interests in its histories, cultures, and people. Unlike Kainuma, whose tactic, much like the radiation experts' top-down ghostbusting risk-communication strategy we saw in chapter 5, was to update the public by providing only "correct" facts about Fukushima, Komatsu approached it less academically and more inclusively. Fukushima needs to become a more accessible topic and place so that more people, locals and nonlocals alike, could feel like they are *toujisha* (an interested party) and *kankeisha* of the ongoing accident. An individual who is interested in discovering and cultivating *en*—both preexisting and new—with Fukushima and its people is what Komatsu calls *kyoujisha,* or a person in synchronicity with others. In this context, visiting 1F is one opportunity for an individual to explore *en*—the invisible thread that connects 3.11 to each of our lives—with Fukushima and remain updated about and interested in Fukushima and its residents' past, present, and future.

A few days before my scheduled tour of 1F in May 2015, I visited Naoko at her temporary residence in Kashima. As we chatted, a TV in the living room was broadcasting a local news program. Behind me, I heard the host reporting the recent progress of TEPCO's decommissioning (*hairo*) project and a new leak site discovered inside the plant property. Glancing over to the TV, Naoko asked why I wanted to visit 1F and warned me that a young person like myself should be careful of radiation exposure. I felt compelled to see the accident site and talk to workers, I told her, because I was an outside researcher trying to engage deeply with the accident. By 2015 I was well aware that an individual in coastal Fukushima could be both a TEPCO employee and an evacuee. I felt that learning about this complex entanglement of roles was essential to understanding the region's ongoing coexistence with nuclear energy and its waste. Unlike in Tokyo, in coastal Fukushima the question was not how one could avoid the risks of nuclear energy and remain distanced but how one might live with its legacy.

All Naoko said to me in return that day was how she was looking forward to hearing about what I would see there: "You go there for us and observe what the accident is all about. We have the right to know." Her comment made me realize my privileged position as an ethnographer and the obligation

I had to report back in order to give residents like Naoko a kind of proxy access to what they might be unable or unwilling to access themselves. After all, outsider and ethnographer or not, I was living the same time with her, synchronically experiencing the continuous unfolding of the TEPCO accident. This was what locals meant by *en*, when they used it to name the invisible threads interconnecting the people, things, and other matters they live with. While the idea seems rather esoteric and surreal, it is not unknown outside Asian contexts. For those who are unfamiliar with *en*, Carl Jung's idea of synchronicity might offer a few insights.

Scholar Laura Kerr summarizes synchronicity as an idea that describes "circumstances that appear meaningfully related yet lack a causal connection."[4] Contrasting the concept of synchronicity with causality, Jung (1984, 85) believes that synchronicity is "the prejudice of the East." While I am not interested in exploring the East–West divide, I observed and experienced locals mobilizing the term *en* to acknowledge and embrace being in the same time (*kyouji*) with others—with the living, the dead, the radioactive, and the environment. When I got on Route 6 to get to the heart of the accident in the Futaba district, about ten kilometers south of the Odaka district, I thought I felt the presence of invisible threads. How, if not by some surreal working of *en*, had I come to coastal Fukushima and met and learned from various residents, like Hatsumi, Naoko, Saeki, Sachiko, Miura, Oshima, and others not named here?

At the southern edge of the Odaka district and visible from Route 6 about two kilometers inland, a tall metal pole stood right by the coast. In Odaka, unlike in Haramachi and Kashima, the ocean becomes more visible from Route 6, making the cold Pacific wind (*yamase*) more palpable. Miura, from the agriculture cooperative in Shinchi, whom we met in chapter 3, used to live in this area. He once told me about this pole marking the site between Odaka and Namie town, which was devastated by the tsunami, radiological contamination, and long-term evacuation. In 1968, Tōhoku Electric Power proposed to build a nuclear power plant (NPP) at this site. Miura belonged to a local activist group that had been resisting nuclear energy in Odaka and Namie since the 1970s. "If the TEPCO accident had not happened, Odaka could have been hosting an NPP by 2021," Miura commented.

Although the plan was never realized, the phantom NPP benefited the region.[5] Since 1986 Odaka town—and then, after the 2006 merger, the city of Minamisōma—had received around $4,000,000 total in subsidies from the national government based on the Three Electric Power Development Laws of 1974, or Dengensanpō (see Aldrich 2005, 2008, 2014; Fujigaki 2015;

Lesbirel 1998; Onitsuka 2011).[6] The state established these laws to provide incentives for the local government, cooperating municipalities, and neighboring communities to host power plants. The subsidies were used to promote social and cultural aspects of community building, such as fixing infrastructure, supporting the operation of the annual Nomaoi festival, and organizing 1F tours for residents.[7]

Following the 2011 accident, the city refused the subsidies. With this decision, Mayor Sakurai wanted to assert Minamisōma's disapproval of nuclear energy. Noticing the radical shift in the national and local sentiment on nuclear energy, Tōhoku Electric Power eventually halted its forty-five-year plan in March 2013. The TEPCO accident was a wake-up call for many, including the tsunami-inundated coastal community in Minamisōma and the Namie town officials, who had been actively recruiting the NPP.[8] In March 2015, the city of Minamisōma declared itself the first anti–nuclear energy city in Japan.[9] The TEPCO accident in 2011 had propelled Minamisōma to cut its quarter-century *en* to nuclear energy and its associated benefits.

Back on Route 6, I passed the bowling alley—Namie Bowl—that marked the entrance into the Futaba district. At 865.71 square kilometers, the Futaba district is slightly larger than San Diego and consists of six towns and two villages: from the north, Namie town, Futaba town, Ōkuma town, Tomioka town, Naraha town, and Hirono town; and to the west, Katsurao village and Kawauchi village (map 4). Unlike Minamisōma, the Futaba district still preserved much of its early disaster-stricken landscape even in 2015, four years after 3.11. Heavily contaminated and guarded with metal fences and security guards, most tsunami-damaged buildings had been left untouched, the salt damage caused by sea freeze having further damaged the already precarious structures. Despite the reopening of the entire Route 6 in September 2014, no access beyond the route was allowed without proper permission.[10]

On Route 6 in Namie, I primarily saw disaster recovery and reconstruction trucks and decontamination vehicles as well as tour buses with people visiting the coastal area to learn about 3.11. Minamisōma used to get these tour buses in 2013, but as the city gradually cleaned up its disaster-stricken buildings and rebuilt its infrastructure, the tourists' attention moved to the Futaba district. Here, it was easy for tourists to intuitively understand the effects of 3.11 through spectacular sights of destruction, unlike in Minamisōma, which was dealing with the invisible and messy consequences of radiological contamination. This is exactly the reason that TEPCO decided to open 1F to more visitors: seeing was believing.

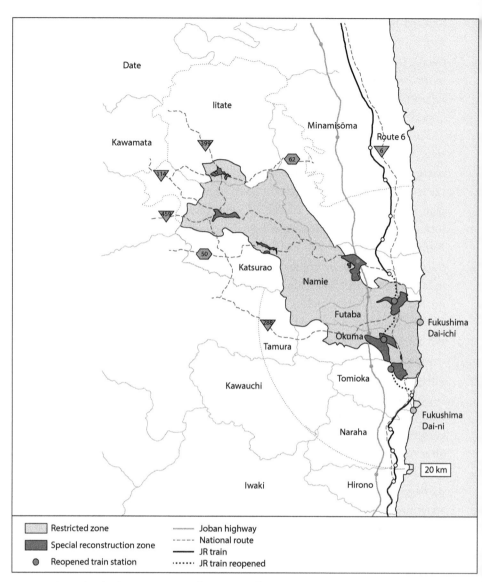

**Date**

**Iitate**

**Minamisōma**

**Kawamata**

Route 6
⑥

399

114

459

62

50

**Katsurao**

**Namie**

288

**Futaba**

**Fukushima Dai-ichi**

**Tamura**

**Ōkuma**

**Kawauchi**

**Tomioka**

**Fukushima Dai-ni**

**Naraha**

20 km

**Iwaki**

**Hirono**

| | |
|---|---|
| ▢ Restricted zone | —— Joban highway |
| ▓ Special reconstruction zone | - - - National route |
| ● Reopened train station | —— JR train |
| | ······ JR train reopened |

MAP 4. Coastal Fukushima in 2020, with an image of Route 6 along the coastline. The map also shows that the JR train between Namie and Tomioka reopened in January 2020. Taken from the Ministry of Economy, Trade, and Industry, https://web.archive.org/web/20220629142426/https://www.enecho.meti.go.jp/about/special/johoteikyo/fukushima2021_01.html, and translated by the author.

Passing by multiple tour buses, I thought about Hatsumi's nuclear ghost and how anyone, especially outsiders, could experience these messy and rather strange postfallout social and political landscapes. If the residents' experiences of the accident were so atomized and constantly shifting that they themselves could not fully articulate them, who among the disaster tourists could believe—let alone remember—their individuated experiences? I asked myself this question as I drove past gated Namie town, Futaba town, Ōkuma town, and Tomioka town to get to a multipurpose facility called J-Village at the southern boundary of Naraha town and join the TEPCO bus tour of the 1F site.

### SURVIVING GROUND ZERO

The TEPCO 1F tour originated at J-Village, located within the twenty-kilometer zone in Naraha town.[11] TEPCO built and donated this sporting facility to the local community in 1997 as a gesture of its dedication to mutual prosperity and coexistence. A national soccer-training center before the accident, with two hundred hotel rooms and eleven soccer fields, J-Village served as a critical off-site center for self-defense forces, firefighters, cleanup workers, and experts as well as TEPCO and its related companies during the nuclear emergency in 2011. TEPCO gradually moved its main operation back to 1F, where they built new facilities, including a meal center for its workers in May 2015, a convenience store in March 2016, and the main administration building in October 2016.[12] TEPCO eventually ended its off-site operation at J-Village, which resumed some of its original functions on July 28, 2018.

On the TEPCO tour I took, there were three guides. Overseeing the tour for the day was Hoshi, a man in his fifties from TEPCO's Fukushima Daiichi Decommissioning Company, established in April 2014 to optimize the organizational structure for the safe cleanup and dismantling of 1F. Accompanying Hoshi, his two younger male assistants were responsible for taking photographs and surveying radiation. From the front of the room, Hoshi welcomed around twenty-five participants. Everyone in the room looked nervous and no one was speaking to the others. Instead, everyone busied themselves with reviewing the document that TEPCO had provided for the tour. TEPCO had prepared a name tag for each participant with an explanatory title. Mine said that I was a researcher from the United States, but most of the others in the room were Naraha residents. I watched as some people underlined words in the document while others stared at it blankly. Hoshi did not seem to care about the participants' behavior, and, after checking the microphone a few times, he started the tour by introducing the general plan of the day. "Welcome to our tour," he said. "First, we would like to apologize sincerely for the accident, and all the difficulties and concerns we have caused as a result. Today, we hope to show you various sites for you to observe the progress of our decommissioning efforts."

Listening to Hoshi's eloquent narrative, I wondered why his announcement started with an apology and how this group of evacuees from Naraha town, within the thirty-kilometer zone, received it.[13] Later, through my repeated participation in various 1F tours between 2015 and 2019, I discovered that this apology was part of TEPCO's routine protocol. They articulated their position as the company with legal and moral responsibility for Fukushima's recovery. On the company's website, they announce that "Fukushima Decommissioning and Revitalization Are Our Top Priority."[14]

After relating some highly technical details about the current state of 1F, Hoshi mentioned, "As you will see, the condition of the power plant has become increasingly better. We have come a long way and are still making our best effort to improve working conditions for plant workers." He continued by describing that day's tour route, which could change depending on what was happening inside the facility. The tour's aerial map of the 1F site flattened about 3.5 square kilometers of the facility, slightly larger than Central Park in New York City. It rendered 1F like an amusement park, with notable attraction sites, instead of as ground zero of the TEPCO accident. Hoshi stated that inside the facility there were two rules concerning nuclear safety and security: "During the tour, one of us will be monitoring the ambient radiation level and will report it to you constantly. Inside the property, you must carry a personal dosimeter we provide, and at the end of the tour, we will check your accumulative radiation dose from being inside the plant. One thing we ask you is not to bring any recording devices inside. My assistants will take pictures on your behalf."

The tour started with a drive northward on Route 6 from J-Village to Ōkuma town. On the way Hoshi shared his memories of various locations and comments that residents on previous tours had made. Pointing at what used to be rice paddies along the route, Hoshi commented, "A resident told me that particular plant [by the road] did not exist before the accident, but it has taken over. I again apologize for how the accident made the area inaccessible. Our hope is for all the evacuees to be able to return and resume their lives as soon as possible." About twenty minutes later we made a sharp right turn on Route 6 toward the coast, a turn that was usually forbidden. Soon after, the bus arrived at the gate of 1F, where guards checked its license plate. After reading the license plate number aloud twice, they opened the gate. Now we were entering the heart of the accident.

From inside the bus on TEPCO's fifty-minute tour, it was surreal to think that the causes of community disintegration and the chronic, shapeshifting suffering of individuals I had witnessed in Minamisōma were nearby. I was in close proximity—less than fifty meters —to "ground zero" of the TEPCO accident. Despite that, I did not see much. It would be more accurate to say that I did not see anything beyond what the media and TEPCO had been portraying in their daily reports on decommissioning: countless temporary tanks of contaminated underground water, multiple facilities processing contaminated water, construction sites for building frozen soil walls intended to block the flow of underground water from contamination, and the broken Reactor Units 1, 3, and 4, which hydrogen explosions had shattered in 2011.[15]

From my view through the tour bus windows, I saw only suggestions that something dangerous and unknowable was nearby. The plant workers outside wearing their anonymizing protective gear and heavy-duty masks disoriented my senses. I was near the sources of contamination and danger, yet I was also far removed from the threat in a way that those nameless workers were not. After all, they were the ones directly encountering the high-level radiation emanating from the melted fuel rods. After the tour, I talked to one Naraha resident wearing a suit and tie who looked to be around sixty. He commented how the tour made him feel "*kankakuga okashikunatta*" (disoriented). For him, the experience of being at the accident site and his continued displacement from his home were confusing. "My residence is about seventeen kilometers away from here," he said. "How is it possible that the radiation levels there are not so different from here?" He shrugged his shoulders.

This reported sensory disorientation resonates with Joseph Masco's analysis that the discrepancy between the presence of potentially harmful agents and their imperceptibility renders "everyday life strange" and shifts "how individuals experience a tactile relationship to their immediate environment" (2006, 33). Notably, the tour seemed to produce and highlight this type of disorientation for its visitors. We reached the pinnacle of the tour as the bus went down the hill to approach Reactor Unit 4, whose 1,331 spent fuel rods and 202 new fuel rods had been successfully removed in April 2014.[16] It was when we collectively had this brief exposure to high-level radiation that Hoshi became most animated.[17]

He announced excitedly, "This is one of the most irradiated sections of today's tour," and signaled his fellow employee to show him the readings of a radiation monitoring device. "The reading right now says 84, 128, 239, and 300 μSv/h. Wow, it is very low now!" Calling what seemed to us visitors to be very high levels "low" grabbed our attention. Hoshi justified his claim immediately: "This area used to get readings of over 3 mSv/h [3,000 μSv/h], even after we tried to decontaminate it a couple of times. As you can see, the color of the soil on the hill looks darker as the workers decontaminated and applied shielding mortars on it." He continued, "Once we pass this hot spot, we are getting about 15 μSv/h right by Unit 4. It is one of the only units we can approach because it's been emptied [all the fuel rods had been removed]. As you can see, TEPCO and its related company quickly built a large-scale structure annexed to the unit to complete this great operation."

For a TEPCO employee like Hoshi, the ambient radiation of 300 μSv/h was a sign of concrete progress rather than a problem. He had insider knowledge about the past conditions of the plant. On the other hand, for many visitors, these few seconds of high irradiation (about three thousand times higher than the average ambient-radiation levels in areas like Naraha and Minamisōma) were the numerically translated experience of the ongoing disaster itself. To our senses, this experience was doubly mediated. Radiation was insensible, and the TEPCO worker held control over the tool that relayed its activities. The tour participants had access to neither the knowledge of radiation's presence nor its meaning. As Hoshi described it, the readings in sieverts were objective; the numbers were tangible signs of the progressive containment of the damaged reactors.

After we survived the high exposure, I overheard a few Naraha residents on the bus commenting how they were surprised to see the "real" progress inside the plant since the news media presented the situation differently. One person sitting behind me even mentioned that the media might have been exaggerating to make Fukushima look worse so that people who lived in other places could feel safer. However, Hoshi did not clarify that the radiation levels they reported were taken inside the bus. The readings did not accurately reflect the radiation levels outside the bus, where the fully geared plant workers were. Visitors could only access a "safe" place, protected by the containment structure of the bus. In September 2015, the area around Reactor Unit 3 measured about 220 mSv/h, and inside Reactor Unit 1, a probe robot measured 9.6 Sv/h or 9,600 mSv/h, a fatal radiation dose. A few seconds of exposure to such a high level would kill any individual.

For the participants, the bus tour produced two successive experiences of radiation exposure. The first was the experience of sensory disorientation; one could be near the accident's ground zero and still be safe. At the end of the tour, everyone handed back their dosimeters to get their external exposure from the tour, which indicated 0.01 mSv, or 10 μSv. Hoshi explained that the amount of radiation exposure from the one-hour tour was similar to that received during a dental X-ray.

Second, the tour produced photographic records. As I later found out, the pictures TEPCO provided to its visitors turned out to be disappointing. One could find more detailed and informative images of 1F on TEPCO's home page or through various news reports.[18] Only the tour, however, provided photos that would prove that one had been physically present inside 1F. Among many out-of-focus, blurry images, there were a few clear photographs of each visitor, busily taken throughout the tour by one of Hoshi's assistants. Participants could take away a memento of their firsthand experience, photographic proof of "having been there" (Barthes 1981), right at the heart of the nuclear catastrophe. Those pictures helped to solidify TEPCO's claims that one could witness the catastrophe and safely return from it.

Even though the tour did not provide further insights into the psychosocial consequences of the accident, it helped me experience the unsettling sentiment among Minamisōma residents I had observed over time. They often felt that their continuous presence at the edges of evacuation zones had come to justify the state and TEPCO's claim that the region was safe and livable. My presence on the plant's property became yet another sign and headcount supporting the radiation-centered claim that the accident had been well contained and was indeed under control. Like other postaccident policies and discourses, the TEPCO tour focused on radiation and radiological contamination as the primary issue—*the* sign of the accident. This half-life politics enacted by the 1F tour emphasized radiation exposure and visitors' collective survival as a possible formation of ties, or *en*, to Fukushima, making Fukushima an undesirable place to relate to. At the same time, the tour also overshadowed other, less visualized lives in coastal Fukushima, like the nameless workers outside the bus windows.[19]

When I returned from the tour, I stopped by Naoko's house to share my experience. I told her honestly that the tour was disappointing and inspired more questions than answers. She was happy that I had come back safely. Naoko was worried, she said, because she watched the evening news report of an incident of contaminated water leakage inside 1F shortly before I was on the tour. Her comment disoriented me for the second time on the same day, following the overall experience of the tour. Living thirty kilometers away from 1F, Naoko knew what was happening inside the plant. On the other hand, I had been unaware of it, despite my proximity to the incident. I wondered if Hoshi had known of this incident as he was telling visitors that the situation at the plant was improving. In my disoriented state, the TEPCO accident felt ghostly. "It is there and not there," was Hatsumi's forewarning in 2013. After immersing myself for almost two years in Minamisōma, all I could do was acknowledge and affirm that the accident and its polyvalent consequences played an incessant game of hide-and-seek.

Overhearing my disappointment about the tour, Naoko's son, Tengo, shared why he thought it would still be meaningful for me. He did not believe that I could get to "the heart of the accident" by visiting 1F. Still, he thought it was good that I had gone there, because "being there and not understanding the extent of the disaster should tell you something fundamental about the nuclear disaster and the way we have been experiencing it." Tengo disclosed how he had been frustrated with people who tended to assign a locality to the accident. "I think many people believe that radiological contaminants are in some specific place, unmovable like a piece of a gigantic rock sitting in one place. They think believing Fukushima is the only contaminated place would save them from accepting the fact that the radioactive debris traveled everywhere."

Tengo was pointing out that half-life politics produced discrimination that he experienced as a resident and revealed the structural inequality between the center and periphery.[20] For him, the name "Fukushima accident" signaled inequality. "While we have to deal with the accident here and its name," he said, "there are people out there who talk about it, own the accident like it is a card they use to play a game. The nuclear accident is an opportunity for them to change whatever they see as problematic in the country or even in the world." Tengo felt that he had been caught in an ideological war in which residents like himself were not the stakeholders but mere exhibits for both sides of the argument: for or against nuclear energy.

Tengo admitted that the accident had altered his attitude toward TEPCO, the company he had regarded highly before the accident, yet things were not straightforward. He told me that anyone could identify that TEPCO had failed and complain about them. It was much harder to try to think about the present and future of coastal Fukushima while dealing with its messy past; the power plant had existed in the region and coexisted with the community for more than forty years. "To me, nuclear energy is not a black-and-white matter. Sure, I do not want it! But I know many people who work there [1F], exposing themselves to contain the accident. Do you question those workers if they are for or against nuclear energy? Before any discussion to be possible, somebody has to contain the mess first. Don't they, like all of us, need to be included in the discussion?" With those words, Tengo pointed out the structural and relational issues that the NPP mediated between the center and periphery; locals, TEPCO, and nonlocals; and past, present, and future.

Tengo's theorization of the complex local relationship with the accident, TEPCO, and nuclear energy reminded me of my brief interaction with Jyunko at J-Village. Wearing a blue uniform with the red-and-white TEPCO logo on her chest, Jyunko, a woman in her late forties, was one of the public relations workers who lived in Ōkuma town but evacuated to Iwaki city. After the bus tour I had asked her if I could explore J-Village, and she kindly offered to show me around. She was a friendly person, and I felt comfortable enough to ask how she felt about working for TEPCO after what had happened to her home. Jyunko thought for a second and said that as a resident of Ōkuma, she felt victimized by the TEPCO accident. Still, as a TEPCO worker, she felt a sense of duty to continue serving the company because someone had to take care of the mess.

Jyunko asked me, "If the people in this region do not keep on working for TEPCO and attending to the damaged plant, then who is going to do it?" Observing my silence, she continued, "I certainly do not think many people outside Fukushima or even outside coastal Fukushima would commit to it. What is sad is that you hear people speaking negatively about TEPCO." She mentioned that she experienced discrimination in Iwaki and felt compelled to hide her status as one of the more than eleven thousand evacuees from Ōkuma and as an employee of TEPCO. "In reality, there are many of us from this region who have been working hard not just because TEPCO caused the accident. More importantly, our homes and land have been ruined as a result of the *Tokyo* Electric Company generating electricity for the center." She put an emphasis on Tokyo to signal the dislocation of cause and effect. Jyunko continued, "What I am disappointed in is not TEPCO itself but how the decisions are made in Tokyo, far away from *genba* [the site where the accident occurred]."[21] What Jyunko found annoying was not only the company's diffusion of responsibility but also its dislocation, which forced the residents, not the authorities, to work at the *genba* to combat the issues. Even though the safety of the residents and the country depended on the safe decommissioning of 1F, many TEPCO employees I interacted with between 2015 and 2019 repeatedly talked about the importance of saving money. One employee commented, "People think TEPCO is an evil company. We are constantly being watched, and we cannot be asking for too much money from the state."

Just like Tengo, Jyunko elaborated on the burden of being a resident. For residents, TEPCO was not just an evil profit-seeking company that had done nothing to prevent the accident from happening. Even if the company and its management had made a series of terrible mistakes, it did not reflect how responsible individuals within it felt. As Jyunko put it, "We local workers are willing to face the challenge. Outside, people might think we are not genuine because we belong to TEPCO, though. Sometimes I feel discouraged from decommissioning the plant because trying to make the situation better here might support people in forgetting the nuclear disaster quickly. People seem to care if, and when, things are failing and dangerous, something is leaking, or someone is exposed. Here we are exposing ourselves to reduce those things happening."

Jyunko's comment suggested her concern that the current efforts to contain the nuclear catastrophe in coastal Fukushima might be antithetical to the collective remembrance of it in the present and in the future. Similarly, the residents I interacted with expressed their concern about *fūka;* by 2015, the country and the world seem to have already forgotten about the nuclear catastrophe in general and locals' persistent and ever-changing struggles on the ground in particular. At the same time, they did not want to participate in the radiation-centered "nuclear victimhood" narrative, which overemphasized the role that radiation exposure played in their lives and living environment, just to make sure people remember the accident. *Fūhyō* and *fūka,* the two winds, made life in coastal Fukushima challenging to navigate.

In this sense, Hatsumi's nuclear ghost of Minamisōma speaks to the series of disorienting local experiences of the TEPCO disaster. In the context of the radiation spectacle, residents had become a medium and message of the danger of radiological contaminants as they were caught in the oscillating, collective negotiations of distance between radioactive contamination and containment in society. In this sense, the nuclear ghost is not merely the experience of the generalized sense of disorientation caused by the radiological danger. For Tengo, the experience of disorientation was psychosocial; he was a victim of but not a stakeholder in the postfallout politics and policies. For Jyunko, the sense of disorientation was spatial; the *genba* of the TEPCO accident was not in coastal Fukushima, but in Tokyo. For Naoko, it was temporal; the accident caused her to be out of sync with her land and ancestral spirits, with which she desired to live and die.[22]

The nuclear ghost is a category of experience standing between the insensible radiological danger and the spectacle of nuclear victimhood. It is a surreal experience that embodies the social, cultural, political, and historical contradictions at play between nuclear containment and contamination. Tengo called this experience discrimination and structural inequality afforded by nuclear energy and its failure. His experience resonates with what Indigenous communities in Native North America have experienced and named "radioactive colonialism" (Churchill and LaDuke 1986; Kuletz 1998).

When I reported my 1F visit to Hatsumi, she described how the TEPCO accident made her remember the neglected NPP nearby. She said, "It is bizarre how we knew a lot about the NPP as it came to coastal Fukushima in the 1970s but never cared to think much about it until 2011. I, for one, all of a sudden remembered that I took a trip to the plant when I was young and was impressed by its scale and technology. It is like some spectacular magic show where something disappears in a way that seems impossible. After a while, you forget what it was that disappeared, or that something was there to begin with, and then you move on to another act in the magic show."

### EPHEMERA OF LOSS

"What objects can we preserve to communicate the impacts of the nuclear catastrophe to people who have not experienced it?" a leader of a Fukushima disaster heritage preservation group in his fifties asked its members.[23] At the end of June 2014, I joined a group of scholars, museum curators, individuals from a local teachers' association, and Futaba municipality staff to survey different parts of Futaba town to identify material objects to preserve. As one of the towns hosting a nuclear power plant in coastal Fukushima, Futaba initially had evacuated 2,200 residents, except for the 20 residents killed by the tsunami, in a group promptly on the morning of March 12, right before the first hydrogen explosion at 1F. The accident transformed more than 95 percent of Futaba's territory into an exclusion zone. As a result, most of its residents (around 7,000) remain evacuated, and as of August 2021, 179 had died due to the long-term evacuation. In the 2020 prefectural survey among the evacuees, over 62 percent of Futaba residents answered that they had no plan to return.[24] The majority of those respondents stated that they had already established new livelihoods where they had evacuated. Meanwhile, the state continues to decontaminate the highly contaminated town for the roughly 10 percent of residents who are still hoping to return and the 24.6 percent who are still unsure. The state plans to reopen a small segment of Futaba in the summer of 2022.

At an abandoned junior high school, we saw a room in the same state it had been in since early March 2011; haphazardly placed water bottles, blankets, chairs, and papers suggested that people had used the room to shelter in before they suddenly disappeared, as if some supernatural force had magically spirited them away all at once (figure 7). Even though I knew, in reality, that the state had ordered people to evacuate and they had boarded buses to get to twelve different locations, including Kawamata town to the west, something about the disheveled state of the room was chilling. We all stood in the room speechless for a while.

Breaking the moment of silence, one of the group members in his early forties voiced his desire to somehow preserve the whole scene, since he thought it "convey[ed] the hectic nature of nuclear emergency." He argued that the scene represented urgency and confusion, the two raw feelings that had overwhelmed residents after the accident. He himself was in that state, he admitted, though he lived in the western part of Fukushima, about eighty kilometers from 1F. Another member chimed in, "But how would that be different from, say, a war zone where people just have to run?" Their debate exemplified the sheer impossibility of representing, let alone capturing, the uniqueness and extent of catastrophe.

Collecting materials from and about a large-scale catastrophe generates moral, ethical, pragmatic, and institutional challenges. Reflecting on the role of museums in light of the unprecedented levels of urgency and care required for curating material objects of 9/11, James Gardner (2011, 287), the associate director of curatorial affairs for the Smithsonian Institute, writes that despite

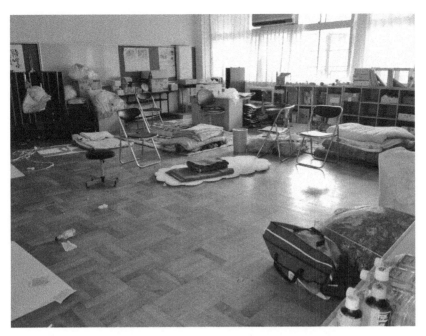

FIGURE 7. An evacuated room in Futaba town. Photograph by the author.

the issues, not collecting anything was not an option, since if the museum did not collect, then "we would have to be prepared to explain why." Just like the curators of 9/11, this group felt compelled to preserve something before the state's decontamination work removed as many things as possible in order to lower the place's ambient radiation level. A monitoring post placed outside the school showed close to 10 μSv/h reading, about nine times higher than the average level in Minamisōma in 2014. Staying here for an hour would expose us to a level of radiation similar to that I had received during the TEPCO 1F bus tour.

None of us knew how to preserve the whole room as it was. Some suggested using 3-D modeling, while others thought of transferring it all to a museum.[25] "A larger question is this," the leader said to the group while walking around the room, taking care not to disturb anything. "Let's say we preserve the room, but what is it for and for whom?" I did not know, and neither did they.

To assess the entire school as efficiently as possible, we each explored different areas. We had to work swiftly to cover the three-story building, since it was not desirable for us to be inside the exclusion zone and receive more exposure than was necessary. Our plan was to stay for about two hours. The leader told us to consult him when we found potential items for preservation. We would evaluate each item and then survey them for radioactivity to follow the state's safety procedures. The state prohibited transferring any item (including a car) from the exclusion zone when the item had a higher count per minute (CPM) than 13,000 or the contamination level was higher than 8,000 Bq/kg.

As I walked around by myself to find materials, I also observed what others were doing. Some were taking pictures of a bulletin board with TEPCO posters promoting the company. Those materials indicated the embeddedness of the company in the town. I wondered how schoolchildren might have perceived those posters in their everyday lives. Maybe they saw them with a sense of pride, as many of their parents probably worked for TEPCO. Somewhere at the far corner of the floor, one elderly member raised his voice, reporting that there was a mess inside the bathroom. "They probably did not have any water. There are still remains, so I suggest no one go there!"

While others were busily identifying objects to possibly take back with them, I did not find anything. Nothing spoke as loudly and powerfully as the various accounts shared by the residents or the presence of the exploded and unmovable reactor buildings at 1F. Gardner (2011, 286) called the material residues of TEPCO accident that the group was after "the ephemera of loss," spontaneous objects that were produced by a catastrophe and constantly modified through its aftermaths.

Feeling at a loss, I found myself in a collective teachers' office where books and files were spread all over the desks and floor. By one of the desks there was a pile of books. One in particular drew my attention: a large blue book with TEPCO's company logo and an intriguing title, *Symbiosis and Mutual Prosperity—with the Community—The 45-Year Chronicle of the Fukushima Dai-ichi Nuclear Power Station*. Its front cover showed contrasting images of a color aerial photograph of 1F and a black-and-white photograph of children in 1968 painting Reactor Unit No. 1 being assembled (figure 8). I did not know why the book was in the teacher's office. Maybe the school used it as a textbook? I grabbed the book and presented it to the group to be preserved. In my presentation, I pitched its significance as a material object that could make all of us reflect on our collective ignorance of local history and the broader history of Japan's engagement with nuclear energy. The book was a kind of document to help outsiders and locals alike be *kyoujisha* of the accident.

共生と共進
——地域とともに——

福島第一原子力発電所45年のあゆみ

TEPCO

FIGURE 8. The front cover of a TEPCO history book. Photograph by the author.

Despite my plea, most of the group members did not think the book suf-
ficiently evoked the raw emotions they were hoping would be inspired by the
objects they gathered. They collectively decided not to include the book in
their collection, but the leader told me that I could keep it. Produced
by TEPCO for the surrounding communities, the book archived the local

history of 1F. It featured multiple historical photographs and TEPCO's interviews of the residents of Ōkuma and Futaba. The book was published in 2008 but marked as "not for sale" (*hibaihin*). According to the Futaba town staff in the preservation group, the TEPCO book was circulated only among residents in a small segment of the Futaba district and a few local libraries. Using a series of interviews with elderly residents, the book details their memories of the rapidly changing Futaba district as it was modernized with the arrival of TEPCO's nuclear power plant. What becomes clear in this book, as in other historical records of the region, is the importance of developing a sense of mutual trust and bonds (*kizuna*) between the residents and TEPCO in making coastal Fukushima nuclear.[26]

## TRUST IN SYMBIOSIS AND MUTUAL PROSPERITY

TEPCO began to develop the plan for its first nuclear power plant in the early stages of Japan's radioactive nation building in the 1950s (Tokyo Electric Company Fukushima Dai-ichi Nuclear Plant Operation 2008). By 1955, TEPCO had established a department dedicated solely to researching and developing a nuclear power station.[27] Immediately after the release of the national government's nuclear energy development plan, in 1960, TEPCO started selecting a suitable location for its proposed reactor. By October of the same year, the state, TEPCO, and local government had chosen Ōkuma and Futaba towns to be the home for TEPCO's new facility. Both towns endorsed this decision in 1961, pledging their complete support of the plan to the prefecture. Importantly, unlike the Odaka-Namie NPP, which met with local resistance based on land disputes, the situation in Ōkuma was ideal. The location covered a former military facility and nationally owned land,[28] making it easier for the state and the nonlocal company to step in without having to participate in too many local negotiations.[29]

In 1966, TEPCO signed a turnkey contract with General Electronic (GE) and began to prepare for the power plant's construction. To do this, TEPCO had to bulldoze the region's thirty-five-meter hills, first down to ten meters and then down to seven, so that the placement of the reactor's base structure would fit with GE's layout plan (Leatherbarrow 2022; Toyota 2008). According to some disaster reports, such as one by the Independent Investigation Commission on the Fukushima Nuclear Accident (2014), this modification of the previous landscape was one of the structural vulnerabilities that led to the 2011 accident. In December 1970, the first reactor began operation. By 1979, the TEPCO Fukushima Dai-ichi Nuclear Power Station—or 1F, as it came to be called locally—had become one of the highest-generating power plants globally, with six reactors in total, four in Ōkuma and two in Futaba.

Despite its smooth development, according to the Ōkuma Town History Editorial Association, "There was indeed an unbridgeable gap [between TEPCO and the townspeople] at many levels on the understanding of what was being built" (1985, 833). For local landowners and others without much technical knowledge of nuclear energy, safety was the prime concern.[30] The local fishing community raised similar concerns upon the agreement to release fishing rights in 1966,[31] which resulted in fishers receiving a lump sum of one million yen.[32] The amount was framed as the compensation for the unlikely, though still possible, damages that might occur because of the potential change in the oceanic current and marine ecology resulting from the release of coolant water from the power plant. At that time, there was not much explicit discussion of the potential damage in the event of an accident contaminating the region's water. Ironically, in 2022, more than fifty-five years later, the disposal of contaminated water into the ocean is the biggest topic of debate between the state, TEPCO, and the local community.

Trust was what the locals asked for and sought from TEPCO and its engineers—the cultural outsiders who arrived with the NPP. Although locals worked for the NPP during and after its construction, the newcomers, all of whom had college degrees, tended to occupy higher positions in the institutional hierarchy. Thus, locals were not exposed to the working of the machinery itself. One resident describes the arrival of the NPP in Ōkuma accordingly: "When I think about it now, not just myself but many of the residents in the town did not understand what it meant to build the nuclear power plant in Ōkuma town" (Tokyo Electric Company Fukushima Dai-ichi Nuclear Plant Operation 2008, 57). The residents had to overcome their obvious lack of knowledge with the cultivation of interpersonal relationships and trust in the TEPCO workers. For the residents, nuclear energy became imaginable and manageable because they chose a relationship of trust with TEPCO.

Trust in the company, if not in the technology itself, promised the residents the sustainability of their region, culture, and history—a happy life. In an interview, Takakura, a prefectural staff member working for the Isotope Center in Ōkuma town, recalled residents' strange behavior, suggesting the residents' general awareness of the risk of radiological danger and their obvious lack of knowledge about it: "A memorable episode of those days [the 1980s] is how the residents looked down and walked hastily when passing in front of the isotope center. Finding it very strange, I asked my superior about this behavior. He told me that they walked past the center without breathing because they believed that radiation was leaking from the center. Also, since TEPCO's dorms and company-owned houses were built on the hilly side of the town, people spread rumors that radioactive particles gather in the lower area and do not travel high up" (Tokyo Electric Company Fukushima Dai-ichi Nuclear Plant Operation 2008, 92–93).

Instead of learning the science of radiation, the locals sought reassurance from TEPCO and its workers that nuclear energy was safe, that it was multiply contained, and that an accident was impossible. As Kainuma discusses (2011), trusting TEPCO—acting as if the power plant were safe—was a feature of residents' lives in coastal Fukushima. In the postfallout period, various scholars have critiqued the safety myth (*anzen shinwa*) as the prominent characteristic of the Japanese nuclear industry and its failure.[33] Here the myth was not only about people's unconditional trust in the safety of the power plant as a technological structure. It was, as Roland Barthes (1972) would put it, the enacted spectacle of societal hierarchies; the residents' trust in TEPCO affirmed and naturalized the existing social, economic, and political configurations, such as center-periphery, rich-poor, and technocrat–blue collar.

This sense of trust became internalized over time as more locals became TEPCO employees themselves. In the decades preceding the construction of 1F, more and more locals began working for TEPCO and its related companies (Ōkuma Town Editorial Association 1985). While in the past, less knowledgeable residents had participated in the initial stages of the plant's construction, a newer generation became technical staff who worked inside 1F after earning degrees and obtaining the knowledge necessary to maintain the plant's functions. Some climbed the ladder of the corporate hierarchy. Almost fifty years after the first reactor became operational in 1971, Ōkuma town had become a TEPCO town, with more than half of its working population (3,600 of 6,000) working directly for TEPCO or its related companies (Kowata and Kowata 2012). Over time, trust had shapeshifted to the bonds (*kizuna*) that shackled TEPCO to the local residents.[34]

Nationally, however, 1F's legacy was the way that it physically displaced the potential risk of the nuclear energy production from the center to the periphery and from urban areas to rural regions. The power plant imposed a double standard, or the logic of what historian Hisato Nakajima calls the "development of underdevelopment" (*teikaihatsu no kaihatsu*) (2014, 82), in the rural region.[35] The "development of underdevelopment" had several layers. It was an economically rational option for a private company pursuing nuclear energy, since land and labor would be cheaper than in more urban, populated areas. For the national government, which was concerned with developing nuclear power during the country's postwar modernization, the goal was to balance fulfilling its political ends with protecting the majority of its population from a potential radiation threat.

The country was dealing with the traumatic memories of radiation exposure in Hiroshima and Nagasaki (Yoneyama 1999). For most Japanese people, putting a nuclear power plant in the Futaba district about 250 kilometers away from Tokyo was the safest way to alleviate the fear of nuclear things from past experiences.[36] The development of underdevelopment reflected the national-local negotiation of risk. The majority's interest in displacing the risk of nuclear things far away met with the poor rural communities' need for economic stimulus to survive. The TEPCO book and other historical records from the two towns hosting nuclear power plants reveal how the relationship between TEPCO and the locals was not one of mutual prosperity but more one of codependency (Kainuma 2011). Trust and ignorance were what bonded the two parties and fueled nuclear energy in coastal Fukushima.

Furthermore, the local history of 1F suggests how the general public's ignorance of nuclear energy policy contributed to the disappearance of the power plant from the collective memory, as Hatsumi put it, like a spectacular magic show, until its failure in March 2011. *Fūhyō* and *fūka,* the two winds blowing in coastal Fukushima, have broader implications for the national and collective memory of the TEPCO accident. While focusing on radiation and its potential danger, its related *fūhyō,* to human bodies is essential to advocating for the residents' right to live where they feel they belong, the sole emphasis on its biological effects and scientific (half-)life make it easy for radiation's more-than-biological impacts on the residents to be forgotten—the *fūka.* If, as Komatsu contends, radioisotopes are the agent of memory, then their removal could diminish what it is possible to remember about the accident. Against the radiation-centered approach, I suggest that navigating the two winds in coastal Fukushima requires our attunement to what residents consider to be their ephemera of loss, which the state and TEPCO have been removing with radioisotopes in the name of environmental remediation and human security.

In the winter of 2015 I accompanied Tengo to his evacuated house in Odaka. He had to talk to the state's decontamination agents to discuss the plan for cleaning up around his residence. The workers showed us one area of Tengo's property and told us how they would remove soil from the ground to capture contaminants. Tengo said sarcastically to me, "No matter how much the state decontaminates, and how well it tries to hide contaminants from us, it cannot erase the history. We will not forget our miseries."

For residents, the TEPCO accident has not been just about the radiological danger that began in 2011. It is also about a longue durée of struggles for the land. "What the accident taught many of us," Tengo continued, "is we cannot let the state and TEPCO take over our land. We can no longer trust what they would do with it." Tengo's comment indicated the emerging issue with nuclear waste in the region. Unlike the disaster ephemera the preservation group tried to rescue, no one wanted nuclear waste, but it kept accumulating every day from the ongoing decommissioning at 1F and the surrounding communities where the state worked to decontaminate the environment to make it possible for the residents to return. Although coastal Fukushima's accumulating nuclear waste helped to visualize the lingering and broader effects of the TEPCO accident, it also came to contain and render local histories and individual memories invisible.

# SEVEN

# Frecon Baggu *and the Archive of (Half-)Lives*

People's memories are maybe the fuel they burn to stay alive. Whether those memories have any actual importance or not, it doesn't matter as far as the maintenance of life is concerned. They're all just fuel.

HARUKI MURAKAMI, *After Dark*

## "*FRECON BAGGU* ARE EMERGING!"

In early 2015, Saeki from Kashima attended a decontamination inspection at her parents' property in central Odaka; a few days later, she sent me pictures of the progress with the message *"Frecon baggu* are emerging!" Many locals referred to those large, opaque polyethylene bags of decontaminated waste as *frecon baggu.* Each bag holds about one ton of materials. Usually, the bag is used for construction-related work. Immediately after that, she sent another message; she wanted to convey how the emergence of the black bags made her remember something crucial she had forgotten during the last four years since 3.11: the TEPCO accident had indeed contaminated her living territory. Observing their steady accumulation throughout the city, Saeki was worried about how outsiders might consider Minamisōma to be an even more dangerous place than before. Her concern was spot on.

As Saeki experienced, the emergence of the *frecon baggu* illustrated the paradox of decontamination: it simultaneously contained and revealed the ever-changing consequences of the TEPCO accident. Even though the dread of fallout had initially stemmed from the invisibility of contaminants and their potential health effects, the accumulating black bags provided a new experience of the accident. The progress of decontamination—the removal of radioactive materials—rendered the contaminants visible. With the emergence of *frecon baggu,* radiological materials' ghostly presence had become the most noticeable stuff in the region. The accident and its aftermaths

FIGURE 9. Accumulation of *frecon baggu,* flexible containers, in central Odaka. Photograph by the author.

gained a color, form, shape, and feel for the first time. *Frecon baggu* became the cloud chamber of the TEPCO accident (figure 9).

As much as decontamination is essentially a process of cleaning that might offer a sense of order to locals who were experiencing chaos, it also selectively materialized the TEPCO disaster in the process.[1] Due to the lack of scientifically validated decontamination procedures, the state described its *jyosen*/除染 (decontamination; *jyo* = eradicate; *sen* = contaminant) policy vaguely. The Act on Special Measures Concerning the Handling of Radioactive Pollution issued in August 2011 defined decontamination as the removal of as many radioisotopes as possible in the environment: "In this Act, 'measures for

**Before**

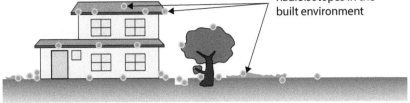

Radioisotopes in the built environment

Contained & isolated radioisotopes

**After**

FIGURE 10. Before (*top*) and after (*bottom*) decontamination, taken from the Ministry of the Environment website. English subtitle added by the author. https://web.archive.org /web/20220627040323/http://josen.env.go.jp/about/method_necessity/decontamination .html.

decontamination of the soil, etc.' means measures taken for soil, vegetation, structure, etc., contaminated with radioactive materials discharged by the accident including the removal of the contaminated soil, fallen leaves and twigs, sludge accumulated in ditches, etc., preventive measures to keep the pollution from spreading, and other measures" (Article 2.3).[2] However, the eradication of the radiological materials required tangible materials.

For any space to be decontaminated, the invisible radioisotopes must first be "captured" using material objects in the vicinity. In the decontamination framework, dirt or any other tangible materials in the environment are not pollutants on their own. They become toxic insofar as they are carriers—nonhuman hosts, if you will—of invisible contaminants. As the Ministry of the Environment's graphical representation indicates (figure 10), the main function of decontamination was to spatially carve out and (re)articulate the geographical and psychosocial boundaries between the safe and the unsafe.

*Frecon baggu* served a critical function in this process. They enabled the bundling of radioisotopes and objects in the environment and produced a new hybrid material—decontaminated waste—that could be moved from one location to another. Importantly, the state defined what was being removed from the zones to be *contaminated* regardless of their actual degree of contamination. As the container of the traces of invisible toxins, *frecon baggu* visualized the regional contamination. At the same time, their generic appearance homogenized what was being removed: local histories, personal memories, and meanings residents attached to various materials. Exploring the flattening visual effect of containers, cultural theorist Alexander Klose (2015, 341) notes that they "decompose [material objects] into at least two parts, of which one part means nothing, apart from containing something . . . and the other part bears a plethora of meanings but cannot contain them for long."

I personally experienced how *frecon baggu* had come to stimulate sociopolitical concerns about the invisibility of radiological materials. Those bags fueled the public imagination of the TEPCO accident and the desire to see and feel it. Since 2014, people outside Fukushima have asked me to share pictures of *frecon baggu* in Minamisōma more than ever before so that they could use these images in presentations or workshops about Fukushima. Previously, I had shared images of ruined structures or overgrown vegetation. Those images, however, somehow did not convey what the public imagined to be the Fukushima-like quality, and they were indistinguishable from other disaster-stricken areas in northeast Japan. Around the same time, many aerial images of *frecon baggu* in coastal Fukushima emerged in the media and on online platforms, providing a panoramic view into the extent of contamination throughout the region. Artists increasingly used *frecon baggu* in their artwork to incite public reflection on energy issues and the relationship between nature, culture, and technology. Takashi Murakami's 2014 *Power to the People,* Haruya Nakajima's 2015 mixed-media work *Odyssey,* and Haruya Nakajima's 2015 *Penelope's Border* are iconic examples.[3]

The appearance of tens of thousands of *frecon baggu* conveyed the fallout's extensive though invisible ecological consequences throughout Fukushima. Nevertheless, not everyone interpreted *frecon baggu* in the quite same manner. While for the general public, these bags seemed to represent the real and ongoing danger in Fukushima, residents like Naoko interpreted the same bags as their sacrifice, desire, and hope for returning to their alienated homes.

The residents' relation to the bags suggested that *frecon baggu* contained various contradictions of the TEPCO accident: past and present, memories and contaminants, hopes and despairs, and the visibility and invisibility of its harm. As such, they are an archive not only of the everyday consequences of the TEPCO accident but also of the accident's broader effects on the sedimented local ecologies, histories, and personal memories with which decontamination workers mechanically stuffed those generic black bags.

A nuclear ghost played hide-and-seek in and around *frecon baggu*. As Hatsumi told me, "You might not see it ever and forget that it is there or could be there, but some other people can see what you do not see and tell you it is there." Understanding what *frecon baggu* signified locally and individually requires an act of reading against half-life politics, the collective desire to uncover and see the invisible.[4] For residents, what had gone inside the bags—their misplaced and displaced livelihoods—mattered as much as, if not more than, the imperceptible risk that the bags rendered visible.

## PLOTTING A FUTURE

Naoko did not question the specifics of decontamination. She seemed unconcerned, for example, about the particularities of *how* to decontaminate an irradiated environment or *whose* standards of decontamination (the national government, the local government, a certain ministry, etc.) needed to be met. Instead, for her, the burning question was *when* decontamination would happen. It was not that she did not care to keep the land she had cultivated along with her other personal belongings. She was willing to do whatever it took to get her land back. Naoko was concerned that her clock was ticking, and she was increasingly aware of her less able body. "So long as I have my land," Naoko often said, "I can recultivate it." She knew that the land needed someone to care for it constantly.

During his stay in coastal Fukushima, journalist Hideyuki Miura (2020) heard about the state's use of the term "white space" (*hakuchi*). Bureaucratic jargon, *white space* described any space that fell outside the state's reconstruction plan or was considered to have no use value. In the postfallout context, it also meant that a specific area was highly contaminated and would be nearly impossible to decontaminate. As early as 2012, decontamination signaled to evacuees that, at the very least, the state thought their homes held some value and were worth saving. Evacuees, in turn, had to keep on proving that value to the state by expressing their unwavering willingness to return home whenever it was possible.

By early 2014, with the reopening of part of Tamura city, southwest of Minamisōma, where 121 households and 381 residents fell inside the twenty-kilometer zone, it had become clear that decontamination signaled the state's intention to reopen the evacuation zones. With the emerging association between decontamination and reopening, decontamination became for the locals not only a technoscientific measure to redress the fallout's environmental consequences but also a symbolic process, something that must happen for the region to recover. If, as Erikson (1991) argues, fallout's dread stems from the general uncertainty generated by the lack of plot in its aftermath, decontamination provided a sense of structure for the evacuees who were caught in their present anxieties and future anticipations.

Ryōji Aritsuka, a clinician in Sōma, conducted a survey in 2019 among 513 evacuees of the highly irradiated Tsushima district in western Namie. Due to the wind direction in March 2011, western parts of Namie were heavily contaminated and remain so today. Aritsuka's study revealed heightened risks for PTSD symptoms caused by the long-term displacement, a weak sense of belonging at the places to which residents evacuated, and a loss of ties to one's home and land. The risk increased for those evacuees who left Fukushima Prefecture. Furthermore, in his interviews with the evacuees of Iitate village, just north of the Tsushima district, Ōkubo (2019, 26) encountered their strong desire to return home despite the contamination. In 2020, Iitate village, unlike other municipalities in coastal Fukushima, requested that the state reopen the exclusion zone without decontamination. Finding their yearning for home to be almost involuntary, Ōkubo notes that "the desire to end life at their home might be the case of [biological] 'homing' [kisei honnō]."

The idea of home is not reducible to biological impulses.[5] Indeed, many long-term evacuees claimed that the loss of their overall livelihoods (furusato no soushitsu) from the TEPCO accident to be the central harm. Many legal battles between the evacuees, the state, and TEPCO centered on the idea of what constitutes one's home, or furusato (Yokemoto 2021; Yokemoto and Watanabe 2015). Unlike life itself, and the potential biological damage to individual bodies from radiation, the loss of livelihood, the evacuees claimed, amounted to the loss of a broader network of relations, or en—the interconnected webs of individuals, families, the landscape, the community, and the environment (Yoshimura 2018). Similarly, sociologist Katsuhiro Matsui

(2021) observes that the gradual thinning of *en* from long-term displacement led to the evacuees feeling like they were in a constant state of suspension.

If the fallout and its environmental consequences had denied the evacuees' past livelihoods and compensation forcefully translated them into legal and economic values, decontamination demanded a total reset. From evacuation to compensation to decontamination, the accident, in its shapeshifting consequences, ripped apart the preexisting ways of life and intangible *en* that supported it. The state, though, would claim that they had done their best. Unlike what has happened in other postfallout regions like Chornobyl, the Japanese government has gradually reopened portions of land that had initially been deemed uninhabitable (Kawasaki 2018a, 2018b). As of August 2021, more than 31 percent of currently registered residents in the former evacuation zones (more than 14,000) had returned. In Minamisōma, after the opening of its zones in July 2016, around 6,000 residents had returned within its twenty-kilometer zone by March 2022, and 3,812 residents remained evacuated.

In contrast to the evacuation-led centrifugal movement of people outward from 1F, decontamination was a centripetal process of moving contaminants in the environment toward 1F. Since 2012, the national government has budgeted more than $40,000,000,000 for decontamination and the construction and maintenance of radioactive-waste storage sites. In Minamisōma, the state declared the completion of the decontamination of the city's more than 61 square kilometers in March 2017. Inside Minamisōma's 20-kilometer zone, the state decontaminated 4,500 households, 17 square kilometers of farmland, 13 square kilometers of forests, and 2.7 square kilometers of road. The results materialized as a spectacular display of more than 965,000 cubic meters of decontaminated waste in black bags, distributed throughout the city's twenty-two temporary waste-storage sites (figure 11). But it was not the presence of bags themselves that revealed the work of decontamination.

In April 2015, four years after the TEPCO accident, the state-operated decontamination project was finally operating around central Odaka. Decontamination work proceeded from the east of the region to the west, presumably coinciding with regions that were more contaminated—those at a higher elevation and with more vegetation—to areas near the coast that were less contaminated. At that time Saeki invited me to observe the initial inspection of her parents' house in Odaka. As we drove from Kashima, we first encountered an assortment of black bags along Route 6 and, in Odaka, around people's properties.

FIGURE 11. Decontaminated waste in a temporary storage facility in Odaka in 2014. Photograph by the author.

"Four years after the accident, Odaka became a graveyard of *frecon baggu*," Saeki cynically commented. "Doesn't the view indicate how the region has been transforming into the land of waste?" As she lamented Odaka's current state, we arrived at her parents' house, where we met with an elderly woman, Tae, from the neighborhood. Saeki's parents, who had evacuated to outside Fukushima Prefecture, could not come to the initial assessment, so Saeki and Tae attended on their behalf. I was there as a third-party witness to the exchange between the residents and the decontamination workers.

The state-hired six workers from Chiba Prefecture (about three hundred kilometers away) were there to take an initial radiation survey and learn about items the property owner wanted to spare from decontamination. The leader, a man in his late forties in a uniform, mask, and helmet, told us that this part of the inspection process was the most important. Otherwise, he said, their job was to discard everything around the house to achieve the state-promised 1 mSv/y exposure standard. According to the Ministry of the Environment, the party responsible for decontamination, its goal was to achieve the ICRP-mandated 1mSv per year or a standard of 0.23 μSv per hour within a specific locale.[6]

While looking at the layout of the property displayed on the clipboard, the leader told us, "We are going to decontaminate the property by cleaning up around the bushes in front of the house and weeding the whole area, clearing up drains and taking out any sludge, and throwing away any items left outside the house." Based on his description, decontamination sounded more like deep cleaning a house than removing radiological contaminants. Keiji Kogure (2013), a geoenvironmental technology specialist, argues that there is no silver bullet for decontamination. Currently, the only known method to decontaminate an irradiated environment is through "washing, scraping, and diluting" (179) its radioisotopes.

As the leader suggested to us, decontamination, in principle, treats all material objects in a specific area as belonging to the homogeneous category of "contaminants," or what the state defines as pollution.[7] The language of pollution reflects the state's approach to the TEPCO accident; it is not a technological disaster but an environmental disaster like the one in Minamata city, Kumamoto Prefecture, where the Chisso corporation's local chemical plant contaminated the nearby bay with methyl mercury between 1932 and 1968, causing the bioaccumulation of poisons in fish and neurological disorders among the residents (see, e.g., George 2001; Walker 2010; Ui [1971] 2006). As an effort to remediate environmental pollution, the end game of decontamination is nothing but the overall reduction of ambient radiation levels according to the numbers that represent technoscientifically detectible levels of background radiation. As such, the effectiveness of the decontamination of a given area was measured by comparing the pre- and post-decontamination levels of background radiation after the removal of what was deemed "waste."

Under the Act on Special Measures Concerning the Handling of Radioactive Pollution, the state defines "waste" produced through decontamination as "refuse, bulky refuse, burnt residue, sludge, excreta, waste oil, waste acid, waste alkali, carcasses, and *other filthy and unnecessary matter,* which are in a solid or liquid state (excluding soil)" (Article 2.2; emphasis added). Notably, the state's definition of waste does not specify who determines what is or is not "filthy and unnecessary matter." This lack of detail became a point of contention for residents, since the state and decontamination workers had the authority to decide what fit these criteria, but the residents did not.

Because of its goal of removing as many objects as possible, the decontamination policy enacted half-life politics, where the reduction of the background radiation levels was the state's sole concern. From this perspective, the more objects that become waste in each space, the better the outcome of the decontamination. The state was there to protect residents' present and future lives from exposure while also plotting out a possible future inside the zones. To meet this end, residents should oblige by sacrificing their past livelihoods, family histories, and memories embodied in various material objects. When the radiation levels mattered more than anything else, residents' past livelihoods and anything that stood for them became good to waste.

### NONCONTAMINABLE STUFF

Although in theory decontamination is a nondiscriminatory overhaul of a specific space, in practice, it was a negotiation between the state and residents to strike a balance between the reduction of life risks (contaminants) and the preservation of aspects of their past livelihood. By 2015, almost three years after the beginning of decontamination, the state had learned that residents cared a great deal about what the state considered "waste" and that the total alteration of people's lived environment discouraged their return. If the evacuees did not come back, decontamination was pointless for the state. Unlike the state, however, for the evacuees, their livelihoods before the accident mattered as much as, if not more than, the present radiation levels.

At Saeki's parents' house, the leader of the inspection team told her, "Before we start decontaminating, please mark anything you want to save around the house so that we can tell what not to dispose of," and then he handed her yellow and pink ribbons. Pink ribbons were for items to be kept, and yellow for those to be removed. While the leader instructed Saeki how best to mark items so that workers wouldn't miss them, I walked around with my Geiger counter to measure the background radiation present at the property. It gave readings of between 0.14 and 0.22 μSv/h. According to the state's equation, the area had already met the international standard for exposure, but if the radiation level was already lower than the standard, what could be the purpose of decontaminating it?

The question prompted me to continue looking for hot spots, but there were none to be found. As I was surveying, I witnessed workers at the property next door taking multiple photographs of that property from various angles (figure 12). Saeki had told me that the neighbors' house was newly built

FIGURE 12. A decontamination worker documenting the work in progress at the property next to Saeki's family home. Photograph by the author.

at the end of 2010. The family had used it for only about four months before 3.11. When I questioned the workers, they clarified that they were documenting every step of their work to avoid a potential conflict with the house owner. A worker in his late fifties with a digital camera described that, on average, he took a few thousand pictures per household. Then he demonstrated how they took each photo with a small plate placed in the bottom right corner of the camera frame to indicate the date and location of each photograph.

When I rejoined Saeki and the others, the leader described how it was common for residents to criticize their work. They would, for example, accuse decontamination workers of discarding valuable items against the owners' will. Usually, he explained, people had forgotten to mark the items during the initial inspection. The workers just did their job, which was to get rid of everything that was unmarked. Since residents were not allowed to live in the houses, and many had evacuated far away, photographs were essential to making the decontamination process as transparent as possible. For those who lived far away and could not come to the inspection session, sometimes they had to do the entire inspection process through photos. "I wish many residents knew how to use a video conference, but many elders don't," the leader stated.

Responding to the leader's comment, Tae spoke on behalf of other Odaka residents. "We, Odaka people, have been very frustrated and angry because of the long waiting time for decontamination and even longer time of evacuation." She continued, "Our mindset right now is such that we want to find any excuse to complain. Decontamination workers are good targets because, unlike state officials, you guys come to our territory and work directly with our properties." The leader nodded and said that a significant part of the decontamination work involved listening to homeowners' complaints. According to him, residents most often vented about how slow the decontamination process was in Odaka, how demoralizing the evacuation was, and how the entire country had forgotten about the accident and its victims. "To tell you the truth, interpersonal communication skills are something I gained through decontaminating many houses for the last few years," he commented with a practiced smile on his face.

The inspection process provided evacuees some sense of control over their past livelihood, but exercising power involved making difficult decisions. Although prior to the appointment her parents had told Saeki which items they wanted to keep, she had to call her parents to discuss the fate of many forgotten things around the house. Saeki mentioned that the last time her parents had visited their home was about a year ago, and their memory of the place and their belongings was fading. "Are you serious that you are keeping that old blue bucket?" Saeki said loudly into her smartphone while examining items in the alleyway with the workers. "It is a good opportunity to throw it away. . . . . Yes, I used that in the past as a pool for my kid, but we do not need it anymore." As she fought with her parents on the phone, I saw her directing one of the workers to throw away the bucket.

Saeki and her parents went over several other items—like vases, a wooden sculpture, and gardening and farming tools—and it was clear that her parents won some battles, and a few items got the pink ribbons. Their conversation indicated that decontamination was, at least for Saeki, an opportunity to throw away unnecessary things from the past. Saeki later shared that she was trying to convince her parents to discard as many things as possible, because if they did not throw things away now, it might become difficult to do so later inside the twenty-kilometer radius from 1F, which the state had designed as the special decontamination area (SDA)—one of the additional geographic divisions that the act introduced within already-divided coastal Fukushima. It was only within the SDA that that state claimed responsibility for the collection, transfer, monitoring, temporary storage, and final disposal of radioactive materials resulting from TEPCO accident. The other areas, which included prefectures other than Fukushima, were designated as intensive contamination survey areas (ICSA), where decontamination fell under the responsibility of each municipality (map 5).

Despite her persuasion, Saeki's parents did not want to throw away many of their personal belongings because, unlike Saeki, they did not approach the selection as a pragmatic choice to be made in the present for the future. For them, each item evoked a nostalgic feeling and invited them to contemplate their past and now-lost livelihood in Odaka. For them, those items felt necessary to keep their evacuated home homey. The decontamination workers were on Saeki's side and encouraged the removal of as many objects as possible in order to increase the chance of reducing sources of radiation from the property. To me, their narrow emphasis on radiation was puzzling. While the workers stressed the importance of decontamination to lower the ambient radiation levels, the place already met the standard. They must have been aware of this. I noticed the decontamination workers had put up a sign at the property that indicated the results of the pre-decontamination radiation survey.

The dynamics between the evacuees and the decontamination workers illustrate how decontamination is, first and foremost, about negotiating the meanings of postfallout life in Odaka. For evacuees, decontamination entailed several things. It could mean the juncture to determine the continuity or discontinuity of their belongings, the preservation of mementos of a specific event, and an opportunity to reassess what might be essential for

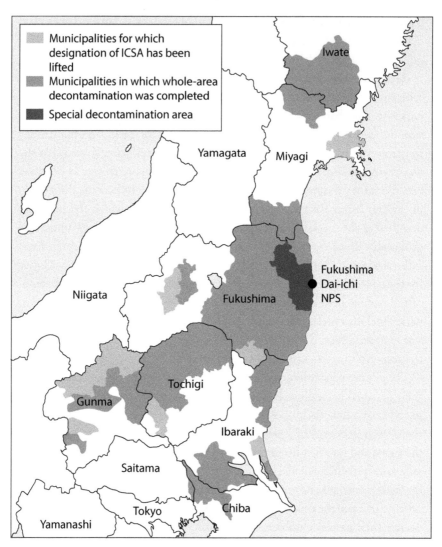

MAP 5. Decontamination zoning by the Ministry of the Environment, March 2018, from "Environmental Remediation in Japan," https://web.archive.org/web/20220423211330/http://josen.env.go.jp/en/pdf/progressseet_progress_on_cleanup_efforts.pdf.

them to return. The question Saeki and her parents discussed was not whether a particular item was contaminated. Instead, they negotiated what each thing meant or what way of life each item might make possible or impossible when they returned to Odaka. Like the preservation group I worked with in Futaba, they were looking for a narrative that helped them connect their presence and absence in Odaka.

By focusing purely on the removal of as many radioisotopes as possible, decontamination threatened to homogenize the unique biographies of their objects. Beck (1999) points out that modern risks like radiation exposure are egalitarian, and this idea of egalitarianism applies perfectly to how the decontamination policy frames material objects in coastal Fukushima. Under the regime of decontamination, all objects in a specific space-time were and are contaminated equally from the fallout and thus are subject to removal.

While Saeki was busy talking on the phone, Tae was kneeling by the small garden in front of the house looking for alpine leeks. Finding them, she carefully wrapped their stems with pink ribbons. I asked Tae why people cared so much about alpine leeks. Tengo had also made an effort to save the alpine leeks growing on his farm in Odaka and had transferred them to his temporary residence in Kashima. Tae elaborated, "It takes five to seven years for alpine leeks to grow fully and become edible. They are challenging to take care of and require thorough attention. The amount of time and labor that need to be invested in them makes the leeks especially valuable and tasty too." The historical value of mundane items like leeks further illustrated the significance to the evacuees of predisaster time, which the accident had ruined and the decontamination threatened to reset.

Navigating the radiation-centered logic of decontamination and the threat to halve the meaning of their preaccident livelihoods, the residents had to emphasize the historical significance of various belongings, or perhaps sometimes the residents needed to "reinvent" their significance (Hobsbawm and Ranger [1983] 2009). They used personal memories to counter the generalizing logic by which everything in the space belonged to the category of pollutants. Overhearing our conversation on the alpine leeks, the leader chimed in to promote the value of the inspection work: "This is why we have people come for the inspection, since we would not know what is valuable and what is not, and sometimes not-so-precious-looking things turn out to have some personal value attached to them."

On the way back to Kashima, Saeki voiced her doubts about the decontamination paradigm. "Decontamination is just a fancy name for ordinary cleaning around a house," she said. "It is nothing special, no high-tech machinery, but just like good old low-tech household tasks we know how to do. How are we supposed to believe that it is a tested, effective scientific procedure?" Here Saeki was complaining about the Ministry of the Environment's much-debated claims about the effectiveness of decontamination.[8] Long-term environmental-survey data suggest that decontamination efforts, along with the expected decay process of one of the two most prominent radioisotopes in the region, have reduced the ambient radiation levels around the residential areas by 40 to 70 percent since 3.11 (Saito et al. 2019). Nonetheless, the reduction is usually assessed by using the metrics of external radiation to humans in a specific space (i.e., sieverts) but not the overall contamination levels in the same space (i.e., becquerels) (Kawasaki 2018a, 2018b).

Regardless of the cost-effectiveness of the efforts to remove radioisotopes in the environment, decontamination stimulated both the national and local economies for a select few. Huge construction companies that were responsible for building nuclear plants, such as Kajima Construction, Shimizu Construction, and Taisei Corporation, have also been responsible for the decontamination project, although they have often subcontracted smaller, local companies to do the actual work. Whether they were building or cleaning, the companies benefited.[9] Much as described in Naomi Klein's *The Shock Doctrine: The Rise of Disaster Capitalism* (2008), political and economic elites used the crisis as an opportunity to maximize their profits while exploiting the needs and wants of vulnerable people like Oshima, who lost his job due to the TEPCO accident and worked a precarious job as a decontamination worker to survive. The data from the Ministry of the Environment indicate that there were more than 13.6 million workers in the SDA and more than 180 million workers in the ICSA by the end of January 2018. Those workers together transformed around fourteen million cubic meters of soil and personal items in Fukushima into decontaminated waste.

Saeki expressed another concern. "A more annoying thing is that the people in front of my parents' house might decide to demolish their house. Don't you think radioisotopes fly with dust from demolition? I wish that the state had coordinated its decontamination team and demolition team. My parents' house gets cleaned first, and then it might get contaminated again." As she expressed her frustration with the state policy, I shared my finding that the background radiation at her parents' house was already below the 1 mSv/y standard. I asked her why, considering the data, residents think decontamination is necessary, especially near the center of Odaka and in the coastal region, where the radiation levels were lower than they were in western Minamisōma.

As a resident of Kashima, the area that fell outside all the zones and was determined to be less contaminated than neighboring districts in the city, she thought about this for a while and then answered, "Getting your place decontaminated is kind of like a ritual. I mean, people like my parents feel better to know that their house has been decontaminated even if decontamination is just a joke, if you ask me. At least from cleaning around the house by gathering fallen leaves, weeding the overgrown vegetation, and getting rid of old tools and items, decontamination makes the place looks more habitable." The production of the sense of habitability rather than the creation of an uncontaminated space, Saeki pondered, might be the key work of decontamination.[10] She continued, "Maybe the visible change between before and after is what's convincing for people who have been living away, who have not visited and seen their houses for a while, and who only have the image of their not-well-maintained houses they saw from years ago. My parents are one example. No one wants to go back to a house in a mess, but if it gets clean and looks livable, many might be tempted to return."

A survey of Marumori town in Miyagi Prefecture, fifty kilometers north of 1F, conducted by Murakami et al. (2018) suggests decontamination's psychological effects. Regardless of decontamination's effectiveness in lowering the already-low radiation levels, Marumori residents reported that decontamination helped reduce their anxiety and improve their sense of well-being. Saeki's view illuminated that the evacuees' focus might shift from the invisible-though-omnipresent contaminants to the fact that their houses that have been made to look cleaner through decontamination.

Saeki's observation that decontamination produces a sense of order or purity follows Mary Douglas's classic conception of purity and danger ([1966] 2002). As in Douglas's theory of dirt, radiological contaminants are essentially considered by the public to be a disorder and thus "eliminating [them] is not a negative movement, but a positive effort to organise the environment" (2). In addition, decontamination staged an opportunity for residents to reflect on the memories of the past that they had to, but could not, leave behind. In making decisions about which of their personal effects should be classed as waste, the decontamination process forced evacuees to contemplate their inalienable history and memories archived in the place they called home.

Goffman (1971) observes that individuals use objects to articulate their sense of self and to delimit their territory from that of others; one's body, for example, is a personal and intimate territory that can be used to distinguish oneself from another person. Personal possessions such as clothing, furniture, and tools help to articulate the boundary between one's self and other in social worlds through a person's privileged access to these personalized items. From this perspective, making decontamination the necessary condition of evacuees' return posed a threat to their sense of self and territory by prioritizing the removal of contaminants in the "environment," a depersonalized framing of space and the very personal things inhabiting that space. Saeki's parents' desire to save the useless blue bucket spoke to the affective value they attached to it. They did not need the bucket to live, but for them, it was a necessary marker of their territory and the kind of livelihood they had and wanted to regain. Wrapping the pink ribbons around their belongings, the evacuees claimed that some memories and objects were noncontaminable. However thin and weak the connection might be, each personal effect contributed to the network of *en* with which they resisted their postfallout life from being suspended in the air.

## THE PROOF OF LIVING IN THE WORLD

Between 2013 and 2016, before the state decontamination project demolished her house, I often accompanied Naoko into the evacuation zone to tend to her house in Odaka and till her land. We used Route 6, the former Route 6, and sometimes a mountain road in the west, which the locals call Sanrokusen, to observe different locations in the city as we drove to her residence. The drives made visible the city's emerging wastescape, located either on the coastal side along Route 6 or on the mountainside along Sanrokusen.[11]

While radically changing the local landscape, decontamination achieved the two radiation-safety principles of distance and shielding, leaving the residential areas in the central part of the city largely unaffected by the visible presence of *frecon baggu*. Since 2019, special vehicles have transported black bags one by one to the newly established Interim Storage Facility around the 1F property between Ōkuma and Futaba towns, so that it is further removed from residents.[12] By the summer of 2021, all decontaminated waste within Minamisōma had disappeared from sight. Once inside the storage facility, *frecon baggu* also disappeared. There, workers at the state-owned Japan Environmental Storage and Safety Corporation ripped each bag open, incinerated burnable waste to reduce its size, and buried soil-based waste in ditches. Although currently there is no concrete plan to host nuclear waste for perpetuity nor any municipalities willing to do so, the state has promised to move all decontaminated waste out of Fukushima by 2045, thirty years after the first *frecon baggu* arrived at the facility.[13]

Upon each visit to her house, Naoko would enter her former bedroom, open the dresser drawers, and quietly contemplate the items within. She would remove a few things from a drawer and place them on the floor, as if to indicate her willingness to transfer them somewhere, and then she would clean up around the Buddhist altar in the adjoining room. The shiny appearance of the golden altar and its large size indicated her family's faith in Shin Buddhism. As she offered incense, the smoke would rise to the wall where memorial portraits of her deceased family members hung. Dusting the photographs, Naoko joked how she would soon join them but added that for that to happen, she first needed to return to her home. When she first joked about it in 2013, she was already in her late seventies.

Since the TEPCO accident, she had thought about how her evacuated home had become not just an important place to live her life but also the only place where she wanted to rest after her death. Naoko never doubted that she would eventually join the cremated remains of her deceased family members resting in the family grave on the property, a few hundred meters from the house. The altar, memorial portraits, and the nearby grave together articulated Naoko and her family's multiple relationalities, temporalities, and territories archived in the place she called home. The accident threatened to sever her tie to it as it had denied her the right to choose where to live and die.

Before leaving, she would return to her bedroom and place the items laid out on the floor into their respective drawers once again. Her temporary residence was too small to hold the dresser in her bedroom or the family altar, she would tell me. And it was only "temporary"; soon she would be able to resume living in her old house, in her old room, surrounded by her clothes and her family, both the living and the dead. The family left the calendar from March 2011 hanging up. It was not because they resisted the vicissitudes of time, nor did they feel nostalgic for the lost time. Instead, there had not been many activities in the house. The calendar appropriately marked the lack.

Although, for the moment, the deceased represented in the altar had to remain in the alienated residence, Naoko expressed how important it was that somebody take care of them. As material and semiotic proof of the groundedness of her personal lineage, the unmovable altar and grave were one of the main reasons she felt that she must return to her home. With the family altar still in place, the dead continued to dwell at her evacuated home and haunted Naoko in her sleep at the temporary residence.

In late 2013, more than two years after the initial evacuation, the state began decontaminating the region around Naoko's home. Eventually, state officials proposed to demolish Naoko's home at no cost, and her family consented. The earthquake that had catalyzed the tsunami and the series of nuclear meltdowns had partially damaged their house. Thus, in 2014, as she turned eighty, I returned with Naoko to visit her residence. This time her task was to identify which items would be preserved from the demolition and which would be disposed of as contaminated "waste." The plan was to transfer the valuables to a warehouse next to the family's house, where they kept farming tools and machinery.

During this sorting process Naoko became more vocal about the items I had seen her unpack and repack many times before. Contemplating a kimono she had not worn in ages, she asked me, "What do I do with this? Even if I keep this kimono, no one is going to wear it, including myself." She touched the faded gray fabric. "To tell you the truth," Naoko confessed, "I might have worn this only a few times. It was from sixty years ago, when I married in this house. It was the first and last good kimono I got from my parents."

Still holding the kimono, she explained how in the old days, long before the 1F development plan was taking shape in the 1950s, people wore kimonos every day. When I asked her if she would throw away her kimono, Naoko replied, "The demolition date has not been set yet. I could keep it for the time being. I mean, isn't it a nice kimono?" She stroked the kimono before folding it with care and returning it to the dresser, closing the drawer with deliberate force. It sounded heavy, as if the dresser carried all the sedimented memories of her life in the house. When she opened and closed the drawers, the chest gave off a musty, though familiar, fragrance of old clothes, of lost time.

On the way back to her temporary residence where her family waited, Naoko spoke about the difficulty of letting things go. We passed a few temporary waste-storage sites along Sanrokusen, which stored more bags than they had a year before. She said, "My family said to throw away any unnecessary things in my life. But the more I thought about each item, the more I felt that my life was being taken away by the accident and reconstruction." Preserving her personal belongings was Naoko's way of resisting the force of a particular kind of plot that the state had initiated with its reconstruction through decontamination. Even though she wanted to return to her home— and, by extension, for the place to be decontaminated—Naoko was not ready for her personal belongings to be "cleaned up" as carriers of contaminants and transformed into waste, along with the house she had lived in and the land she had cultivated. "It's not that I did not want my home to be less contaminated," Naoko described. "It's just that whenever I go there, my home looks less like how it used to be."

Although decontamination, which involved the careful cleanup of a house, might provide a sense of order to evacuees, decontamination could also take on the appearance of a large-scale destruction of residents' familiar environments, as Saeki observed. I vividly remember visiting Hatsumi's temple in early 2015 and noticing its empty appearance. Usually, Buddhist temples are surrounded by tall trees to protect them from fire. They are also used for eventual restoration when necessary. As a result of decontamination, most of the trees had been trimmed and some had been cut down. This reduced the ambient radiation level, but at the same time, as Hatsumi put it, "It feels like the temple lost its dignity. Now it feels like it was built only a few years ago. The only positive thing is there won't be as many cicadas buzzing in the summer." A sense of dignity was not the only thing lost; it took a long time for Hatsumi's beloved birds to come back.

When I visited Naoko's house with her in March 2015, I noticed that there were many *frecon baggu* in the house's backyard, on the field east of the warehouse. She told me that the state had partially decontaminated her property in December 2014 to experiment with the effectiveness of various techniques. The purpose of this experiment was to evaluate ways of decontaminating forests, which was a considerable challenge, since more than 42 percent of Minamisōma consists of forests, compared to slightly less than 6 percent of residential areas. Forests cover around 68 percent of Fukushima Prefecture. In December 2015, the state determined that decontamination of forests would not extend past a twenty-meter radius of the outskirts of one's residence (Taira et al. 2019) due to the cost that would be incurred. This meant that for some evacuees like Naoko, who lived by forests and whose livelihood depended on encultured natural resources in and around forests (*satoyama*), they had to wait for the local ecology to be less contaminated after the passage of time.[14]

Pointing at the field with *frecon baggu,* Naoko said that the bags were from cleaning parts of the forest, the persimmon tree, and the bamboo bushes on the west side of her house. What might have been the forest that surrounded and overwhelmed Naoko in her recurring dream now looked less somber and eerie. About fifty or so *frecon baggu* were piled up in one space, which she called "temporary temporary storage" (*kari kari okiba*) to hint at the dilemma of decontamination. Locals called decontamination *isen*, or moving of contaminants, to signal its actual work—not removing, or *jyosen*, but moving contaminants from one location to another. The radiological contaminants did not disappear. Instead, through each movement, they only become further distanced from people and less visible.

To my eyes, the bags she pointed to looked exactly like any other *frecon baggu* I had seen in other parts of the city. As I shared my impression about how *frecon baggu* made everything look the same, she told me otherwise. "I think many people do not understand how it takes a long time for a farmer to cultivate land and to farm. For us, the quality of the soil is the foundation of every activity we do." As she talked, she showed me her fingers, which were crooked and darkened from working with the soil for her entire life. "I spent sixty years of my life, since my marriage, in Odaka cultivating the soil, and my in-laws before me spent more years before that. The soil is the history of this house, and the soil is what they [decontamination workers] take away and put in those black bags." Each *frecon baggu* captured the soil that the TEPCO accident had contaminated. In the process, each concealed what Naoko had cultivated throughout her life in Odaka. The soil was not only hers; it was the intergenerational patrimony of the house she had married into sixty years ago, which she wanted to pass along to future generations like Toru.

For a mere spectator like myself, what was visible—the bags themselves—seemed to signify something general; *frecon baggu* were, to me and increasingly to others, a visible sign of the TEPCO accident, regardless of what might have gone inside that bag. Naoko's comments powerfully emphasized that what had become imperceptible inside those black bags were what made Odaka *her* place. Decontamination produced multiple *frecon baggu,* which made what was previously visible invisible, and vice versa. "When I see my soil removed and stuffed in those bags," she said, "I feel like my history has been stolen from me. I know the state and people say that is what is necessary in order to make the place livable again, but how I am supposed to come back to a place where everything is cleaned up, and there's almost nothing I care for left?"

In the face of the outcry of the evacuees like Naoko, the state took measures to preserve the fertile soil. If the level of contamination in a location turned out to be not too high, the decontamination workers used a method called reverse tillage (*hanten kou*). They flipped the soil sedimentation using agricultural machinery first to scrape off the topsoil and then to plow up the field and bury the topsoil in a lower layer. Since cesium tends to stay on the topsoil by binding tightly to the soil, this measure helped to shield residents from radiation while saving farmers from completely losing their connection to their land. It also benefited the state by reducing the amount of waste. The decontamination act states that the method is "expected to reduce the radiation dose via shielding and to inhibit the diffusion of radiocaesium by shielding it with soil. This has the advantage of not generating removed soil since the topsoil is not being removed" (2.33). Despite all this, decontamination took things away from Naoko. The evidence was undeniable; the decontamination work archived her losses in the hundreds of *frecon baggu*.

In June 2015 I once again visited Naoko's house with her. She took me to the warehouse to show me how she had moved items from her house there. I noticed her chests among these items, and I knew that one of the drawers contained the gray kimono that she would probably never wear ever again yet nevertheless embodied many memories of her past life in Odaka. I asked Naoko what she thought would happen to the items she was preserving. She remained silent until we exited the warehouse and she closed its heavy door. Then she glanced at her slowly decaying house and said that she was getting old and soon she would naturally be parting with them. For the time being, she insisted, "So long as I live, I need them as proof that I lived in Odaka."

It was neither decontamination nor the *frecon baggu* as its material artifact that archived the stories of the TEPCO accident. Instead, what was concealed inside the bags and survived decontamination testified to Naoko and others' histories, memories, desires, hopes, and determinations—their atomic livelihoods. They fought and are still fighting to live and die in coastal Fukushima despite the accident, contamination, evacuation, community disintegration, decontamination, and collective imagination of the risk of radiation exposure that has reduced their lives before and after the TEPCO accident to contamination. There was more to their lives than what had been made visible.

# EIGHT

## *In Search of the Invisible*

My basic view of the world is that right next to the world we
live in, the one we're all familiar with, is a world we know nothing about, an unfamiliar world that exists concurrently with
our own.

HARUKI MURAKAMI, *New Yorker* interview, August 2018

### NO ONE PRAYED FOR THE LIVING

In the summer of 2018 Naoko and I spotted an animal bone that we could
not identify inside her warehouse, where she had moved her personal belongings before the state-initiated demolition of her evacuated house. Except for
occasional short-term visits by family members, no one had gone inside the
warehouse, and it was locked. The wilted flowers at the family altar indicated
that some time had gone by since the family's last visit. From the crime scene,
it appeared that something had been chased around and killed. During the
chase, something had knocked over items like chests, tables, and boxes that
had placed in the corner of the warehouse to be later moved into a new house.
There were traces of blood left on the tatami mattress underneath, suggesting
there had been a struggle.

Studying the scene carefully, we guessed from its size and shape that the
bone might have belonged to a rabbit or civet, but we had no idea why it was
ossified inside the enclosed warehouse or what had possibly chased and killed
it. Naoko guessed that it might have been attacked by either a monkey or a
boar, but then she immediately remembered that monkeys are usually herbivores. And if the culprit was a boar, she said, it was strange that any bones
remained. "A boar would have eaten the whole thing. Also, I do not see any
opening where a large animal like a boar could sneak in," Naoko said. Not
knowing what had happened during her absence gave us a chill. It reminded
us that humans were not the sole inhabitants of the evacuation zones, and
radiation was not the only thing invisible.

I documented the scene with my phone to show it to Tengo later, and Naoko reluctantly removed the remains by sweeping them in a plastic bag with a broom. Feeling sympathetic to the dead creature, Naoko offered incense at her family Buddhist altar at the other side of the warehouse and briefly prayed, closing her eyes and putting her hands together in front of her chest.

The purpose of our visit was to observe the condition of Naoko's property. After the state demolished her residence as a part of its decontamination project in early 2016, the family had finally started working with an architect to build the new house. Without anything standing on the land, Naoko's property looked smaller than I had remembered. Decontamination had thinned down the surrounding forest, which had arrested Naoko in her reoccurring dream. I noticed that a tall Japanese cedar tree, which the family had been growing intergenerationally to eventually use to renovate the house, was gone.

"Cesium went inside the tree," Naoko commented matter-of-factly. The state and TEPCO duly counted trees in the evacuation zones as assets to be compensated. Naoko did not even know how many trees the state had "bought out" or for how much. She never checked her bank account. Nothing important in her life was there, she said. Naoko pointed to the empty lot, where there would soon be a new house. "I do not need anything elaborate at this point. It will just be a place to die in," she commented somberly. As she walked around the flattened property, her back looked more bent and she looked even smaller than when I had first met her in 2013.

Despite the quick demolition of the house, it took Naoko and her family a few more years to get any more information about rebuilding it. The local construction companies were all busy, hastily fixing and rebuilding homes in Odaka. Many residents longed for an early return but, like Naoko, they first needed to rebuild their damaged houses. Saeki's parents, whose house was decontaminated in early 2015, decided to renovate their house before returning, and this decision delayed their return to Odaka. It was rare for the evacuees to be able to return to their original residences immediately after the zone's reopening in June 2016, for many reasons. The residents who evacuated to locations far away, like Saeki's parents, could not occasionally attend to their evacuated houses to maintain their condition, while other properties were damaged from the earthquake, its aftershocks, and the tsunami. Like Naoko's family, some others had discovered that wildlife and pests had invaded their places, causing structural damage. The houses deteriorated without a human presence, and years of continuous human absence were enough to change the local ecology in favor of nonhuman others.

In June 2019, more than eight years after the initial evacuation, Naoko finally reunited with her inalienable belongings inside the warehouse. In place of her one-story farmhouse stood a modern two-story house with a shinier and more elaborate Buddhist altar. Naoko complained about the new residence's second floor, saying how it would be harder for her to go up and down the stairs with her weakened legs. Although a home with less measurable contamination was desirable, evacuees like Naoko found their decontaminated and renovated homes rather strange. She found the modern electric stove unfamiliar; it spoke as she turned it on and off, and the new smart toilet opened and closed its lid by itself and flushed automatically. "Everything else but me is getting smarter!" Naoko laughed.

Like many people moving into a new house would, Naoko used her relocation as an opportunity to replace the altar. Naoko was concerned that her ancestral spirits were angry from the long-term displacement and she felt it was necessary to appease them by placing them in a new altar. Luckily, the state compensated her for the cost of the replacement. This, however, required Naoko to carry out a set of protocols to securely transfer the ancestral spirits from the old altar to the new one. "It was challenging to move the spirits," she told me at her new residence in June 2019 as she was cleaning the former altar in the warehouse while she waited for staff from the altar shop to remove it.

She said that prior to this moving arrangement, "a Buddhist monk came and chanted to transfer the spirits from one altar to the other. It is funny how the deceased get better treatment than the living ones. No one prayed for me!" Naoko's sarcasm was a not-so-veiled critique of the state's decontamination policy, their half-life politics, which lacked any consideration for disaster-related damage beyond biomedically significant radiation exposure. For officials, her return to the now supposedly decontaminated region signaled the success of their remediation efforts; Naoko was yet another statistic—one of more than three thousand in Odaka—proving the efficacy of decontamination and the safety of the area.

When the altar movers arrived to dispose of the now spiritless altar, Naoko proudly told them how carefully she had cleaned it as a gesture of bidding farewell to the object that had marked her territory for more than sixty years and her family's territory for even longer. One of the workers voiced his concern that the altar looked *too clean;* it did not look like a deserted altar in an evacuated home. After the staff chatted among themselves, a younger staff member asked Naoko if it was all right to make the altar "look dirtier"; this was their routine way of documenting that an altar

FIGURE 13. An altar mover photographing the condition of Naoko's family altar for state compensation. Photograph by the author.

belonged to a displaced family. The procedure would allow Naoko's family to receive a higher compensation for the altar's disposal, he said. "Is that so?" Naoko responded, pursing her lips together to protest in silence.

Without securing Naoko's consent, the worker grabbed the incense ash holder and spread ashes over the old altar. As he photographed the spoiled altar from multiple angles, he exclaimed, "It looks much better!" He then assured Naoko that the photos of the adequately "ruined" altar would garner her family around $10,000 that they deserved for the time away from their original residence (figure 13). Watching the truck departing with the old altar, Naoko lamented, "The ruined altar is worth something, but what about my last eight years of waiting and more wrinkles on my face?"

Despite her commitment to following various spiritual practices and tradition, I would not call Naoko a religious person, at least not in the sense that the term *religion* is employed analytically to explain a faith in a set of doctrines. Commenting on the ambiguous sense of religiosity among the Japanese, prominent scholar of Japanese religion Helen Hardacre says that "if we were to characterize the way in which people are religious in Japan, it would be closely linked to the family and to tradition, emphasizing the things that people do, rather than their strict adherence to a set of doctrines."[1] Naoko had no obligation to pray for the unknown dead creature in her warehouse. She did so because feeling compassion toward all beings—humans or otherwise—was a way that Naoko had learned to live with the others that happened to coexist with her. However, as unnoticeable and inconsequential as some of those beings were to her, their activities were *kyouji* (synchronous) to the overall ecosystem in which she lived.

More than anything, Naoko, like many other people I met in Minamisōma, believed in invisible things that were copresent with and influenced visible things. Whether ghosts, spirits, gods, nature, or *en,* the name of such invisible things did not matter. Instead, residents acknowledged and respected something larger than themselves, supernatural forces that, like often-invisibilized material infrastructure (Star 1999), did the work of connecting one person and one thing to another. Radiation was not the only invisible force at play in Minamisōma, and it was certainly not the most important thing in their lives. Indeed, this was what Hatsumi warned me in the beginning of my ethnography of fallout. She said that my Geiger counter only communicated the presence of radiation; it did not tell me what other visible, invisible, and invisibilized things mattered to people who had been living in the region before the TEPCO accident and despite the radiological contamination after it. In order to understand what mattered to them, I had to get beneath coastal Fukushima's spatial and temporal surface where radioactive particles had fallen and remained.

## ECOLOGIES OF THE LIVING, DEAD, AND SUPERNATURAL

"Come to think of it, I have not seen any snakes since 3.11," Hatsumi said in early 2016, commenting in passing on the changing local ecology. Situated in the urban area of Minamisōma, Hatsumi had not noticed some issues many others had increasingly brought up since 2013. Those residents, especially the evacuees, reported seeing wildlife such as monkeys, boars, Japanese raccoon dogs, and other pests like mice and civets when they visited their evacuated residences. Hatsumi, however, spoke not of the presence of those creatures but of the lack of snakes. "I do not know why I just thought of snakes. I hate them and am happy if I do not see them ever. Maybe somebody recently mentioned this. I just cannot remember," she said, sounding less cheerful and more disengaged than usual. I promised Hatsumi to investigate this matter and left her temple.

In the summer of 2016, when the accumulation of *frecon baggu* signaled the steady progress of the decontamination project and overwhelmed Minamisōma's visual landscape, I stopped seeing Hatsumi. One day I visited Hatsumi's temple as I usually did, but I was greeted by one of her daughters in her early thirties. She reluctantly informed me that Hatsumi was under the weather and was unwilling to receive visitors. It was the same upon my multiple visits in 2017. Each time Hatsumi's daughter thanked me for coming and apologized for sending me away. When I asked how Hatsumi was doing, she looked down and told me that she was not allowed to say. Feeling apologetic, Hatsumi's daughter shared with me that she had taken over the business at the family temple and had learned what Hatsumi had always done for the temple. Each day, small tasks like taking phone calls, talking to visitors, scheduling funerals, arranging anniversary visits for the families of the deceased, cooking, cleaning, and so on were daunting for the daughter. Growing up, she had observed that her parents were always running around busily, but it was only after she took Hatsumi's place that it finally dawned on her that the temple serves the local community of both the living and the dead.

Now, five years after the sudden evacuation in March 2011, the same daughter shared how she understood why Hatsumi had felt compelled to come back from her evacuation in March 2011 after only a few days. Hatsumi had wanted to assist her husband, who had remained in Minamisōma, to attend to the families of the more than seventy tsunami victims who belonged to Hatsumi's temple as parishioners. In a time of crisis, the community desperately needed spiritual guidance. Without Hatsumi's temple, about 570 cremated remains of unidentified tsunami victims (*muen botoke*) would not have had any place to wait for their families and relatives to find them. Even though it took about three years for all the displaced remains to be reunited with their families, Hatsumi's husband prayed for them every single day to hope for their eventual reunion. Some remains brought about unexpected *en* to nonlocals. They belonged to individuals who were washed away in Ishinomaki city, Miyagi Prefecture, about 120 kilometers north of Minamisōma.

Even in 2017 the crisis was still ongoing. As the state recovery and reconstruction efforts made the tsunami damage throughout the coast less visible, the delayed effects of the nuclear fallout kept Hatsumi's temple ever busier. When I drove past a funeral home between Kashima and Haramachi district during my routine drive on Route 6, I was constantly exposed to signs for funerals that were taking place. If I was driving with Naoko, she would often identify the individuals on the signs and mourn another loss in the city. With the city's high rate of aging individuals and various secondary health issues from the postfallout evacuation, the twenty-six temples in Minamisōma had been bombarded with the responsibility of assisting people in ending their lives peacefully. Temples and monks helped tame their souls and send them securely to the other world (*ano yo*). Proper rituals were needed to guide their souls so that they would not dwell in the world of the living (*kono yo*) as ghosts. In Minamisōma, boundaries between *kono yo* and *ano yo* and the contaminated and decontaminated were intertwined like the figure of the ouroboros, a serpent eating its tail. While some residents were ending things, others were beginning.

In Odaka's coastal region, about ten kilometers southeast of Hatsumi's temple, the chief priest at one of the oldest Shinto shrines in coastal Fukushima, Nishiyama, a man in his sixties, was busy preparing to return to Odaka. By mid-2015 the state had announced its reopening in June 2016 and began letting evacuees stay overnight at their original residences. Like Naoko, Nishiyama never doubted his eventual return, and he belonged to one of the first groups to receive a permit to stay overnight at his evacuated residence. As early as 2013, when I first met him, Nishiyama consistently stated that he had a transgenerational duty to protect his shrine—that is, to serve Minamisōma, its people, and its *kami* (gods). With his charismatic personality and dynamic laughter, he said that he did not wait patiently at the temporary housing in Haramachi.

Between 2013 and 2016 Nishiyama hosted a few community events at his shrine, inviting old neighbors and volunteers from outside the city to sustain a sense of community. One event he hosted featured the history of Odaka.[2] A storyteller recounted the local legend of a monster serpent (*daihizan daijya monogatari*) with a picture storyboard, or *kamishibai*.[3] The tale chronicles various places in Odaka and their connection to the monster serpent who once tried to destroy Odaka with an endless storm. A traveling blind monk, Tamaichi, was visiting Odaka to pray at a holy spot (*daihizan*) to cure his condition. The melody of the *biwa* (Japanese lute) that he played as he prayed drew the attention of the serpent, who shared with him its evil plot to destroy Odaka by causing a series of floods. Learning about this, Tamaichi convinced the skeptical domain lord to stop the serpent while sacrificing his life to prove the maliciousness of the serpent. Thanks to Tamaichi, the disaster was averted, and the domain soldiers defeated the serpent. The legend goes that a few locations in Odaka were named after different parts of serpent's body, such as teeth (Onaba), ear (Mimigai), and horn (Tsunobeuchi). In fact, several such locations were devastated by the 2011 tsunami.

Nishiyama wanted to feature the legend because, as he said to me, it communicated the importance of local knowledge and histories passed down from generation to generation. He believed that in postfallout Minamisōma it was critical to remember, more than ever, how people in the past tried to coexist with nature while allowing the supernatural to mediate their relations. "Science is great unless we foolishly come to believe nature is controllable," he said, often asserting that the problem with nuclear energy was human arrogance: "Science that does not pay any respect to the power of nature is a human disaster [*jinsai*]." The TEPCO accident was the human disaster, Nishiyama asserted, that had threatened to disintegrate local history in addition to local spirituality. "But I believe we still can recover," he said. "For that, we need to put our faith in things unseen, like gods and spirits, or what Buddhism would call *en*, a divine force that brings things together. Guess what? Radiation is not the only invisible thing in Minamisōma."

Nishiyama repeatedly told me that one significant function of the shrine is to mark the beginning of things, like offering a prayer to farms for successful harvests and to the land before building a new house or blessing a newborn or a newly purchased car or motorcycle. As the date of Odaka's reopening approached, he became busier celebrating and blessing numerous beginnings. According to him, there are different roles assigned to Buddhism and Shintoism, the two spiritual practices that have coexisted and coevolved in Japan.[4] "Usually, Buddhism concerns the ending of things," he said, "like parting with the deceased. On the other hand, Shintoism concerns the beginning of things." As a Shinto priest, Nishiyama emphasized the importance of witnessing the beginning, and for that, his physical presence was a must. "As a priest, I need to be the first one to return so that I can welcome others as they restart their livelihoods in Odaka." Successful historical continuity

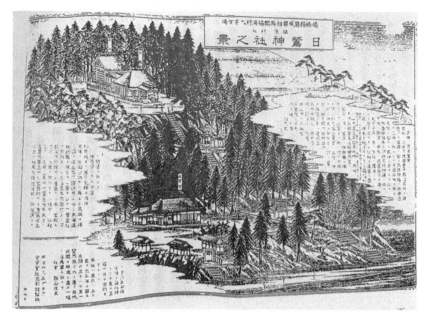

FIGURE 14. A drawing of Hiwashi shrine. Date unknown. Taken from Kowata, Idogawa, and Monnma (1996, 8).

resided in the sustained witnessing of the newness of everyday life in the present.

"The nuclear disaster disintegrated the entire community," he said in 2015 while sharing his estimate of how many neighbors might return to the area. "It forced me ask the question none of my ancestors asked. What can a shrine do in the time of a nuclear catastrophe?" Nishiyama constantly asked himself this question. "Since long ago, Hiwashi has always served the community and the entire domain," he told me.[5] "It was there to protect them from terrible things like disasters and diseases. What Odaka needs now more than ever is a place where people can gather, and a shrine is one such place for the community. That's why I need to return."

He was proud of being the priest of one of the oldest shrines in the region. Nishiyama's shrine, Hiwashi shrine, is about 650 years old (figure 14). Its history is connected intimately to the legacy of the Sōma domain. According to official accounts (Kowata, Idogawa, and Monnma 1996; Sōma History 1969), around 1323, twenty-seven generations ago, his ancestors came to Sōma with the domain lord from the current Chiba Prefecture.[6] In 1364, the domain officially placed Hiwashi shrine in its current location in Odaka. It enshrines a war god, among other essential divine spirits such as a god of fertility and a god of sericulture.

Further reflecting on the weight of history on his shoulders, Nishiyama discussed how the relationship between divine spirits and disaster was another reason to preserve the shrine. When showing me his shrine, he mentioned the Isewan typhoon in 1959, one of Japan's worst typhoon disasters on record, which took more than five thousand human lives from as far west as Hiroshima Prefecture and as far north as Hokkaidō. At the Hiwashi shrine the typhoon damaged the first of its three stone-based torii, the gateway structure for the shrine's entrance, which marks the boundary between the world of humans and divine beings. "But that was the only damage," Nishiyama remarked. "Hiwashi is very resilient."

The events of 3.11 became another testament to the shrine's resiliency. He continued, "Some researchers are now investigating the reason why most shrines along the coast survived the tsunami.[7] Call it divine power or human wisdom, but the tsunami stopped around shrines as if our ancestors knew how far a tsunami could reach inland. I think people in the past knew how things are connected better, which informed them how to live with nature." Nishiyama said that for him, Shintoism is primarily about a way of living with nature and the supernatural entities that surround it.[8] Shintoism's polytheism honors all natural phenomena as divine; gods are present everywhere. According to Nishiyama, this belief helps people to respect and venerate more than just human worlds. "Saving Hiwashi, therefore, means not just preserving the domain's history. It also means passing down the symbol of resilience and the wisdom of our ancestors who lived in harmony with nature. One day people like you find Hiwashi and then relate its significance to someone else. It is a place to mediate that chance transmission."

While the flatter coastal region of Minamisōma escaped much radiological contamination, it was devastated by the tsunami. Nonetheless, the tsunami stopped less than two kilometers inland near Hiwashi, whose main sanctuary is located about thirty meters above the sea level. Usually, a community emplaces shrines on a hill or on high ground surrounded by woods (*chinjyu no mori*). This land allocation adheres to the Shinto conviction that divine spirits reside in prominent animate and inanimate objects like large rocks and distinctive trees, which are often located in high places. It also indicates that people in the past honored supernatural beings and offered them the best locations, which are less vulnerable to natural disasters like earthquakes, tsunamis, and flooding.

In social scientific research on disaster, scholars employ concepts like "resilience" and "vulnerability" to address existing socioeconomic and racial-ethnic inequalities as a broader sociocultural cause of a disaster and its messy aftermath (see, e.g., Bankoff 2008; Barrios 2017; Hannigan 2012; Oliver-Smith and Hoffman 2002). While this political ecology perspective allows us to consider disaster as never purely "natural" and always "culturally" embedded, it highlights the duality between nature and culture and tends to underestimate the role the supernatural plays in how people have organized their space/place. The case of Hiwashi (and the story of Tamaichi), however, suggests how spiritual hierarchy also configured the distribution of people and things across various landscapes. In the present period, those past practices and beliefs resulted in the specific local topography where most shrines occupy naturally disaster-resilient locations, yet neither individuals in the past nor supernatural entities knew of nuclear fallout, which defies all former material, geographical, and spiritual boundaries and the protection they afforded. While accompanying me up the 135 stone steps to the top of the shrine, where I could catch a glimpse of Route 6 to the west and the coast to the east, Nishiyama remarked, "The nuclear accident is something else. No one in the past had anticipated it. Not even gods [*kamisama*]."

Local landscapes like shrines become differently vulnerable to combined natural and technological disasters. While many shrines survived the tsunami's inundation and served as emergency evacuation centers, their fortress-like structure made them particularly vulnerable to radiological contamination. Tall trees collected radiological particles suspended in the atmosphere and prevented them from traveling further. This local historical ecology designed for divine spirits posed many challenges for Nishiyama in the post-fallout period. For one thing, it was almost impossible to maintain the close to two thousand square meters of the shrine's tree-covered land by himself. To grapple with this issue, he had a plan to partner with the local nemesis, TEPCO.

Nishiyama contacted TEPCO for the first time in 2014 to help conduct a deep cleaning of the shrine. About fifteen uniformed TEPCO employees from Minamisōma and the greater Tokyo area arrived on a large charter bus. Since 2013, TEPCO had increasingly invested in repairing its ties to local communities by sending its employees to volunteer in those communities. At the shrine the workers swiftly positioned themselves along the stone steps connecting this world, at the bottom of the shrine, to the other world, where divine spirits rest, at the top. Using chainsaws, handsaws, lawn mowers, and wheelbarrows, TEPCO volunteers thinned out the forest. Birds flew away as they cut down dozens of trees and bamboo to let the ocean breeze circulate. As they knocked down one tree after another, the intense sunbeams of the early summer found a way in, lighting up the property. Next, the volunteers gathered fallen leaves and twigs and put them in *frecon baggu*.

Despite its resemblance to the decontamination work happening throughout the region, the TEPCO employees called this work something else— cleaning (*souji*). As they described it, what they did was to address emerging issues with the abandoned and cluttered properties in the evacuation zones, which had become nests for wildlife and pests. The leader of the TEPCO employees said it was a small gesture to assist in the local community's recovery. "Thank you for giving us the opportunity to help and learn about the importance of Hiwashi shrine," the leader said, concluding the work by expressing his interest in sustaining a new *en* from their chance meeting.

At a dinner he hosted for a few local TEPCO employees after the cleaning, Nishiyama openly admitted his dislike of TEPCO for causing the accident. As a retired high school teacher in Futaba town, Nishiyama said he was always skeptical of nuclear energy. But the *en* he had with the town brought personal connections to TEPCO, where many of his former students worked. "I sometimes regret that I should have been a better mentor for those kids, who must be suffering tremendously now both as residents and as TEPCO employees. I always told my students that nuclear energy is not safe, though." Emptying a glass of beer in one swallow, Nishiyama said that he needed all the help he could muster to be able to return promptly, and this desire outweighed his personal feelings. "Either TEPCO or not," he said, "disaster recovery requires people, who have gone missing in this region. After all, the divine spirits would not discriminate against anyone who would pay proper respect." His loud laughter echoed throughout the busy restaurant.

At the end of the dinner, Nishiyama promised the TEPCO employees that he would tell others in his community to contact them for help. "If the company feels genuinely responsible for the accident, then show us not with your words but through your acts," he told them. "Cleaning Hiwashi was just about doing some landscape work, but I know some of my friends in the western parts have been struggling with wildlife." With this statement, Nishiyama reminded the workers that the TEPCO accident and its aftermaths had threatened the local ecology. His comment made me think of Namie, the desolate town a few kilometers south of Hiwashi that I had visited in late 2013, where nature has threatened to erase all vestiges of human occupation. The residents of Minamisōma described Namie as the "radioactive ghost town."

## RADIOACTIVE GHOST TOWN

At the end of October 2013, I obtained a permit to enter Namie's exclusion zone. In Namie, the TEPCO accident forced its surviving residents (approximately 21,000 people) to evacuate. More than 440 residents have died during their evacuation since 2011. Even though Namie is not one of the nuclear power hosting towns, TEPCO had agreed to notify the people of Namie in the case of an accident. However, the state and TEPCO failed to communicate information about the accident promptly.[9] The result was inhumane. Many Namie residents initially evacuated to the Tsushima region, where radioactive plumes from the fallout traveled and contaminated snow fell. After the state's rezoning in April 2011, more than 80 percent of Namie fell into the highly contaminated exclusion zone, and as of August 2021 it remains so.[10] In September 2021 the state announced its intention to reopen the highly contaminated sections of the exclusion zone before 2030 if there are residents willing to return to them. More than any other place in coastal Fukushima, Namie has revealed the uncontainable risks of nuclear energy across bodies, regional borders, and ecologies.

The purpose of my visit was to observe the reality of the TEPCO accident in the "ghost town." Unlike Minamisōma, where evacuees like Naoko, Nishiyama, and Kobayashi, whom we met in chapter 1, occasionally visited, Namie was rarely visited by its former residents. Before I could enter the gated town, I had to jump through bureaucratic hoops at one of the town's satellite offices located in Minamisōma.[11] When I told the officials I was a student-researcher, one of them reluctantly handed me a form to fill out. From the back of the room, an older staff member with a baritone voice warned that a young person like me should not go in there for too long. However, my age did not prevent me from getting a permit.

A week later the town hall called to confirm that they had issued a permit for me to enter the city at the end of October. On the phone, the same individual with the low voice warned me again that I should not remain in Namie for too long because it was risky. He said, "Please understand that we [the town] are not responsible for anything that happens to you while in Namie. Something still could happen at 1F, so please pay attention to the city's loudspeaker, especially if there is an earthquake. And make sure to stop by a radiation screening station after leaving." That the town was not responsible for any risk I incurred was precisely the agreement I had made by signing the form.

In addition to regulating who could go inside the exclusion zone, the town had to protect themselves from responsibility for the risks a curious outsider or a Namie resident might incur when visiting the alienated place. Various gates in Namie served the double function of preventing the unnecessary exposure of visitors and residents and safeguarding the latter's abandoned property from potential intruders.[12] Once inside, safety became the responsibility of individuals. Different municipalities had different rules for entering their exclusion zones. For example, when I visited with its officials in 2014, Ōkuma town provided protective gear and a high-end personal dosimeter to visitors entering the zone. The difference, according to the one of Ōkuma officials, was the result of how well the towns hosting power plants have been funded by the state and TEPCO.

Being inside Namie felt surreal, as if it were a world apart from any other part of Minamisōma and probably the rest of the country. While there were many construction and decontamination workers in the coastal and central regions of Namie, once I drove toward the mountainside in the west, ambient sounds suddenly vanished. I spotted a few dead animals on the street and one feral cat in what seemed like an old rice field. Many pets were left behind during the sudden evacuation in March 2011. Before 3.11, around 116,000 pets were registered in Fukushima Prefecture, and the tsunami killed about 2,500 of them. According to the Ministry of the Environment's 2012 report on disaster-affected animals, only 637 dogs and 462 cats were rescued, and even fewer of them were returned to their original owners.[13] Naoko and her family were one example of evacuees who fled without their pet, believing the evacuation would be brief. They constantly mentioned their lost cat, which suggests the psychological significance of pets to the evacuees; the separation from their pets was another invisibilized loss in the shadow of radiation exposure, and it points to a need to prepare pet-friendly evacuation centers for a future accident.

As I passed the second and last security checkpoint to the west of the town hall, where an elderly security guard checked my permit, there were no people to be seen except for a few individuals dressed in all-white DuPont protective gear. Their presence, though sporadic, signaled the out-of-ordinariness of the town. In March 2011, many fleeing residents reported having witnessed individuals in white protective gear surveying the region (Sanpei and Futakami 2018), who never answered the residents' anxious inquiries about the purpose of their presence. They gave the residents nothing but a chill; something was terribly wrong at 1F.

Even in late 2013 it wasn't just people that were missing. There was no cell phone signal or Wi-Fi connection in the area. Only my Geiger counter was active, busily sounding out the ghostly presence of nuclear things. The more I drove northwest toward Tsushima, the noisier my Geiger counter became. I had no navigation system other than a physical map to guide me in this unfamiliar town. At times the readings oscillated between 25 and 40 μSv/h inside the car, communicating radiation levels about forty to sixty-five times higher than those outside the vehicle at Naoko's house in Odaka a few kilometers north. In some areas in Tsushima I experienced readings above 200 μSv/h. On March 13, 2011, a monitoring post in the area recorded over 1,200 μSv/h. However, neither the municipality officials nor the residents, who initially evacuated to western Namie to be physically distant from 1F, were informed of this data (Terashima 2021).

In the zone, the dynamic force of nature had encroached upon what had once been streets, signposts, rice fields, railroads, cars, and houses. All the human vestiges of Namie together produced in me a strange sense of awe and creepiness. This radioactive ghost town reminded me of the whimsical and surreal scenes from Hayao Miyazaki's anime, such as *Princess Mononoke* and *Castle in the Sky,* which contest the anthropocentric myth of the symbiosis between nature and technology. Although highly radioactive and strange, the environment also looked pure and impressive. Like the matsutake mushrooms that thrived in the blasted landscape after the atomic bombing of Hiroshima (Tsing 2015), nature seemed to coexist well with the invisible contaminants and the absence of human others. If anything, my presence seemed to disturbed their synergy.

Namie offered one poignant example of what Kim Fortun (2012) calls "late industrialism," where both the infrastructure and the paradigm that produced and sustained it were exhausted. The failure of modern, low-carbon-emission, "eco-friendly" nuclear technology ended up creating a "greener"-looking state of nature in a highly radioactive place like Namie. Not in its brilliance but in its meltdown, nuclear technology had achieved its intended goal of "going green." Postfallout Namie became an idyllic carbon-neutral spot, not because nuclear energy reduced harmful pollutants, such as carbon dioxide, from the ecological system to combat much-anticipated climate change but because it polluted the human territories and subsequently purged people from their built environment.

When I described my raw impression of Namie to Hatsumi a few days later, she thought about it for a second and said, "Nature is powerful and scary. Remember what the earthquake and tsunami did in 2011. Our technologies were so frail and useless against it. Because we are always inferior to nature, we must pretend as if we own it. Because we pretend it so much, we forget that we cannot own nature." If anyone had forgotten it, Namie offered a place of remembrance.

As I hurried out of the radioactive ghost town, I took a wrong turn while trying to get out of the exclusion zone and approach Route 6 in the east, which led me down a dead-end path. The road narrowed very quickly and unexpectedly, and my Geiger counter began buzzing loudly. The sun was starting to set, adding more hues to the already stunning autumn colors, which, Hatsumi later told me, was something that had once drawn locals to visit Namie. Close to 5 p.m., the city's loudspeaker informed people to finish working and exit the town.

As I struggled to turn my car around on the narrow path without hitting the gutters lining the road, I noticed a persimmon tree standing on the property of a deserted house. I got out of my car to assess if the property offered enough space to turn my car around and spotted a wild boar by a storage structure at the far corner of the property, picking up a fallen persimmon. Surprised by the sound of my footsteps, which were amplified by the dried grass, the boar jerked its head in my direction. We stared at each other, eye to eye and several feet apart, and became frozen for a moment. I could hear the boar making threatening sounds.

The boar made the first move. It jumped over a small wooden fence behind the storage structure opposite from where I stood and ran away into the woods. Feeling a chill in my spine, I ran back to my car to escape as well. As I was turning my vehicle around, I kept repeating to myself what had happened until a thought came to me: I could bring back one kilogram of persimmons to Minamisōma to get tested.[14] After double-checking to confirm that there were no more boars, I once again left my car, this time with the Geiger counter in hand. I quickly picked a dozen persimmons, put them in a plastic bag, and ran out of Namie.

As I exited, no one bothered to stop me at the gate. On my way back on Route 6, I stopped on the border of Namie and Minamisōma to be screened. The survey workers duly checked the entire car and my body at the screening station to ensure no item coming from the zone was above the standard (less than thirteen thousand counts per minute). Luckily, the persimmons from Namie were not radioactive enough to violate this rule.

## MEMORIES OF PERSIMMONS

I seized the radioactive persimmons from Namie town not to consume them. Instead, I wanted to measure their contamination levels in order to apprehend how badly persimmons could get contaminated. If Namie's persimmons were safe to eat, Naoko and I reasoned, then we could eat ones from Odaka. Unfortunately, Naoko could not get persimmons from her tree in 2013. When we thought of measuring them for contamination, it was already too late; wildlife had taken over the region and eaten most of them. In 2014 Naoko's tree did not produce much fruit, which she attributed to the lack of care and the impatient wildlife that had harvested them too early.

Just as Dr. Tsubokura and his team increasingly uncovered the relationship between a contaminated environment and human radiation exposure, residents had also learned by 2014 that a contaminated environment did not equal contaminated food. Since the accident residents had not remained passive. Instead, they had grown and tested local foodstuffs to visualize how each absorbed contaminants differently. I visited a local radiation-testing center a few times to test vegetables and fruits Naoko had harvested at her temporary residence in Kashima and her evacuated home in Odaka. She did not say much about the results, but each test seemed to have boosted her confidence; after receiving one result, she wondered about the next thing she could grow to test.

The accumulating local data suggested that green onions, rice, green peas, cucumbers, eggplants, and plums were less likely to be contaminated than wild plants, mushrooms, soybeans, and yuzu. The local testing regime was less formalized and politically driven than citizen science efforts to monitor food security in response to the general mistrust of the state and authority that scholars like Sternsdorff-Cisterna (2020) and Kimura (2016) observed burgeoning throughout the country. Just like local farmers who used the testing device at Miura's agricultural cooperative in Shinchi town, the residents and evacuees tested foodstuff not to reveal lies or to challenge the state but to understand, for their own sake, which pursuits from their prefallout livelihoods were still possible. Although elders like Naoko often complained about the complexity and alienness of the language and science of radiological contamination, through her homegrown vegetables she came to understand more intuitively what the TEPCO accident had done to the local ecology and how to cope with it. Confronting environmental changes was nothing extraordinary; farmers have always had to adjust to survive and deal with the precarity of nature. From their persistent efforts and desire to continue cultivating the local soil despite its contamination, they generated hope for the possibility of a postfallout agrarian livelihood.

Four days after my trip to Namie, in early November 2014, Tengo helped me drop off the persimmons at a city-operated measurement center in Kashima. The test result came back in an hour. As Tengo had expected, the Namie persimmons turned out to be highly contaminated. The result showed sixteen times more cesium-134 and -137 per kilogram present than the 100 Bq standard advised for safe consumption.[15] Tengo complained to me that the staff at the measurement center had interrogated him, since it was extremely rare for them to test something so highly contaminated. As soon as Tengo told them the persimmons came from Namie, the staff stopped the questioning.

Sharing the result with Naoko and me, Tengo exclaimed that persimmons in general would not be good to eat anytime soon, and he told Naoko to give up their persimmons in Odaka. The results seemed to sadden Naoko. Pursing her lips, she remained silent. My persimmon hunt was unsuccessful at stirring hope in the family. I never got to taste Naoko's persimmons. Eventually, the state's decontamination process removed her persimmon tree, along with the house and soil she had inherited and cultivated.

Namie's highly contaminated persimmons were particularly discouraging for Naoko, who had told me many times about how she wanted me to try her Odaka persimmons. I told her that I would eat them without testing since I knew a loophole in the logic behind the state-led food survey. Usually, foodstuffs were tested with a set quantity of one kilogram, and the result indicated the contamination one would receive by consuming a kilogram of the tested item. Even if Naoko's persimmons had more than permissible levels of contamination (higher than 100Bq/kg), it was rare for an individual to consume a kilogram of persimmons unless they were being used to make jam. As the assessment of radiation exposures among Native American communities near the Nevada Test Site reveals (Frohmberg et al. 2000), the standardization of dose calculation is necessary to prevent the broader circulation of contaminated food, but it is not in itself sufficient to understand food contamination.

When living with radiological contamination, local foodways and eating habits mattered.

Despite my reassurances, Naoko was still hesitant to feed me her persimmons; her pride as a farmer prevented her from serving me potentially "contaminated" food. Also, persimmons are valued for more than their taste, she told me. Persimmons are significant to the community; they stand for the local history of migration and land reclamation, which reveals the magnitude of the nameless harms caused by the TEPCO accident. With her persimmons, Naoko wanted to tell me how mundane things like fruit embody the cultural biography of current residents' affective ties to their traveling ancestors, *en* between regions, and memories of migration, discrimination, and the spirit of reconstruction. As such, persimmons tell a story of how the TEPCO accident contaminated not only the physical spaces on maps but also the places where residents' histories lived, their memories persisted, and many of their ancestors shed tears. Those evacuees who desired to return despite contamination all emphasized their inalienable attachment to the land, their ancestral spirits, their intergenerational succession, and the livelihoods they made possible. Persimmons were one proof of that.

During the eighteenth century, coastal Fukushima experienced severe famines.[16] Most notably, the Tenmei famine between 1782 and 1788 dramatically altered the local demography. A series of environmental challenges—a frigid summer (*yamase*), volcanic eruptions, and flooding—led to extremely poor harvests. The famine devastated the Sōma domain (present-day Sōma, Minamisōma, Iitate, Namie, Futaba, and Ōkuma), reducing its population of around eighty-nine thousand people to thirty-two thousand due to starvation, plague, and residents fleeing in search of better living conditions elsewhere (Iwamoto 2013b, 2000; Iwasaki 1970; Ōseko 2013; Sōma History 1983).

With the help of Shin Buddhism, Sōma successfully recruited migrants from different parts of the country in exchange for emptied residences and barren, though free, lands.[17] Since migration by farmers was forbidden at the time, they had to risk their lives to leave their domains for new homes in coastal Fukushima. Sōma survived the hardship, thanks to many able and eager migrants from Hokuriku (in the northwest) and other parts of the country.[18] Migration continued until the end of the nineteenth century. As Japanese historian Fabian Drixler (2016, 25) puts it, "Despite its distance from Kaga [current Kanazawa Prefecture], Sōma attracted a large number of Etchū natives [from current Toyama Prefecture] to its Pacific shores. By 1845, 1,800 households had immigrated from Hokuriku to Sōma. In a survey conducted in 1871, fully 3,000 households claimed descent from Hokuriku migrants, accounting for over a third of Sōma's population." However, despite the benefits and subsidies they received as the indispensable labor force brought in to recover and reconstruct the shrinking domain, migrants struggled in the unfamiliar land, as they were often without friends and family and they confronted discrimination as they reclaimed deserted lands.[19]

Sōma natives found the migrants' religion, language, and culture strange. They often discriminated against the newcomers by ridiculing them, calling them "goofy peasants," and excluding them socially (Toyama History 1983, 1026–27). To deal with the everyday discrimination, the migrants worked tirelessly to reclaim the infertile lands given to them. At home, they cried (*kaga nami*) in the dark. Their faith sustained their spirit. Gathering and praying at the Shin Buddhist temples that were gradually being established throughout coastal Fukushima to support the mass migration helped them deal with the stress and keep working (Ikebata 1995).

Those migrants not only brought their Shin Buddhist faith, which was symbolized by the small wooden statues of Amitabha or the great savior Buddha that they carried with them.[20] They also brought other exogenous cultural materials and practices, including but not limited to persimmons, landscape design (a small forest surrounding a house or *igune*), folklore, and regional cuisine (*benkei*).[21] Plentiful historical anecdotes relate a story centered on the traveling persimmon, which represented migrants' memories of hardship.[22] It goes something like this: Waves of migrants carried sticks of persimmon as they hiked across mountains and hills to get to the Sōma region on the other side of the coast, about five hundred kilometers away. They stubbed the persimmon branches in a daikon radish to keep them moist. The new persimmon trees sustained their past livelihoods in new and unfamiliar territories.

A year after we tested Namie's radioactive persimmons in 2014, Naoko and her family decided to get rid of their persimmon tree. It was one of the first things on their property to be removed and stuffed in *frecon baggu* by the state-led decontamination team. Naoko never suggested that its potential contamination was the reason for the family decision. Instead, she explained that keeping the tree could attract more wildlife to her property, which could cause more damage in the future. But the decision was not an easy one. Naoko expressed her regret with a popular proverb: "*Momokuri san nen kaki hachinen*," or, literally, "It takes peach and chestnut trees three years to bear fruit. Persimmons take eight."

If the TEPCO accident threatened to deny people's pasts and their rights to choose where to live and what kind of livelihoods to engage in, decontamination, the removal of as many contaminants as possible, necessitated the removal of the entire structure and material, semiotic, and affective infrastructures that made a particular kind of livelihood possible and desirable. Purity, in this instance, endangered historicity.

In 2016 the state moved the bags of "decontaminated waste"—or aspects of Naoko and her families' life histories—to a temporary storage site about two hundred meters from her Odaka residence. Although the removal of the decontaminated waste from her property was desirable, it caused more damage to other parts of her property. Naoko told me that she had agreed to loan the state a section of her rice fields by the storage unit. They needed to build a route to enable trucks to bring in decontaminated waste being produced in nearby neighborhoods, and the location of her old rice fields was ideal. She had agreed without any negotiation. It was "for the community," she said.

Unlike decontamination, the construction of a road required her rice fields to be filled in with concrete and covered with asphalt. "They promised me to return the land as it was before," she said. From the tone of her voice, I knew that she did not believe them. I thought of all the similar promises the state and TEPCO had made to local communities as we observed the accumulating *frecon baggu* in the temporary storage unit to the east of her former rice fields. Pointing in the direction of the storage, I asked Naoko if she was hesitant to go near the storage site full of decontaminated waste. She looked at me with a smirk and said, "Which one do you want me to choose? Not going back to my own house and die from stress, frustration, and regret or going back to my own house, take care of my ancestors, farm, and die peacefully? Those bags are not going to kill me, but believing that those bags could kill me at my temporary house would" (figure 15).

Evacuees who desired to get back their irradiated land resisted the fact that the technoscientific determination of radiation exposure and its highly anticipated biological harm is the only valid form of experiencing a nuclear catastrophe. The selective focus on radiation exposure is what I have been calling half-life politics to highlight the narrow emphasis of radiation and its biological effects as the sole consequences of a nuclear catastrophe. This perspective erases, ignores, and invalidates other forms of struggle and losses that individuals may experience as they confront the lingering nuclear disas-

FIGURE 15. Naoko at her farm in Odaka in 2014. Photograph by the author.

ter, which is not a quantifiable, all-or-nothing encounter with contaminants in the lived environments.

The desires and sacrifices of residents like Naoko and Nishiyama to sustain some aspects of their personal memories and collective histories in their postfallout livelihoods show how a nuclear accident is not merely about the radiological danger. This narrative only reveals, at most, half of the residents' postfallout struggles. Even if low-dose radiation did not cause immediate harm to individual bodies, when selectively visualized, it damaged the preexisting social and cultural fabric—what many locals glossed as *en*—that had woven the interconnections, however subtly, magically, and invisibly, among the living, the dead, the environment, and the supernatural.

"Radiation is not the only invisible thing in Minamisōma," Nishiyama had said. Radiation that the state visualized and managed using various technoscientific devices certainly has not been and is not the most important thing through which to understand and remember the TEPCO accident and coastal Fukushima. And humans were not the sole agents whose lives were altered by the TEPCO accident.[23] Inside the local ecosystem, humans, nonhumans, and abiotic factors like water and wind have been moving radioisotopes haphazardly and unexpectedly, coproducing accidental nuclear waste by doing what each must to survive.[24] As I had witnessed in deserted Namie, living beings like wild boars moved around and consumed contaminated persimmons and other irradiated things. Where would radioactive boars go, and what might happen to the contaminants they unknowingly embody?

---

# NINE

# A Wild Boar Chase

Violence does not always take visible form, and not all wounds
gush blood.

HARUKI MURAKAMI, *1Q84*

## YOWAI TSUNAGARI

"Contrary to what outsiders fixate on about Fukushima," Tengo told me in
early 2019 in Odaka, "the practical obstacle of living here is the wildlife, not
radiation!" As a state-recognized evacuee from Odaka, Tengo had spent the
last eight years making a series of damage claims to TEPCO regarding the
land's contamination and his loss of livelihood. Tengo's sustained efforts had
distilled for him, more than ever, the psychological attachment he had to his
natal home, its physical upkeep, and his spiritual ties to his ancestors.
Observing Naoko, his mother, visibly aging since 3.11, Tengo wanted her to
be able to go back before the end of her life.

According to Tengo's family history and the Japanese legal system, the
land was registered under his name, and he owned it. Even before 3.11, the
family had transferred Naoko's rights to Tengo to proactively deal with the
issue of inheritance. However, in multispecies worlds, the land always belongs
to anything that roams it or finds in it the necessities of life. In this region,
macaques, masked palm civets, green pheasants, and wild boars have long
coexisted with the human locals. The city museum's exhibitions of stuffed
animals celebrated the diversity of local habitats.

In coastal Fukushima, wild boars are particularly emblematic of the ecological upheaval (Walker 2005) that evacuees experience as they return to
find their predisaster residences overtaken by animals that once abided by the
anthropocentric divisions between home and wilderness (cf. Cram 2016b).
In Minamisōma, the city reports that the number of wild boars captured
in the city limits jumped from 39 in 2011, to 1,300 in 2014, and to more
than 1,700 in 2016. In July 2018, the city launched an online hazard map for

visualizing locations where residents have previously spotted wildlife.[1] In some other municipalities, like Tomioka in the south, the town office printed out a map, placed it at a local shopping mall, and asked the returned residents, construction workers, and visitors alike to mark the locations where they had witnessed wild boars.

Upon his return to his Odaka residence in June 2019, Tengo was confronted with the mismatch between his newly reconstructed, modern-looking two-story house and the desolate wilderness surrounding it. Odaka is now, more than ever, the quintessential Japanese *inaka,* or nostalgic countryside. Odaka officially reopened in July 2016, and by June 2022, more than 50 percent of its currently registered residents (6,600) had returned to the area. While most returnees live in Odaka's central region, Tengo's residence is in western Odaka. Here, it has become more common to encounter wildlife than any of the 300 or so, mostly elderly, returnees. As a result, Tengo's daily routine included patrolling his property with an air rifle to chase after wildlife, like the wild boars and monkeys that roamed around and stole vegetables from the family garden (figure 16).

By 2019—the Japanese zodiac year of the boar—wild boars had become the prime resource competition for returnees like Tengo. The Japanese wild boar, *inoshishi* (*Sus scrofa*), has a long, intertwined history with humans, ranging from coexistence in the Jōmon period (ca. 6,000 BCE) to boars becoming farm pests in modern times (Niitsu 2011). Excavations from the Urajiri shell midden in Odaka illustrate the intimate local human-boar relations of the past. According to a special exhibit at the Minamisōma city museum to celebrate the year of boar in 2019, the Jōmon people used boars' fangs to make accessories and knifes. Archaeologist Takeshi Niitsu argues that representations of boars in Jōmon earthenware and the evidence of carefully buried boars suggest that boars served as an important symbol of fertility and successful harvests.

In the modern Japanese cultural imagination, the wild boar is figured as *chototsu moushin* (recklessness); individuals born in the year of boar are believed to possess attributes such as directedness, determination, and adventurousness. Against this cultural and historical backdrop, the unruliness of boars—their movement across and beyond coastal Fukushima—has come to signify the fallout's protracted and boundary-crossing impacts, and the social and material harm caused by the state's efforts to contain the radioactive environment. In and of themselves and in relation to their human cohabitants, the boars' biotic lives and deaths have challenged the fundamental separation of

FIGURE 16. Tengo patrolling around his property with an air rifle to chase wildlife away. Photograph by the author.

humans from a radioactive environment that the state-imposed evacuation zoning and the subsequent decontamination project attempt to hold steady.

By 2019, eight years after 3.11, state efforts to recover and remediate the region had radically changed coastal Fukushima's environment and human and nonhuman geography on a granular scale (O'Neill 2019). Nonhuman animals that once roamed on the outskirts of properties like Tengo's had been taking advantage of the sparse and sporadic presence of humans (Lyons et al. 2020). The eight-year retreat of humans made places like Odaka safer for many nonhuman species despite the persistent presence of contaminants. Dwelling for years in the evacuation zone with only a few decontamination workers to contend with, wildlife in the region seems unthreatened by the human others' gradual homecoming, yet their presence had increasingly posed a threat to the evacuees and returnees throughout coastal Fukushima. Hearing numerous complaints from the returning residents, TEPCO attempted to solve this issue by using drones.

In early 2019, the time Tengo shared his struggles with wildlife upon his return to Odaka, TEPCO invited me to its Fukushima reconstruction head-quarters in Tomioka to discuss coastal Fukushima's boar issue. Even though I first approached them to learn about their new drone technology for boars that TEPCO had announced on its website, they somehow thought I was an expert on the wild boar issue.[2] In a meeting room, eight TEPCO employees of various ages and genders welcomed me. From their business cards, I learned that they were a group of engineers and public relations experts.

Menshiki, a TEPCO employee in his fifties with the highest status in the room, proudly told me, "We have invented a drone with EAMS Robotics that releases ultrasonics at boars in the field." He explained that the drone project was a part of TEPCO's new decontamination-support initiative started in 2018. Commissioned by the state, TEPCO has been supporting the evacuees and returnees by conducting post-decontamination follow-up activities such as offering their technical expertise on environmental moni-toring by performing group-radiation surveys. The use of drones to chase boars was a new but important monitoring challenge they were undertaking to assist coastal Fukushima residents.

From their video and PowerPoint presentation, it became clear that TEPCO had learned about the boar issue when locals increasingly com-plained about wildlife roaming in the area. Just as Tengo had experienced upon his return to Odaka, residents in the Futaba district reported that wild-life's presence, not radiological contaminants, made it difficult for them to return or to live safely at their former homes. "It is our responsibility to ascer-tain the safety of the residents," the same employee said, explaining why TEPCO was using the drone to chase wild boars. An engineer in his forties continued, "First, we attached an infrared camera to investigate boars' habi-tats at night. While this proved helpful in identifying where they lived, the question was how we could deter them from staying in the area. That's when we thought of using ultrasonics, which has proven to be effective for chasing away deer."

In their first field tests in September and October 2018, eight employees from TEPCO conducted an experiment using two drones, one with an infrared camera to locate boars from seventy meters above and another with an ultrasonic emitter to chase them away. The result was favorable. It appeared that boars were bothered by ultrasonics ranging from the upper limits of human hearing of 4 kHz to the human-inaudible 50 kHz. The video showed boars reluctantly moving away as the drone approached them, which they interpreted as a success. Based on this initial achievement, they told me their plan to continue with field experiments in different parts of the region and refine their technologies for chasing boars away from the evacuation zones. However, their confidence was shattered when I asked where those chased boars went. The room became silent, and no one spoke. Impatiently, the lead engineer in his sixties said, "I do not know what you mean by that."

From my own experience, I told them, I knew that many boars in Minamisōma and Sōma in the north came from areas like Namie, where TEPCO had conducted the experiment. A female employee interrupted and asked how I know those boars were from Namie. A fair question. I answered honestly that I had no proof. Nonetheless, by talking to city officials I had learned that boars captured in the north were highly contaminated, which seemed strange given that those areas had been already decontaminated and, on average, had lower ambient radiation levels than places like Namie.

The tension between us made it clear that the TEPCO employees and I approached the boar issue differently. While they were interested in spotting boars in a specific place and relocating them elsewhere, I approached the same topic more ecologically. I asked in what ways residents had come to be in touch with wild boars in the region. In order to understand the extent of the boar issue in the area, I not only talked to evacuees, city officials, hunters in Sōma and Minamisōma, and ecologists, but I also consulted the city hall in Date, Fukushima Prefecture, about fifty kilometers northwest of Minamisōma, where they have been repurposing the skins of captured boars to craft leather goods like purses, card holders, and key holders, among many other items.[3]

To convey the boars' significance beyond one specific location and their boundary-crossing movements, I explained the following to the TEPCO employees: decontamination has produced an ecology of harm in coastal

Fukushima that has rendered locals, wild boars, local food chains, and the land itself complicit in the production of nuclear waste. Intrigued though skeptical, the same female employee asked me to share my research findings. I asked if I could share stories that chronicle the ecopolitics of contamination and containment. I warned them that I had no fancy video, like those captured by TEPCO's drones. All I had was low-tech storytelling, a collection of ethnographic stories that uncover messy interrelations of invisible and surreal threads, or *en,* among people, boars, contaminants, and state decontamination policy in coastal Fukushima.

They asked if I had a title for my presentation, but I did not. On the spot, I decided to tentatively name my account "A Wild Boar Chase." The title, I explained, is an homage to Haruki Murakami's novel *A Wild Sheep Chase,* in which he writes about the almost impossible task of spotting a unique star-marked sheep that possessed superpowers among tens of thousands of sheep.[4] "A Wild Boar Chase" is a similar story of a futile attempt. Still, it renders the local *en* palpable, bringing together loosely connected webs of the living, the dead, the supernatural, and nonhuman others, however arbitrarily and magically such connections are woven. What mattered to me as an ethnographer was the fact that the residents believed such connections to be real.

The key to seeing invisible threads, philosopher Hiroki Azuma (2014) argues, is exploring weak connections (*yowai tsunagari*). Unlike stronger connections, which are more visible and palpable—like the link between Fukushima and contamination or the historical bonds (*kizuna*) between coastal Fukushima and TEPCO—weaker connections are full of noises, uncertainties, and loose ends that do not necessarily confirm our beliefs and taken-for-granted assumptions.[5] Azuma claims that such connections are not ready-made nor searchable on the internet, where different keywords are already programmatically linked as related. Weaker connections, or what I am calling *en,* can surprise us, allowing us to be exposed to things that challenge our habitual way of thinking and seeing. Philosopher Charles Sanders Peirce calls such a mode of operating in the world "abduction" (see, e.g., CP 5.417, 5.51) and says that it is only through this that a new idea can emerge.

"A Wild Boar Chase" as I told them then and now retell it here, is a story of *yowai tsunagari.* It interlaces preexisting and emerging connections in the postfallout period to show the futile assumption behind decontamination: the separability of humans from the rest of the environment. It also is a story of unspectacular and distributed violence. The decontamination policy, motivated only by the reduction only of human exposure at all costs, had forced human and nonhuman victims of the fallout—locals, wild boars, and everything else—to coproduce nuclear waste for their survival. The danger of the wild boar chase lies in the process through which boars became the unruly vehicles of contaminants, resisting decontamination's intended removal of contaminants away from people. In transforming the wild boars into nuclear waste by incinerating them, the decontamination policy forced residents to cut their new and old relations to boars so that they could reclaim their territory as safe.

The chase began as early as 2013.

## THE RADIOACTIVE FOOD CHAIN AND SURROGATE BODIES

"Ever since we evacuated Odaka, I learned how expensive buying vegetables is," Tengo complained in the summer of 2013 as he was running errands at a nearby store. "Look at these expensive, spoiled bamboo shoots!" he said. "Before the disaster, I would harvest them from my backyard. Now I cannot because of contamination. Since I don't take them, boars have been eating them!"

For Tengo, who had grown up eating the vegetables, wild plants, fruits, and mushrooms growing on his family's property, buying produce at the store felt very strange. What was even stranger to him was that one could buy produce regardless of the season, since for farmers, each season is marked by what they harvest from their farm. "Things have changed, and life seems to be more convenient, but I like my old way. Maybe this is one of the few points the boar and I would agree on," Tengo laughed.

Despite Tengo's cheerfulness, the wildlife outbreak in and around the deserted regions was a dire problem, especially in coastal Fukushima, where many people once farmed their food. Frequently, evacuees would visit their abandoned houses to find that boars had harmed their farming grounds and other parts of their property. In April 2019 the city built an incinerator on the tsunami-inundated land to address the ever-accumulating dead boars that had been hunted or killed in collisions with a car, boars that were suspected to be "radioactive" because they fed off local produce like Tengo's bamboo shoots and Naoko's beloved persimmons.[6] Until the arrival of the incinerator, a few thousand dead wild boars within the city's boundary were buried at several undisclosed locations in Minamisōma. One of the locations turned out to be within walking distance of Naoko's temporary housing in Kashima. When I disclosed this information to Naoko, she was disturbed by this peculiar *en* and said remorsefully, "I cannot escape from boars."

What makes wild boars particularly harmful is their indiscriminate appetite, size, and high reproductive potential. In addition to the potential physical threat that their size and tenacity pose to humans, boars reproduce rapidly. A single gestating boar gives birth to four to seven offspring on average, up to two times per year. Although boars are predominantly herbivores, they have also been known to consume roots, seeds, mice, rice, fruits, earthworms, snakes, frogs, and insects, all depending on the season and the availability of other food sources. Nicknamed "the cleaner of the forest" (Knight 2006, 49), some boars can grow to a hefty 180 kilograms. Although boars maintain a nocturnal schedule when living near human activity, they are typically diurnal (Ohashi et al. 2013). Since boars prefer to dwell in human-disturbed environments rather than uncultivated land (Ralph and Maxwell 1984), deserted farmland is one of their favorite dwelling sites. Hence, not unlike in the Chornobyl exclusion zone (Carver 2019), places in postfallout Fukushima like Odaka became a boar's paradise.

In late 2013, upon one of his occasional visits to maintain his abandoned house in Odaka, Tengo found that something had broken into the warehouse. It had smashed open the wooden boards barricading one of the doors. Inside the warehouse, it appeared that an intruder had been eating and spreading old rice and other grains kept in bags that the city was eventually supposed to dispose of out of concern that the grains were contaminated. "Not only that," Naoko, who came with us, complained, "boars went for the persimmons by the entrance of my property!" When I asked her if it was really boars and not monkeys that had eaten the persimmons, she replied that it must have been boars: "Maybe monkeys ate them first, and whatever they dropped from the tree, boars cleaned up." For Naoko, boars were somehow more problematic than monkeys, another local pest.[7] When she initially mentioned the persimmon-eating boars, I had a hard time imagining it. However, my encounter with a boar eating a radioactive persimmon in Namie later taught me more about the local food chain in which wildlife played a part.

One evening over drinks I asked Tengo to elaborate on the dangers of wild boars. "It is my drunken theory," he mused, "but I think that the boar is eerier than monkeys because many of us used to eat boar." He continued, "The accident taught us how we could get exposed internally by eating contaminated foodstuffs, which is more dangerous than being exposed externally. But boars don't care about contamination. They are very ignorant, and because of that, they can thrive in the area where we aren't supposed to live and farm. In a way, they are the experimental bodies for us to know the reality of exposure from eating tasty things like mushrooms and wild plants, which we gave up to avoid the risk of exposure." Compared to the radiation monitoring posts emplaced throughout the region, Tengo felt closer to the boars, which, like the region's residents, did not just sit still in an environment but interacted dynamically with it.

Tengo implied that boars are not only harmful to the built environments but can be a potential contaminant as well. With his theory of a chain of internal exposure through ingestion, Tengo argued that precisely because boars and locals eat similar things and locals categorize boars as consumable food, it was both easier and more intimate to imagine the ecological transfer of radioactive materials from their bodies to his. Unlike other wildlife that are pure pests, the boar, as a potential food source, is a liminal creature that crosses the boundary between what is outside humans—the environment— and what can become a part of human bodies via ingestion. "If the state did not tell us about the accident," Tengo continued, "I could have been the boar." Tengo imagined the boar as a surrogate body, or an environmental monitor of Odaka that consumes contaminated foods on behalf of the risk-avoiding people.

Tengo's narrative offers a view into how people entertain the idea of *en*. For Tengo, the presence of contaminants in the postfallout environment made the boars relatable on multiple levels. Observing the boars eating fresh local foodstuffs reminded him of his severed connections to the land, the sense of seasonality, and the joy of farming. Tengo also realized that boars could help him understand the postfallout environment, since, unlike Tengo, they would continue consuming the local harvests and thus make changes in Odaka's contamination visible.

Tengo experienced the increasingly frequent presence of boars in his community as pointing to the changes that the TEPCO accident had caused to his livelihood, the environment, and their interconnection. Despite the changes to the texture and valence of that relationship, however, *en* sustained their connectedness. They happened to be in the same place at the same time because of some ineffable reason. Naoko did not particularly appreciate her newfound connection to boars, some of whom ate her persimmons and others of whom were buried secretly near her temporary residence. The city staff at Date city, who were cleaning the skins of captured boars for craftwork, commented, "When we first started selling boar goods, many farmers were angry, saying that they did not want to have anything made from their enemies, farm pests." Yet these existing and emerging recognitions of residents' *en* to boars have been threatened, as the state identified boars not only as resource competitors to be captured but also as potential contaminants to be isolated from locals like Tengo.

## WILDLIFE INCINERATOR FOR THE AGING SOCIETY

The outbreak of wild boars has been an issue not only in and around the evacuation zone but also in Fukushima Prefecture as a whole. According to the 2018 prefectural government's annual report on wildlife management, the number of captured boars in Fukushima increased from approximately three thousand in 2011 to twenty thousand in 2017, causing severe agroeconomic damage ($500,000 and close to $1,000,000, respectively). This surge relates partly to the establishment of a boar-hunting reward system in various municipalities in Fukushima. In Minamisōma, before the operation of its incinerator in 2019, hunters could present the tail of a boar they had killed and receive $200 in exchange (figure 17). Although initially people wanted to control the number of boars to prevent them from damaging local habitats and the local economy (since they destroyed the farmland), once the boars are captured and killed, their bodies have posed an additional and unexpected problem.

"I am not a specialist in wildlife," said Nakata, a structural engineer who works for Sōma city. "I do not like them, but now, for some strange *en,* I work with them." Nakata, the chief operator of the $1.6 million wildlife incinerator that Sōma city built and started operating in April 2016, explained why Sōma became the first location to host a wildlife incinerator. The boars' increasing presence became a serious threat to the Sōma residents and their farms, and the boars created a problem there much earlier there than in other regions in Fukushima. Sōma is located outside the state-mandated evacuation zone, and the majority of its thirty-five thousand residents, in addition to outsiders who evacuated to the region, remained in the area following the TEPCO accident.

"But now things are better because, with this machine, we can incinerate up to 120 kilograms of boars at a time, so we usually pile up a few for a single

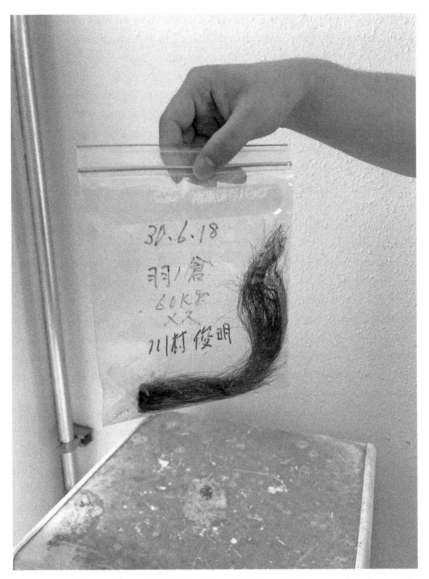

FIGURE 17. A boar tail stored at Odaka district hall in 2018. The name of the hunter, the location of capture, and the weight and sex of the boar are written on the plastic bag. The tail in the image belonged to a female boar of sixty kilograms caught on June 30, 2018, in Hanokura, Odaka. Photograph by the author.

operation. This innovation is one of a kind in the world, I know!" Nakata declared proudly. Previously, the city had cremated killed boars at a much smaller incinerator originally designed to cremate pets, like dogs and cats. Hunters and staff had to manually break down the boars' bodies to fit them in this smaller incinerator. As a result, for hunters, catching boars involved not only capturing and killing them, all of which took time, skill, and patience. Due to the government's ban on eating boars, capturing boars also entailed processing the boars and not consuming them (as they once had), all to simply destroy them. "This was very problematic," Nakata explained, "since there were not many hunters who were willing to put in that much effort, not to mention that the number of licensed hunters in this region and the country has been declining." Also, he noted, "most of them are getting old like myself."

Located in Sōma city, on the property of the waste-management center that processes all household waste produced in the city, the incinerator looked new and clean when I visited it in the summer of 2018. The city hoped that the larger, newer incinerator might mitigate the overpopulation of boars and disaster-exacerbated issues in the region, which include population decline and an aging society. The incinerator provides a means of streamlining the strenuous labor involved in eradicating wild boars, which no one, including Nakata, had imagined would be a persistent problem.

"We thought that the number of boars would decline as time went by," Nakata explained, showing me a chart enumerating the boars processed at the facility. "But even in June 2018, we had already burned 41 boars, and the year before we cremated 440 of them. The number is still high considering that we only process boars captured in Sōma and Shinchi town in the north." When I asked why that was, he guessed that it resulted from how decontamination had been progressing in previously untouched evacuation zones. Nakata observed that as time went on, more people began returning to former evacuation zones, areas that people had completely abandoned immediately following the 2011 TEPCO accident. "To be honest," Nakata lamented, "some of the boars are highly contaminated. I mean, they must be traveling from near the nuclear plant in the south, but if we capture them in our city, then they become our responsibility to dispose of."[8]

Just because boars were captured and processed within Sōma's geographical boundary did not mean that they had dwelled only in Sōma or had eaten things only in Sōma, where contamination was less severe than in other parts of coastal Fukushima. This is what I emphasized to the TEPCO employees. I was concerned that their plan to "chase" boars around 1F using their drones might contribute to the haphazard movement of contaminants that the boars embodied.

According to Nakata, the incinerator was designed to contain the spread of radioisotopes embedded in boars in order to remove contaminants from the ecosystem. Since there was some chance that the boars brought to the facility had passed through highly radioactive localities and eaten organisms and vegetation in those places, it was considered crucial to locate and distill any radioactive material that the cremation process produced. "All burnt particulates go through a pipe connected to an air filter and then into a bag in a steel can in a separate room. Whatever remains after the operation, like some leftover bones, our workers suck those things up with a vacuum. These vacuumed residues go through another pipe, which is also connected to another steel can in the separate room," he said, sounding more like a nuclear power plant operator than a city official.

The boar outbreak has beleaguered the national government, which aims to signal its successful containment of the nuclear disaster to Japan and the rest of the world by gradually reopening the evacuation zones. The boars have posed an additional hurdle to the evacuees, who would like to restart their former, more familiar lives. In the 2016 Japanese National Television documentary *Radioactive Forest* (*Hibaku no Mori*), a local organic farmer in Odaka, Koichi Nemoto, expressed how parts of Minamisōma have transformed into wilderness after the fallout: "Boars don't run away at all. They look at me like I am a visitor. . . . Wild animals that normally live in nature have taken over our world. I wonder if we are the ones now living in the cage." In response to the evacuees' concerns and to assist the state, TEPCO's drones flew the sky. It looked to me, I told the TEPCO employees, that their boar chase seems to assume that boars are isolated from the rest of the local ecology.

Whether working for TEPCO or not, humans did not accept life in the cage for long. Instead, locals and the state have been chasing and eradicating boars so that coastal Fukushima can fully recover and become "safe" again. In this process, the wildlife incinerators encapsulate the complex interdependence between the state remediation policy, which chases radioactive boars away from the evacuation zone, and the locals' desire for regional recovery, which marks wild boars as "obstacles" to be removed and isolated.

"I feel sorry for boars," remarked Gotanda, a Sōma resident in his sixties who works under Nakata. Gotanda is a specialist not in mammals but in fish; he is a retired fisherman who now farms the land he where lives. "I know boars are problematic. They destroy our farms and eat everything they can find, but we always coexisted. The fact that they are contaminated, captured, killed, and burnt like this with no mercy now is not their fault but the fault of us humans." Gotanda showed me inside the industrial-size freezer near the entrance of the incinerating facility where had placed about twenty boar carcasses wrapped in a translucent bags in green bins (figure 18). Their stiff legs were sticking out of the bags, and some were piled on top of the others uncomfortably. Approaching the frozen boars of various sizes, he put his hands together to pray for them. Looking at them, Gotanda recollected how he used to enjoy eating boars and treated them with great respect. According to him, in an impoverished region like coastal Fukushima where meat is a luxury, boar was a special treat.

Like many other local farmers I talked to, Gotanda saw boars as vermin (yūgai chōjyū), especially since the state banned the consumption and circulation of boar meat throughout most of Fukushima and its surrounding prefectures in November 2011. Even in 2022, boar meat remains one of the few items that the state still regulates, along with mushrooms and wild plants. These regulations continue to deflate the enthusiasm of local hunters. Typically, each municipality's local hunting association is responsible for hunting wild boars. "Some of my friends are hunters," said Gotanda, "and one time I was joking to them how the nuclear disaster made them rich because they can get over $200 per boar. Then they got angry at me, saying that they no longer enjoy hunting boars."

A few hunters I spoke with in Minamisōma confirmed that they no longer took pleasure in hunting boars. They no longer hunted to test their skill and technique, to share the meat with their family and friends, or to confront the vicissitudes of nature. Instead, once the state imposed the boar bounty, hunters began to see the boars as mere targets to be killed and disposed of, assigned an arbitrary monetary value. One hunter conceded that although some hunters may enjoy the few extra bucks they earn for capturing as many boars as possible, he was not a killer. He did not enjoy "murdering for money." His boar hunting was motivated solely by a desire to contribute to the region's recovery, which boars thwarted. He was aware of this issue as an evacuee

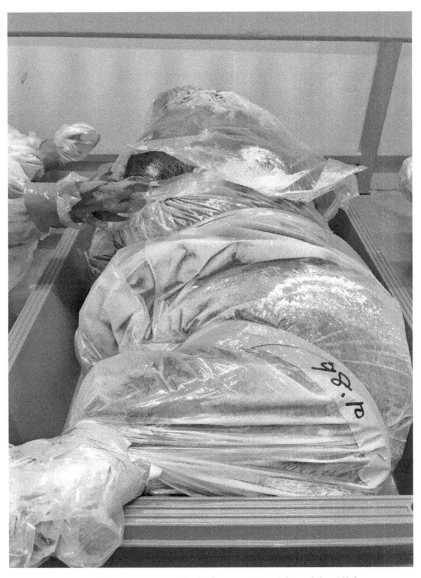

FIGURE 18. Frozen boar carcasses inside the freezer at Sōma's harmful-wildlife incinerator. Photograph by the author.

from Odaka himself. The pleasure and pain of hunting boars, he explained, was not only about exercising a set of skills; it also depended on having luck, which "is beyond one's control." Lighting a cigarette, he continued, "if you don't have *en,* you don't encounter boars. They are intelligent animals and know how to avoid humans. I happened to have *en.*"[9]

In the postfallout context, however, his *en* with boars took an ironic turn. By chasing boars to reduce their number and allow for the safe return of evacuees and for local agriculture to thrive again, he has unwittingly contributed to the production of radioactive waste. The TEPCO employees I spoke to, who had learned about the science of radiation through their work, seemed to be aware of this possibility without my elaboration. While boars may consume radioactive things in the environment, they are not dangerous to human beings unless humans consume them. Incineration, however, transforms boars into nuclear waste by obliterating everything but the contaminants. Gotanda, who had been cremating one boar after another, was keenly aware of this. As he prepared to show me the whole process of incinerating the animals, he shared his feelings of guilt. By producing nuclear waste just to get some extra money from the city during his retirement, Gotanda felt like he was betraying his fellow residents who, like him, wish they had never been victimized by the threat of contamination.

On the day I visited them, Gotanda and his work partner, Watanabe, a man in his sixties, smoothly loaded a couple of boars into a green container and rolled them to the side of the incinerator's bed. They then picked up each boar with a crane and carefully placed each one on the bed. They repeated this process a few times, loading four boars to get the weight to approximately 120 kilograms (figure 19). Once the bed was pushed inside the incinerator's chamber, the door was shut and the burner was ignited. The monitor panel to one side of the chamber door showed the temperature inside jump to eight hundred degrees Celsius. Watanabe explained that the temperature must stay higher than eight hundred degrees so that it does not produce dioxins, which are less discussed environmental-toxicant by-products of the remediation of coastal Fukushima. "Radiation is not the only invisible thing in Minamisōma," I remembered Nishiyama saying.

Once the boars were put inside, Gotanda told me that he was not as skilled as Watanabe, who had been cremating human bodies for a long time, and he left to rest inside a small office space. As the fire burned, Watanabe remained close despite the sweltering heat and intently watched inside the incinerator from a peephole. Stepping back a few inches, Watanabe wiped his face with his sleeve. "Cremating boars is much easier than cremating people," he remarked. "I do not have to stand here for hours, making sure I am not over–burning the body since we are not interested in keeping boars' remains intact like we do for humans." After people's bodies have been cremated, it is the Japanese practice for families and close friends who had *en* with the deceased

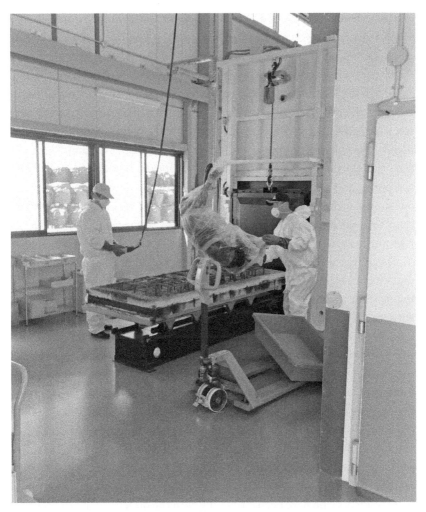

FIGURE 19. Workers piling up boar carcasses to prepare for incineration. Photograph by the author.

to pick out the bones of their loved ones to be placed in a grave later. Despite the difference, the technology used to incinerate the boars was made by Miyamoto, the same company that produces human crematoriums. On their website, the company promotes how their technology simultaneously respects the dignity of the deceased and the environment. They do not specify to which category wildlife belong.

In his comparison of the different mortuary practices for humans and boars, Watanabe suggested that since no one should have *en* to the radioactive boars, there was no need to preserve physical traces of their bodies. Yet, as he returned his gaze to the boars turning into bones and then dissolving altogether, Watanabe knew that even the $1.6 million machine could not make the radioisotopes the boars embodied disappear from the world. Although the incinerator achieved the physical separation of boars and radioisotopes, the boars' cremated remains and their radioactive potential were stuck in the air filter and transferred to a bag in the separate room at an even-higher concentration of particles. Indeed, 2021 data by Minamisōma indicates that the total radioactivity of the ash produced by incinerating wildlife was over 135,000 becquerels.[10] The level is 1,350 times more than the postfall-out food-safety standard in Japan and close to 83 times more radioactivity than the persimmons in Namie that the boar was eating and I picked in November 2013.

This physical death of the boars—and the resultant production of accumulative, accidental nuclear waste—brings to light the ecology of harm lurking beneath the collective recovery projects of coastal Fukushima. What I witnessed was wildlife's confrontation with humans "in relation of shared precarity" (O'Neill 2019, 94). It was also an enactment of half-life politics, in that the boars are destroyed to separate humans from radioisotopes in the name of radiation safety. The state's efforts to decontaminate coastal Fukushima have led to chasing boars away from the evacuation zone, only for them to be copresent and in competition with residents and returnees beyond the zone. This was essentially the problem I posed to the group of TEPCO employees. Where they were chasing boars to and from mattered significantly to the postfallout ecologies of coastal Fukushima. My point was to remind them of the residents living at the edges of evacuation zones. It was these residents who have been struggling with the presence of boars and have been coerced into the production of accidental nuclear waste. TEPCO's actions inside the evacuation zones impacted people outside of them.

After Gotanda and Watanabe successfully cremated the four boars, Nakata invited me to check out the back room, where the ashes of radioactive boars are kept isolated. "I work for the state, so I should not be saying this to you, but," Nakata said, lowering his voice, "when this bag of ashes gets to a certain height, an expert from the prefecture comes to measure its radiation level. If the level is lower than the regulated 8,000 Becquerels per kilogram, then it is the city's responsibility to take care of it as household waste. However, if it is higher, then it is the national government's responsibility to process it as nuclear waste. It might not be appropriate for me to say this, yet

I cannot stop hoping that the ashes are highly contaminated, because if they are, we do not have to keep them in our city."

Here, Nakata struggled to deal with the two competing roles the TEPCO accident and its remediation staged. His responsibility as a state employee was to incinerate boars, but because this act produced nuclear waste, it threatened Nakata's *en* to the land, its people, and the environment. By wishing the boars' remains to be highly radioactive and thus the responsibility of the national government, Nakata also denied Sōma's association with contaminants, as if Sōma were separate from the broader ecology of coastal Fukushima.

Free-roaming boars that elude the decontamination-reopening policy have been transferring radioisotopes across geographical boundaries. Their movements reveal the limits of the state's highly localized approach to the postfallout ecology. Decontamination had demolished Tengo's and Naoko's house, forced residents to sort through their belongings and choose what memories and connections to save and sever, and destroyed the persimmon tree and its fruits that Naoko and her family—and later monkeys and boars—once consumed. At the same time, decontamination and its half-life politics had removed the soil passed down through generations and chased "radioactive" boars out of the exclusion zones. It had also forced increasingly disinterested hunters to kill boars and for Nakata, Gotanda, and Watanabe to cremate their carcasses and disavow their remains. Through this sprawling chain of harm, we can observe multiplicities of loss. As Murakami puts it in *A Wild Sheep Chase* (1989, 20), "Some things are forgotten, some things disappear, some things die." What remains are leveled, "decontaminated" land, ultrasonic-emitting drones, wildlife incinerators, and "radioactive" boar ash.


NUCLEAR FALLOUT, AN ULTIMATE DENIER

In the summer of 2018, I visited radioecologist Thomas Hinton at Fukushima University, who had previously researched the irradiated environment around Chornobyl using wildlife as radiotracers. Inspired by Tengo's provocations and Gotanda's anxieties, I wanted to learn about boars' potential for modeling an irradiated environment. As I started describing my research, Hinton interrupted me, shocked. "Why are people killing boars?" he gasped. For him, the existence of boars was fundamental to his research uncovering the state of the radioactive environment and its changes.

Hinton oversees a group of scientists who follow the movements of wild boars to track the ecology of the contaminated environment in coastal Fukushima. A collar-like device they clasp around the boars' necks monitors the boars' movements through GPS and the ambient radiation through a dosimeter. This device enables his team to use boars as a tracer to help them understand the transfer of external radiation levels to a mammalian body. According to Hinton, the incorporation of radiation from the environment into mammalian biology is scientifically understudied, since the global scientific community, including the International Commission on Radiological Protection, has conventionally treated contaminated environments as something to be modified, decontaminated, and quarantined.[11] In this paradigm, the environment is an object of intervention, an entity that can and should be conceptually and geographically partitioned into safe zones and radioactive ones.

In a 2004 article, Hinton and his collaborators (2004, 333) criticize this prevalent anthropocentric framing of radiation safety: "The fact that humans are among the most sensitive mammals, and therefore the most sensitive species, has led, in part, to the dogma that if we protect humans, then other biota are protected as well." Their critique emphasizes the central issue of anthropocentrism underlying the Japanese state's decontamination policy. Its narrow focus on minimizing the risks *only* with respect to human exposure fails to account for the fact that humans exist within an intertwined ecological system that includes nonhuman biota, with which fallout residues could travel haphazardly. As I have discussed throughout this book, this separation of people from the ecological system also halved the meaning of their lives by undermining not only the ways people had lived but also their hopes and desires to sustain their familiar relationship to their lands.

Locally in Fukushima, this failure manifested as the residents' confrontation with the scientific model of relationality that draws a physical division between humans and contaminants. Just as the radiation maps, compensation, and *frecon baggu* had created the spatial and perceptual divisions between safe and radioactive spaces, the state approached the contaminated environment as something that needs to be detached from the residents to enable them to sustain their livelihoods. While the reduction of radiation exposure was necessary to safeguard the health of individuals, it had impacted people's quality of life in the region. The scientifically driven decontamination had modified and continues to alter the kinds of livelihoods possible in coastal Fukushima. This delimited model justified the half-life politics that destroyed the residents' prefallout livelihoods and killed the wild boars while also severing the pervasive sense of connectedness of people, things, spirits, and the environment. More than just biological monitors, wild boars represent the accident's protracted aftermaths, including the banal distribution of radiological harm and human efforts to distance themselves from those harms. The boars' death "is not their fault but the fault of us humans," Gotanda had said.

At the lab Hinton showed me new research he had begun conducting with graduate students. In this project they were assessing the usefulness of snakes as ecological tracers of the contaminated environment. Due to snakes' habitats (the soil) and their carnivorous preferences, their bodies incorporate environmental contamination more than other animals (Campbell and Campbell 2001). As Hinton and his team write in their 2021 article, "Compared to other animals that travel multiple kilometers per day, limited movement rates of snakes suggest their contaminant burdens should more closely reflect localized radiation levels, meaning snakes could be useful indicators of local environmental contamination" (Gerke, Hinton, and Beasley 2021, 553).

Recalling Hatsumi's fleeting comment two years earlier that she had not seen snakes since 3.11 (see chapter 8), I asked if it was harder to find snakes in the evacuation zones. Although he was not sure about the baseline, Hinton did not think that snakes were that rare—at least they were not rare enough to prevent his team from pursuing the idea of using snakes as radiotracers. In October 2021, Hinton and his collaborators (Cunningham et al. 2021, 12) published new research on the biological effects of chronic low-dose radiation exposure on wild boars and snakes in the evacuation zone. The research concludes that "the results provide critical evidence supportive of wildlife's resilience to the effects of chronic LD-LDR [low-dose, low-dose-rate] exposures, even in areas with sufficient radioactivity to force human evacuations." Their research also suggests that the wildlife inside the zones experience lower levels of stress, as indicated by cortisol levels, which is due to reduced disturbance from human activities, or "anthropause" (11).

The purpose of Hinton's research, whether it involved boars or snakes, was clear: there needs to be a better model to account for the relationship between livelihoods and local radiological ecology, a model that considers broader interconnectedness. Contaminants are not immediately perceptible, but they move incessantly between and across humans, nonhumans, and abiotic factors. The connection was inevitable, and no one could keep them apart. All we could do was to pretend that we could separate them. This pretense had left a severe scar in Minamisōma, as many Kashima residents whose land fell outside the state-regimented zones experienced; the invisible walls the state policy had emplaced in the aftermath of the TEPCO accident only stopped the flow of resources but did not protect the remaining residents from the stigma and ecological circulation of radioisotopes.

Following "radioactive" boars' lives and deaths, the wild boar chase upends the anthropocentric framing of the environment and of contamination itself wrapped up in the state's decontamination policy, in which concerns for human exposure condemn anything deemed potentially radioactive to be the target of removal and disavowal. Per this half-life politics, radioisotopes are the agents of contamination, external to humans and capable of being isolated. This radiation-centered discourse fails to consider the embeddedness of radioisotopes in the broader ecologies of *en* that entangle human, nonhuman, and supernatural in a given environment, however arbitrary, surreal, and unscientific such connection is felt to be.

Those who live in postfallout Fukushima critique this radiation-centered approach for its failure to honor the historical conditions that subjected coastal Fukushima to the fallout. As a lifelong resident of Minamisōma, a city that did not host the nuclear power plant but was nonetheless harmed by it, Tengo reiterated the importance of talking about the structural inequality wrapped up in the fallout that decontamination cannot redress: "It is called the '*Fukushima* nuclear disaster.' With the name, no one takes any accountability for the fallout aside from us who have to live with the name. People only care if we are exposed and do not see our struggles, which certainly cannot be measured by a Geiger counter."

Since the 1970s, the *Tokyo* Electric Power Company has operated the nuclear power plant in coastal *Fukushima* in order to supply energy to the greater *Tokyo* region exclusively (Akasaka and Oguma 2012; Iwamoto 1994, 2013a; Kainuma 2011). By underscoring this asymmetry, Tengo pushes outsiders to consider the TEPCO accident and its decade-long remediation as more than a series of discrete events following an unprecedented radiological crisis. Rather, the TEPCO disaster is a latent condition of radioactive colonialism, through which certain groups of humans and nonhumans, in their *en*, bioaccumulate radiological harm and are rendered exposable.

Like the uranium mines and nuclear complexes located on Indigenous lands throughout North America (see, e.g., Brugge, Benally, and Yazzie-Lewis 2006; Erickson and Chapman 1993; Kuletz 1998), in coastal Fukushima, wildlife incinerators and radioactive boar ash are the material legacy of half-life politics' anthropocentric delimitation of human-environment relationality. Tengo disagrees with this model, claiming that "radioactive materials are everywhere, and we are always already contaminated regardless of what we sacrifice or destroy to try distancing ourselves from them." Echoing Tengo, I suggest that the discourse on Fukushima and the TEPCO accident ought to move beyond concerns about individual radiation doses and address its expansive distribution of harm, across species and abiotic factors and through the coercive severing of their *en*.

Willingly and unwillingly, wild boars have been cutting across the artificial boundary between purity and danger that state decontamination policies have articulated. In order to protect humans from exposure in a narrowly defined way, the wild boar chase will continue, disrupting human and nonhuman *en*. The chase makes local people, boars, and any fallout residue they consume accidental collaborators in nuclear waste production, which will be provisionally dislocated and deferred to another "timescape" (Creager 2013b) for future generations and other species of coastal Fukushima to confront.

The wild boar chase suggests that even though, in theory, nuclear fallout "formally links human actions, technological capabilities, atmospheres and ecologies in a new formation of contamination" (Masco 2015, 140), this is hardly the end of the story. In practice, humans try to undo such links and disavow contamination while disregarding existing *en* among the living, dead, nonhumans, and environments. Confronting this type of violence, Naoko once commented how the TEPCO accident, the public discourse on the accident, and its remediation had been "an ultimate denier of my livelihood." Expressing her frustration, she often wondered if the state's emphasis on radiation exposure only served the public whose fear centered around radiation but not the loss of livelihood.

Photographer Jun Nakasuji explores the ruined landscapes of postfallout coastal Fukushima, or "the other side of the wall outlets" (*konsento no mukougawa*) of central Japan. In his photographic collections "Swabs: Fukushima the Silent Views" in 2016 and "The Other Side of the Wall Outlets" in 2021, he attempts to connect Fukushima and the rest of the country with the concept of *kyouji*, or synchronicity. At the periphery of the peripheries, Nakasuji has encountered people and cultivated *en* that helped him witness and imagine the other side, where people lived despite the accident and its ongoing aftermaths. In his desire to understand Fukushima and its residents, Nakasuji is not alone. A 2021 online Brand Research Institute survey of more than twenty thousand Japanese people investigates various ties using the concept of relational population.[12] *Relational population* signifies the number of people who either already have some biological, cultural, or social connections with a specific location or feel interested in making a connection to the place. As a construct, it expresses the level of interest in a place. In their 2021 survey Fukushima Prefecture scored the highest of all forty-seven prefectures of Japan, indicating that many Japanese people reported having a connection with Fukushima already or hoped to discover a *yowai tsunagari* and cultivate it.

If the nuclear ghost of Minamisōma had a message, what would it say to those interested in making connections with and learning more about Fukushima ten years after the accident? If anything, it would be that while nuclear fallout and radioactive debris materially interlink differently positioned people, the ensuing social, political, ideological, and moral discourses construct and cement divisions among them. By approaching the accident not as an event but as process, we see that a nuclear accident is not merely about radiological danger.

The residents' atomic livelihoods—their diverse lives with radiation that have been rendered invisible, with thriving wildlife and against prejudice and discrimination—teach us that it is not just the state and TEPCO that are to blame. We are also complicit in the construction and exacerbation of divisions produced by half-life politics and its postfallout world. We have contributed to the production of the nuclear ghost of Minamisōma through our efforts to distance ourselves from radiation by creating boundaries between safe and unsafe, Fukushima and the rest, and the self and other.

What I want us to ask is this: if we cannot attend to the residents' present struggles and their narratives of their experiences in the gray zone, how can we attend to the potential long-term effects of low-dose radiation exposure that might not materialize until years from now? There is no easy answer, but because we are complicit in this process, there is hope. If we can reevaluate our relationship with Fukushima and the TEPCO accident and remember our relation to them, not as spectators but as *kyoujisha*, we can help the residents navigate the two winds. What happened to them and what is still happening in Fukushima is undeniably and ineffably connected to us. We have all been living in the same radioactive world. Only in our lack of imagination are there invisible walls dividing us.

# Epilogue

Think about it. Isn't it weird that we met? There must be some supernatural cause, don't you think?

HATSUMI

## THE AGE OF DECOMMISSIONING (*HAIRO NO JIDAI*)

"Whenever an earthquake strikes now," Saeki in Kashima district confessed in 2013, "3.11 has conditioned me to pay attention to the state of 1F." Explaining her disaster response routine, Saeki emphasized the impact the TEPCO accident had on her perception of the risks of living in coastal Fukushima. "An earthquake always meant a potential tsunami warning. But now we have an additional threat to be mindful of."

On July 12, 2014, at 4:22 a.m., a magnitude 7.0 aftershock struck the coast of Fukushima, leading to a tsunami warning. Hearing the eerie sounds of the alarm for the first time, I jumped out of bed and frantically contacted many residents to ensure their safety. They sounded relatively calm. However, like Saeki above, they commented on their newly developed habit of checking the power plant. By 2014, the remaining and returned residents were less wary about radiation exposure from the TEPCO accident. At the same time, however, they all knew that the precarious state of the damaged nuclear reactors nearby continued to pose grave risks in their everyday lives. "It is depressing to think that we must live with 1F. It won't go anywhere unless *somebody* decommissions it," Saeki lamented.

Although Saeki knew that the laborers were likely to be locals, she still hoped that no young residents would be forced to expose themselves to complete the decommissioning. "Maybe the future is we elders become cyborgs ourselves to join the cleanup!" Saeki half-jokingly imagined the future of this graying region where, in 2021, there were only 284 newborns compared to 579 in 2010.[1] Not too long after her remark in 2014, the state selected Minamisōma to cohost the Fukushima Robot Test Field (RTF) with Namie town. With the RTF, the state envisions furthering their techno-optimistic vision of society from the current information-driven "society 4.0" to beyond. "One of the largest research and development bases in the world," the RTF partially opened in 2018 and then fully opened in the former tsunami-inundated region of Minamisōma in March 2020. Out of 3.11, futuristic "Minamisōma 5.0" was born in the gray zone.

Initiated as the state-led Innovation Coast Framework, the RTF aims to attract scientists, engineers, and policymakers from around the world to Minamisōma to develop next-generation robotics for disaster response, mitigation, nuclear decommissioning, and elder care.[2] The city hopes, and the state promises, that the RTF will bring young talent to Minamisōma and help sustain the city's history and culture that survived 3.11. However, unlike in the eighteenth century, when coastal Fukushima offered barren land to enthusiastic migrants, now the land might be irreversibly contaminated. Moreover, the other coastal Fukushima municipalities and rural regions in the country are competing with Minamisōma to attract young migrants.

Even though Minamisōma never hosted a nuclear power plant, the post-fallout reconstruction of Minamisōma as the robot city has come to strengthen its ties to the failed reactors. The demographic and economic survival of the rapidly aging city now depends on the TEPCO accident's futurity and the long-term decommissioning project of 1F, and reactors elsewhere. The TEPCO reactors at 1F represent just six of the fifty-six nuclear reactors in the country. As I write this epilogue in November 2022, Japan plans to eventually decommission twenty-four reactors in the country while restarting many others. Currently, ten reactors are in operation, and seventeen more are in the process of resuming. The 2022 Russian invasion of Ukraine has fueled the urgency to restart the production of nuclear energy to curb the country's resource dependency. The growing global concern over climate change positions nuclear power yet again as a cleaner and more sustainable energy option. What is selectively forgotten here is that 1F came to coastal Fukushima with the same rationale in the 1970s.

Less than thirty minutes before midnight on March 16, 2022, just five days after the eleventh anniversary of 3.11 and amid the global COVID-19 pandemic, two large earthquakes of magnitudes 7.4 and 6.1 struck off the coast of Fukushima. The officials reported about twenty centimeters of a tsunami—enough to make people immobile—on the coast of Sōma. These earthquakes caused the most significant material damage in Kashima district (where Saeki lives), the area considered the least impacted by the TEPCO accident in 2011 compared to other parts of the city. Many Kashima district and Sōma city residents just north of it had to live without electricity and water for a few weeks, reliving the nightmare of eleven years prior, this time wearing masks and trying to maintain social distance. "Seeing how my house got damaged from the constant assaults from the precarious earth," Saeki reported, "I am not sure how the broken TEPCO reactors could still stand!" Saeki's concern notwithstanding, Japan has recorded 109 earthquakes of magnitudes over 5.0 since 3.11.

What my long-term fieldwork in postfallout coastal Fukushima taught me is that the danger of nuclear power is not merely the technical risk of if and when there will be fallout. The other risk is that the failed structure and its unpredictable afterlives will continue to haunt the region with existing and new vulnerabilities, such as frequent natural disasters, an aging population, and the global pandemic as well as climate change.[3] Stopping the generation of nuclear energy does not mean that the power plant structures, their contamination, and residual radiation magically disappear. Someone must labor to decommission them, and land needs to be sacrificed to isolate the waste. As the TEPCO accident has demonstrated, it is inevitable that more bodies will be exposed in the process. This will likely involve local people I have interacted with during my fieldwork, who stayed, returned to, or moved to the region, as well as their communities, families, friends, relatives, and neighbors who hope to preserve their inalienable lands, histories, and cultures.

The TEPCO accident has shown what and who it takes to undo our collective nuclear legacy. It requires a community of people and their commitment to staying in a contaminated environment, maintaining their land, and making the place better for the next generation. It requires a cycle of people and land to be exposed to protect others. Nuclear power is a social technology that produces and reproduces a community of the exposable. It also requires, I suggest, that public and political discourse pay attention to individual stories and local histories behind exposure and not be complicit in this process.

More than eleven years after 3.11, Japan is still under the Nuclear Emergency Situation declared on March 11, 2011, at 4:36 p.m. JST. Around four thousand workers are laboring to decommission the damaged reactors each day.[4] Despite this, TEPCO (optimistically) expects the plant's cleanup to take thirty to forty more years. The global pandemic has already delayed the process, and continuing aftershocks from the 2011 earthquake incessantly threaten the precarious structures and surrounding communities. The question is not "Will 3.11 ever end?" but rather "How can we stop a future 3.11 from happening?" One pragmatic approach is to remember those residents who did not give up or have not given up their homes despite the accident. Doing so allows us to acknowledge that the TEPCO accident is still ongoing and accept that its lingering effects, however spatially and temporally dislocated and shapeshifting, will remain with us and impact all of us for a long time. The TEPCO accident is not somebody else's problem. It is our collective problem.

More than eleven years after 3.11, some evacuees are still willing to return to the exclusion zones, desiring to recultivate their lands and live and die at home with their ancestors.[5] Their atomic livelihoods—like that of Naoko, who took the radiological risk to pass her land down to future generations— sow the seeds for the better future that they want to follow, the age of decommissioning (*hairo no jidai*).

## *SAIKAN* AND THE FUTURE PAST

As of November 2022, over 57,600 residents are living in Minamisōma against the odds and despite the hypervisualized risk of exposure. Over the last decade, social, political, and scientific uncertainties surrounding radiation have challenged people's trust in each other as well as their confidence in their perceptions of the risk and the world. The residents' and outsiders' sense of uncertainty about low-dose radiation exposure has persisted despite (or as the direct consequence of) the state and the nuclear industry's persistent efforts to use scientific reasoning to objectify radiation's impacts on human bodies and contain the fear of the invisible threat as a knowable, avoidable, or compensable risk.

The science of low-dose radiation has become more sophisticated over time, in part due to the use of coastal Fukushima as a real-life laboratory. Nonetheless, the 2021 Ministry of the Environment national survey of 4,200 Japanese citizens shows that more than 45 percent of non-Fukushima residents believe that there will be adverse intergenerational hereditary consequences of the TEPCO accident, compared to around 27 percent of Fukushima residents.[6] The survey result indicates how the state and the nuclear industry have failed to regain the trust of the residents and the broader public and to mediate their epistemic and affective divide using scientific facts.

I have witnessed that much of the residents' postfallout lives amount to recovering and reconstructing disintegrated social connections with their families and relatives, communities, homes, soil, land, and ancestors by choosing where to live and die. Indeed, the instability of state policies and expert opinions has made residents suffer from a series of social meltdowns more immediately and acutely than they have suffered the biological effects of radiation in the environment.

However, this finding does not suggest that contamination in coastal Fukushima is negligible. Even if the physical risks of living in the area are lower than anticipated and there are no medically and statistically significant cases of chromosomal damage, radiation is still extremely harmful socially, politically, and psychologically. Such harms have made the residents' humble hopes and desires to pass down their intergenerationally inherited lands, histories, and cultures to future generations more challenging, if not impossible. From this ethnography of atomic livelihoods in coastal Fukushima, I offer one lesson: a not immediately visible risk like low-dose radiation produces shapeshifting science, politics, laws, and media reports, and the biomedically driven, exposure-centered approach to visualizing the shape of such risk delimits individuals' ability to articulate and share their situated experiences, not only of the risk itself but also of their overall lives.

This lesson is not specific to the region nor relevant only to a nuclear catastrophe. Our collective experience of the global pandemic since March 2020 shows an uncanny resemblance to the diverse atomic livelihoods I have documented in postfallout coastal Fukushima. Journalist Satoru Ishido (2021) has contemplated the link between the two by comparing his coverage of the various facets of 3.11 and his personal experience with the pandemic in Tokyo. Having witnessed the manifold and often-incommensurable struggles of 3.11 survivors on the ground, Ishido writes of experiencing how the pandemic repeats old traumas: traumas of divided communities; mistrusted science, authority, and others; and atomized experiences, voices, and stories that cannot be safely shared publicly or domestically without meeting backlash and moral and personal judgment. Living the pandemic as the return of 3.11 traumas, Ishido retrospectively considers 3.11 as the future past—a past that had anticipated the future—and asks if 3.11 experiences can generate an alternative approach for all who are living with the various divisions occurring in the present.

I have been living with the pandemic in Princeton, New Jersey, and Cambridge, Massachusetts, far away from coastal Fukushima. Despite the apparent differences between the pandemic and 3.11 in the scale and extent of the harm, the structure of pandemic public health policy and enforced protocols, and the reactions of the affected population, the pandemic similarly caused a general loss of confidence in the state, experts, science, and personal perception.

Confronting an invisible threat, no one seemed to know the right thing to do in the United States, Japan, or elsewhere. Only the numbers of the infected bodies and dead keep changing, and the numerical abstractions gradually become out of sync with one's reality. Do I need to wear a mask? If so, when, where, for how long, and what kind? Who is more vulnerable, and who can resume "normal" life? How can I relate to others or accept their subjective experience when their risk perception radically differs from mine? What might be the long-term consequences of the virus, even for those of us it does not kill?

These questions are strikingly similar to those asked by coastal Fukushima residents while confronting the TEPCO accident. They have discovered that the collective imagination of their invisible and ghostly risk walked around independently of their lived experience of radiation and at times spoke louder than the growing body of empirical data that suggested otherwise. With the COVID-19 pandemic, those patterns of multilayered meltdowns seem to be repeating themselves even among diverse populations and across greater geographical and cultural reach. Living through the pandemic, I feel that the nuclear ghost Hatsumi has encountered since March 2011 has reentered the stage yet again. "Once the ghost reveals to you or is revealed itself through other's determination, there is no turning back," said Hatsumi. In this sense, contamination-related disaster—nuclear or viral—manifests itself in a society like an incessantly shapeshifting figure of the Japanese ghost.

My ethnography puts forward the premise that disaster always occurs in a place with a specific history and culture. Thus, it is no surprise that even a globally shared threat like the COVID-19 virus causes dissimilar damage to differently organized societies. Nonetheless, I have shown that a disaster leaves its traces and residues, however invisible, invisibilized, and inconvenient they are. Hatsumi's nuclear ghost captured a process through which individual experiences get black boxed and swept under the rug as a result of science and policy trying to scale up their analysis and progressively favoring the collective over individuals and objectivity over subjectivity. In search of the nuclear ghosts, I learned how the construction of the grand narrative of the past and the average experience with statistical facts had flattened and silenced those atomic truths incompatible with a greater analytical coherency. Against this, I have tried putting the dominant, radiation-centered, half-life politics in suspension to resist reducing coastal Fukushima to "the land of contamination" and its residents to the category of "the damaged." In search of a new narrative, I have instead foregrounded my inquiry into less visualized atomic livelihoods—residues that science, medicine, and the state left behind for the sake of analytical and statistical consistency.

The 3.11 future past has illustrated that a suspension of damage-centered discourse is critical for our increasingly dividing world to attend to residues of technoscience and analytically elided individual variations. However incommensurable and insignificant experiences among differently situated individuals appear to the gaze of scientifically informed data and policies, their atomic experiences nonetheless feel undeniable and indispensable to each of us who have been experiencing the pandemic synchronically in time. Those systematically spilled-over experiences suggest that we may also exist among the ghosts and in a gray zone. I cannot help but think that the experience of alienation and division that residents like Hatsumi experienced could happen to any of us and could become the residue that exacerbates a future catastrophe.

Residue, Boudia and her coauthors (2018) suggest, is an apt concept for approaching chemical toxicants that are characteristically material, irreversible, unruly, transboundary, and accumulative. Here I extend their idea to suggest that atomized experiences, like chemical pollutants, have a residual quality and could remain latent and resurface later. Ishido's epiphany of the global pandemic as the 3.11 future past is one example of how the experience of past disasters could have a long afterlife and nonlinear temporality. If residues from the past may resurface unpredictably in the present, how have the coastal Fukushima residents experienced the global pandemic?

As a city that has confronted the fallout of social ties since March 2011, Minamisōma residents have been surviving the pandemic relatively well. As of November 2022, more than two and a half years since the beginning of the global pandemic, Minamisōma counted a little over 4,000 positive cases, many of whom were nonresident workers attending to the remediation and reconstruction of the city. Although rigorous research must be conducted to examine the relationship between the two catastrophes, 3.11 seemed to help prepare the residents and the city for the pandemic. After talking with residents about their pandemic experiences, I have come away with a greater understanding of how disaster is not a discrete event but a nonlinear process. In particular, I have learned two preliminary lessons.

First, the global pandemic has provided a fresh opportunity for residents to reflect on the ongoing TEPCO accident. Both the pandemic and the TEPCO accident challenge the existing social network and its maintenance, and the vulnerability of relationality based on kinship or physical proximity. These catastrophes suggest the importance of embracing weaker forms of connection, or *en,* whose legitimacy depends not only on blood ties, social and moral obligation (*kizuna*), or physical proximity but also on a general feeling of connectedness.

As I have elaborated in this book, *en* is a mere possibility with which one can entertain a chance encounter of human and more-than-human kinds as meaningful. You may call it an efficacious fable, make-believe, or expansive form of kinship.[7] In my ethnography of fallout, I took the lead of residents who found our chance encounters in their *en*-inflected worldview. What secured our weaker relationships was not the presence of moral duty or obligation but our mutual commitment to the invisible thread that brought us together, which coproduced *en*'s existence and meaningfulness. Following this local theorization of relationality, when crises are so diverse, not widely separated in time, and planetary in scale while also very personal, synchronicity (*kyoujisei*) is critical to radically reimagining sociality, responsibility, and accountability for our individual and collective actions on our shared planet. Highly individualized experiences do not preclude the possibility of being together, as my ethnography illustrates.

Second, the residents of coastal Fukushima have been experiencing the global pandemic amid the ongoing TEPCO accident. Their unique entry into the pandemic (and experience of it) suggests that, as Norihiro Nihei (2012) conceives of it, we are constantly situated in-between disasters (*saikan*) and never out of their cycle. Interlinking the invisible though meaningful connections in Fukushima's gray zone, I have shown the importance of paying equal attention to, on the one hand, what residues from the past disaster(s) resurface in the present and, on the other, how the future disaster that a specific population anticipates influences the way that they experience emergent risk.

When approaching the TEPCO disaster, I have explored the sociocultural significance of the historical disasters in coastal Fukushima and elsewhere (Hiroshima, Nagasaki, Minamata, and Chornobyl) as comparative frames that different actors used to make sense of the present catastrophe.

While the chronological order of "events" is vital to understanding the present, it is not deterministic nor type-specific. For example, residents and outsiders alike have used 3.11 to make sense of COVID-19 and vice versa. As I discussed in chapter 8, local residents found the regional eighteenth-century famines significant for understanding 3.11. Meanwhile, others, including myself, found some resonance with past industrial disasters like Minamata disease. These are just a few examples of living "in-between" disasters. Importantly, *saikan* is not just a negative space. As Kishō Kurokawa (1977) imagined in the architectonic of Japanese design, in-betweenness is where conflicting things, ideas, and people can find unexpected connections. Out of ambiguities and grayness, *en* emerges.

## A LOYAL SERVANT OF *EN*

Hatsumi died in 2018.

I received the sad news in Princeton from our mutual friend in Minamisōma. In the small, tight-knit community, news travels quickly. I had last seen her six months previously. Hatsumi had looked unwell but made time to chat with me. She knew from one of her daughters that I had tried to visit her a couple times between late 2016 and 2017 to update her about my findings on the wildlife issue she had once reported to me. However, each time her daughter had told me apologetically that "she does not want to show her weakening" and requested that I leave.

My last brief sight of Hatsumi, with her awkwardly positioned white knit hat and her petite physique (she was even smaller than when I had first met her), conjured a memory of her yelling at the director and me to leave her property on my first visit in July 2013. In the five years since then, the state had reopened Route 6, lifted the evacuation order for the twenty-kilometer zone, and almost finished installing 7.2 meters of tsunami walls throughout the coast of Minamisōma. Wildlife roamed the city. Meanwhile, I had graduated, gotten a job, and grown more gray hairs. On that last day with her, there was no hint of the film-worthy fierceness the foreign film director had found in Hatsumi. So much had happened to both of us and the city since.

In the living room that had been converted hastily into her bedroom, Hatsumi acted as cheerful and animated as ever, joking about how much older I looked. "I was doubtful when I first met you, but maybe now you could teach something," Hatsumi teased me in her usual manner. Our last conversation was nothing memorable; we just talked about my new job at Princeton and a place in Southeast Asia she wanted to visit. Over the last five years she had traveled to many places, but she was not interested in visiting the United States.

I knew Hatsumi for only five years, until the end of her life in 2018. Even in that brief window of time—merely one-sixth of the time it would take for cesium-137 to decay in half—she proved that centering my analysis of coastal Fukushima on radiation and its bodily harm could only touch, at most, half of how she lived and died in Minamisōma. The same goes for Naoko, Miura, Nishiyama, Tengo, Saeki, and others I have featured in this book and those I spoke to who are not named presences.

Before we said goodbye Hatsumi jokingly mentioned how the Minamisōma people helped me get my new job at Princeton, so I was forever in their debt. I told her how she was right about that, and she said I was a loyal servant of *en*. "Think about it. Isn't it weird that we met? There must be some supernatural cause, don't you think?" she mused. Trying to be witty, I almost mentioned that the nuclear ghost was the missing link, but I swallowed my words and just nodded in agreement.

Although Hatsumi never told me what the nuclear ghost was, I had met many residents in Fukushima's gray zone who forever changed my perspective on the TEPCO accident, coastal Fukushima, and environmental disaster. In a city where radiation's real and imagined presence haphazardly divided its geography, ecology, people, and community, my search for the nuclear ghost shapeshifted to the act of interlacing the invisible, weaker, and unexpected though meaningful connections with strangers I met in the region.

Why was I attracted to a specific group of people over others in my fieldwork? I easily could have featured people who had dedicated their time and efforts to challenging the state and TEPCO, or I could have focused on those who decided not to come back to coastal Fukushima because they thought it was permanently contaminated and unsafe to live there. I had opportunities to interact with those people and learn about their hopes, desires, and struggles. The stories of their postfallout livelihoods are as real, impressive, and pressing as the ones I have told here. I often met audiences, academic or otherwise, who seemed to expect to hear these stories in my ethnography. At the same time, I often read about and was exposed to their accounts in an overwhelming number of books, media reports, and documentary films, which, in my view, had the tendency to extract others' sufferings to fulfill the expectations of the masses. My *en* guided me to tell other stories.

My ethnography of the fallout started in 2013 as a reaction against the foreign documentary filmmaker who attempted to document the "exposed" population in coastal Fukushima. Standing outside the blind spots of his filmographic gaze and acting as a mere translator, I felt the urge to capture what was left unrecorded by his camera lens. However self-serving my answer may be to the question of why I was attracted to certain people and their narratives, perhaps something about each of my dialogic partners, whose mundane lives with radiation stayed out of the public imaginary, made me feel that they could have been me but were not me. Collaborating with them as an outsider, stranger, and ethnographer, I wanted my ethnography to coproduce a new intersubjective narrative out of "what-could-be-told" that might exist in the gray zone between "what-was-already-told" and "what-was-untold" (Shineha 2021, 16–18) of the TEPCO accident.

Writing an ethnographic monograph, just like making a documentary film, is never an unmediated act (Daniel 1984). I have intentionally selected individuals, scenes, facts, and words to patch together what I witnessed in coastal Fukushima, hoping that this monograph would offer the chance for my readers to place their feet in my interlocutors' shoes. Having spent a significant time between 2013 and 2019 in coastal Fukushima, I have found myself compelled to share as widely I could its residents' physical, psychological, and spiritual attachments to the place they called home. Telling their stories, I wanted to convey that the radiation's biomedical threat, however iconic, is just one harm of a nuclear accident. Our collective imagination of radiation's irreversible biological and intergenerational effects could also produce social and psychological damage among the so-called victims of the accident and rid them of their pasts and futures.

Richard Parmentier (1994, xiv) says that for anthropology, "'truth' is the premise rather than the conclusion of discourse." My ethnography indicates one messy truth about the residents who lived through the first decade of the TEPCO accident and the resulting divisions and low-dose radiation in Fukushima's gray zone. Though the accident and its aftermaths coercively put the region and its residents in a less than ideal state, those facts could not annihilate the residents' desires and efforts to make their homes and land fertile and meaningful for those who might come after them.

"There is nothing in Minamisōma." This was how Hatsumi introduced the city to me in July 2013. Yet, in this supposedly unremarkable place with the hypervisualized threat of radiation, I met people like Hatsumi, her daughters, and her grandchildren who nonetheless have remained there, returned there, and will live there. I found *en* with those who happened to call this gray zone their home—however magical, subjective, and invisible such connections are.

In my last conversation with Hatsumi, she did not say anything about her health, and I did not probe, but she did not attribute it to the TEPCO accident. Only a year later, in 2019, when I visited to properly mourn Hatsumi's death, did I learn from one of her daughters that cancer had taken Hatsumi's life. It did not surprise me. I had witnessed her bodily pain with my own eyes. I had finally learned to pay attention to Hatsumi and not to a Geiger counter.

After Hatsumi passed, the same daughter told me about Hatsumi's survival of a previously diagnosed case of cancer, which she had not disclosed even to her daughters. Assuming her daughters knew about the previous cancer, the doctor casually mentioned it close to the end of Hatsumi's life. Without waiting for my response, the daughter shared that even as inhabitants of a Buddhist temple—a place that attends to others' misfortune—the family was not immune to loss.

"She was a strong lady," said the daughter after informing me about Hatsumi in tears. "And a wise one," I added after offering an incense stick to the family altar with her cremated remains, where her family had placed a portrait of Hatsumi from one of her trips to Europe, her big smile capturing her vibrant character. Although it is a common Buddhist practice in Japan to move cremated remains to a family grave forty-nine days after the cremation—at the end of the soul's dwelling in this world—the daughter told me how the family had struggled to follow the custom. "The family told me to take care of it, but I just cannot make myself do so," the daughter said and started sobbing. Soon, a phone call interrupted her crying. There had been another death in Minamisōma. I bowed once again to Hatsumi's remains and left her temple.

Learning about her earlier cancer survival gave me a new perspective on Hatsumi. She had intimately known what it meant to be exposed to radiation, though of a very different kind, from her past medical treatments. As I drove to the coast where reconstruction workers were busily building the Robot Test Field, I wondered if her personal history had any bearing on her view of the nuclear ghost of Minamisōma.

Hatsumi is no longer here to answer this question or crack a joke. All I have now are the words she left with me. "Tell the world and especially younger generations what *you* witnessed in Minamisōma" was the mission Hatsumi assigned to me when I told her about my job at Princeton. I hope I have lived up to her expectations with this book.

The coastal Fukushima residents like Hatsumi challenged me with facts—their livelihoods—that I had not anticipated, which called me, and us, to think differently. In order to understand how people have lived, struggled to live, and died—and for what reasons—following the fallout, it is critical to see beyond half-life politics, beyond how the state and science have tried to quantify the radiological damage, how the media has narrated it, and how the public has imagined it. My ethnography started with a shock—or perhaps a mistake—of assuming that I knew something about their lives without becoming attuned to the place they called home. The residents patiently corrected my ignorance by showing that their decision to remain in or return to the region did not come from a lack of knowledge or recklessness. Nor did they want to become a convenient data set that the state and the nuclear industry could use to justify more civilian exposure in the future. Participating in and observing diverse atomic livelihoods in coastal Fukushima for more than six years, I now know that the local experts on low-dose radiation exposure—the ordinary residents I met in coastal Fukushima—sincerely hope for a world where no one suffers from both real and imagined radiological contamination and exposure.

When I saw Hatsumi's cremated remains in 2019, I remembered the question that I had failed to ask her during our final meeting in 2018. I wanted to know whether she had led a good life (*ii jinsei*) in Minamisōma. Like many other elderly residents in the region, Hatsumi had often talked about wanting to die well. During my time there, I spent a significant amount of time hanging out with elders as their presence in the city became more pronounced after 3.11. I witnessed them suffering and getting angry, sad, and, at times, depressed. Still, those experiences failed to define who these people were or what it meant for them to try to live and die well in Fukushima's gray zone. When I mentioned the hardships of postfallout livelihood to Naoko—in this instance, the need to reconstruct, recultivate, and renurture soils and land, and to be patient—she glossed over these as unremarkable aspects of farming in the region. As a farmer, Naoko knew it would take time, effort, determination, and luck to create and leave behind something good.

For Hatsumi, dying well meant not burdening those who would remain after she departed this world. Remembering her wish has helped me to understand why she did not want to inform me about her health struggles or attribute them to the TEPCO accident. In their idiosyncratic ways, many residents have confronted death's haunting and universal fate by taking on the moral responsibility of leaving fewer burdens to the generations to follow them. As Murakami (2000, 273) puts it, "Death is not the opposite of life but an innate part of it. By living our lives, we nurture death." They did not want to die as a nuclear ghost or to make others become one after their deaths.

In the grand Buddhist (and Hindu) theory of the transmigration of the soul, *en* is as much about interweaving people, things, land, and nonhuman others in a knot of fate as it is about untangling the same knot so that when people depart to the other world they can be reincarnated as something new. Perhaps detachment is a general and pressing concern among elderly residents, for whom death feels more present than the accumulative biological effects of radiation. For them, the moral reckoning of detachment depends on their commitment to intergenerational succession: creating a good life for others.

The residents I interacted with voiced their regrets about leaving behind the crippled nuclear power plant, a disintegrated community, and the irradiated soil and environment. Underlying their concerns about experiencing a good death was their concern for the good lives of new generations. What their atomic livelihoods suggest is that the fundamental issue of nuclear energy that we need to be attuned to is not limited to the risk of an accident, low-dose radiation exposure, or the longue durée of nuclear waste, which means that technoscientific progress alone cannot address it. Instead, nuclear energy is an unsustainable structural threat. Its intergenerational lifecycle collides with the humble individual and collective atomic livelihoods that work toward making life and land better for those who might follow.

Unfortunately, the desire that elders like Hatsumi and Naoko had to create better lives for future generations might not be actualized anytime soon. While I have no magic solution for undoing Japan's nuclear legacy, I want to suggest something to you, readers of this book. The age of decommissioning requires all of us to learn to respect the land that embodies the cycle of existence in which the intergenerational hopes, desires, and struggles of residents and ancestors like Hatsumi keep dwelling.

This is the message I received from the nuclear ghost of Minamisōma nine years after my first encounter with it in 2013. I wish I could tell Hatsumi about my current interpretation. Knowing her, I bet she would have said to me jokingly that I was wrong, and I needed to stay in Minamisōma longer.

I sincerely hope you will visit coastal Fukushima one day, see with your own eyes how people live, and find *en*. Hatsumi was right. There is a nuclear ghost in Minamisōma, and it may change your past, present, and future.

# ACKNOWLEDGMENTS

I did not want to study disaster, nor did I want anything to do with it. When 3.11 happened, I was exploring a different potential dissertation fieldwork topic. Realizing 3.11's significance to Japan, the department of anthropology at Brandeis suggested that I consider turning it into my research topic. I first resisted the idea, but now here I am. I want to thank the faculty and staff of the Brandeis anthropology department, especially Ellen Schattschneider, for seeing something in me and taking me in when I did not know much about anthropology.

When I was confused about the unknown losses in Japan I had experienced from afar, it was Richard Parmentier who patiently listened to me. He suggested that I just go to Japan with an open mind and see what I could find, whether related to the disaster or otherwise. With his guidance I found the *en* that led me to write this book. Since then, he has retired. I am proud to be one of the last students he trained as a semiotician. Rick sensei, you taught me to think semiotically and write accessibly, and I tried to live up to your ideal here. I hope this book serves as a token of my gratitude, one humble demonstration that your scholarship and teaching live on through me.

With encouragement from people in the United States and Japan, I was able to pursue research on 3.11 for more than a decade and met many people in coastal Fukushima and the greater Tōhoku region. My *en* with them fundamentally changed my worldview. I thank all of you, including those who appear under pseudonyms and those who remain anonymous in this book, for allowing me to dwell in the gray zone with you. Without you, this book would not have been possible. Your words, thoughts, and ideas are the building blocks of this book. Instead of naming you here, I will make sure to visit each of you in person to express my deep gratitude.

I hope that this book will eventually become accessible at Minamisōma's incredible library, where I spent countless hours reading local histories and accounts of 3.11 and feeding coins to the scanner. I sincerely appreciate Katsunobu Sakurai and Kyoko Tanaka and their openness to an outsider exploring the city. I also want to thank my old friend and *senpai* Keisuke Kitamura, whose support was essential for my drive around coastal Fukushima. I want to report that the Kitamura pillows I sold in coastal Fukushima on your behalf helped produce good sleep for those who desperately needed it.

Before 3.11 I wanted to research the culture surrounding Japanese cuisine, its culinary arts, and the process of becoming a chef, which was one of the occupations my parents had imagined I might pursue after witnessing my early life as a *hikikomori* (socially withdrawn recluse). They never would have imagined that I would go to a foreign country without knowing much English and eventually become a professor at Princeton. I did not imagine nor plan this either. If anything, my parents, Shigeru and Kyoko Morimoto, my two siblings, Ayumu and Sei, and my extended family members, Takayo, Uta, and Kou, and their nonjudgmental stance toward me made it possible for me to follow this unconventional route (even if my mom would claim that she had given up on me), so thank you.

I traveled a very long and winding path to arrive where I am today. At the vocational high school that I attended to be "rehabilitated" and then reintegrated back into society, I remember writing a cocky essay criticizing the Japanese education system, claiming that I would prove its failure. I must have been an annoying kid to my parents and teachers! But I want to report that all your efforts and patience paid off. Although I do not claim to have proven the system wrong, I have forged a unique path. In the United States, I attended a few language schools before attending a community college (Arapahoe Community College) and then the University of Colorado, Denver, before I moved to Massachusetts for my MA and PhD training. That was by no means a usual track for academics around me, and I thank the University of California Press for recognizing diverse paths to scholarship and endowing me with their FirstGen Program to help me complete this manuscript.

Disaster seems to haunt my life. Around the time I became a *hikikomori,* the Hanshin-Awaji (Kobe) earthquake and the Tokyo subway gas attack shook Japan. When the mass shooting at Columbine High School occurred, I had just moved to Littleton, Colorado, to attend a language school. My host family lived only a mile from the school and were among those who were profoundly shaken. When I returned to the United States after going back to Japan to earn money to continue my community college education by being a mover for a company with the iconic panda logo, 9/11 happened. My host family at the time was dismayed at my apparent indifference or inability to accept reality. Later, I moved to Massachusetts after going through a personal crisis, which prevented me from finishing my MA training in clinical psychology. Just when I felt like I was able to escape from things I did not want to deal with, 3.11 happened.

As Haruki Murakami put it in *After the Quake,* "No matter how far you travel, you can never get away from yourself." The earlier disasters like the Kobe earthquake cast shadows on my psyche. I was not alone, though. I met people in my generation and those who were older in tsunami-stricken Tōhoku. Many expressed their sense of incompetence as teenagers or young adults during the Kobe earthquake and said that they wanted to do something this time, not just to help others but also to confront their weaknesses and unfinished traumas. I wonder if those volunteers I met on the ground have found some answers and resolution. I thank my hometown, Kurashiki city, for sending aid and its residents to support the recovery efforts far away and enabling someone like me, who initially did not know anyone in Tōhoku, to go there and to reflect on my own life and that of others.

As much as 3.11 shuttered many existing and potential connections and networks, it also opened up spaces for me to discover newer and unexpected ones. Because of 3.11 I met the mentors and teachers who helped me to become a scholar. Looking back, meeting Andrew Gordon and Peter Galison at Harvard was formative to my research. They have shown me not only how to be a scholar who uses one's knowledge and expertise to promote social good but also how to excel at doing so.

Andrew Gordon and I worked together for more than four years developing and managing the Japan Disasters Digital Archive (jdarchive.org). Through this international digital archive project, I was able to apply my research to think about ways of archiving and disseminating the memories and records of disasters to the future. It was a perfect platform to continue reflecting on my fieldwork and visiting Japan to conduct my follow-up research. I could not have thought of a better project to deepen my scholarly engagement with 3.11 and my personal ties to Tōhoku in general and Fukushima in particular. I thank Gordon sensei for the opportunity and mentorship. I was extremely fortunate to be able to work closely with and learn from Gordon sensei, who is one of the most prominent scholars of Japanese studies. He is a hidden interlocutor of this monograph. I could not have written the book without his critical historical eyes toward Japan and his expectation for me to be a good researcher.

During my predoctoral and postdoctoral fellowship between 2014 and 2018 at the Reischauer Institute of Japanese Studies (RIJS) at Harvard, I benefited from various individuals' knowledge, kindness, and invaluable support. I want to thank Mary Brinton, Shinju Fujihira, Ted Gilman, Katherine Glover, Yoshie Gordon, David Howell, Koko Howell, Jennie Kim, Fun Lau, Kathi Matsuura, Kuniko McVey, Ian Millar, Yitsy Ooi, Hannah Perry, Susan Pharr, Yukari Swanson, and Gavin Whitelaw. In particular, I thank Stacie Matsumoto for her incredible support and unconditional belief in my ability.

Working with Peter Galison on his documentary film project with Robb Moss, *Containment,* introduced me to the fascinating world of the history of science and the pressing issue of nuclear waste and its relationship to land and underprivileged populations. To this day, I am shocked by the extent of his generosity and his agreement to serve on my dissertation committee, and I am grateful for his continued support. Because of Peter, I can no longer think of science, scientists, and technology in the same naïve way as I once did. His scholarship has indescribably impacted my own, as well as my inquiry into scientific instruments, infrastructures, language, and scientists. Without his guidance I would not have been able to add historical depth and geographical breadth to this book. I hope this book lives up to his expectation that it not be just about 3.11 nor just about Fukushima.

In the early stage of my research, I benefited tremendously from Tōhoku University and its International Research Institute for Disaster Studies (IRIDeS). I still remember when I first met Professor Akihiro Shibayama at a conference at Harvard in January 2012. After his talk, I ran up to him and said that I wanted to work with him and he casually answered, "Sure." When I showed up with my big suitcase at his office in Sendai months after, however, he looked flabbergasted and said, "Oh no . . . you actually came. I thought you were joking!" and could not close his wide-open mouth for a while. Nonetheless, he kindly created a visiting position at IRIDeS for me to work closely with him. Since then, he has been an outstanding mentor and a leader of 3.11 digital archive research in Japan. His Michinoku Shinrokuden project has allowed me to expand my network with various stakeholders of disaster studies in Japan and elsewhere. I thank IRIDeS faculty Sébastien Penmellen Boret, Yuichi Ebina, Maly Elizabeth, Julia Gerster, Arata Hirakawa, Fumihiko Imamura, Daisuke Sato, Shosuke Sato, and Daisuke Sugawara, as well as earlier members of the archive project, Masahiro Iwasaki, Yukihiro Minami, Madoka Ono, and Tohru Okamoto.

My multiyear fieldwork was made possible by the generous support of the 2013–14 Japan Foundation Doctoral Research Fellowship and the 2015–16 Toyota Foundation's Exploring New Values for Society Research Grant. RIJS and IRIDeS also provided continuous support for numerous trips between 2014 and 2018. At Princeton, the funds from the University Center for Human Values and the Council on Science and Technology allowed me to conduct follow-up research between 2018 and 2019.

Joining Princeton has nurtured my identity as a scholar and teacher. I want to thank my magnificent colleagues at the Department of Anthropology. They have offered me a secure environment in which to grow: João Biehl, John Borneman, Lauren Coyle Rosen, Elizabeth Davis, Julia Elyachar, Agustín Fuentes, Hanna Garth, Carol Greenhouse, Onur Günay, Jeffrey Himpele, Rena Lederman, Serguei Oushakine, Laurence Ralph, Carolyn Rouse, and Jerry Zee. I was fortunate to join the department at the same time as Laurence, who has provided invaluable mentorship and support. I thank Carol Zanca, Mo Lin Yee, and Gabriella Dorinovan for selflessly creating a productive and collegial working environment for me to excel.

I also want to extend my thanks to people outside the department. At the department of East Asian Studies, I have greatly benefited from the in-depth knowledge of the experts in Japan: Amy Borovoy, Thomas Conlan, Sheldon Garon, Kimberly Hassel, Federico Marcon, Setsuko Noguchi, Keiko Ono, Franz Prichard, James Raymo, and Shinji Sato. I have also learned immensely from various scholars at Princeton across disciplines, including but not limited to Aisha Beliso-De Jesús, Patrick Caddeau, Allison Carruth, Michael Celia, Steven Chung, Angela Creager, Natalia Ermolaev, Christiane Fellbaum, Elena Fratto, Maria Garlock, Alexander Glaser, Michael Gordin, Katherine Hackett, Alison Isenberg, Sami Khan, Zia Mian, Erika Milam, Anne McClintock, Rob Nixon, Sébastien Philippe, Kim Scheppele, Robert Socolow, Gabriel Vecchi, Frank von Hippel, Keith Wailoo, and Sharon Weiner.

Teaching and interacting with students have enabled me to reflect on my fieldwork afresh. I want to acknowledge my undergraduate research collaborators on the Nuclear Princeton Project (nuclearprinceton.princeton.edu) for allowing me to expand my horizon of nuclear-related research. With you, I have learned to think more critically and creatively about the unexpected connections among the people in Native North America, coastal Fukushima, and Princeton's engagement with nuclear science and technology. Thank you to Blue Carlsson (Cherokee Nation), Travis Chai Andrade (Kanaka Maoli), Thomas Dayzie (Navajo Nation), Lillian Fitzgerald (Klamath Tribe), Brooke Kennedy (Walpole Island First Nation & Ojibwe), Jessica Lambert (Choctaw Nation), Keely Toledo (Navajo Nation), Ella Weber (MHA Nation), and Hunter and Joshua Worth (Kanaka Maoli). You all are the best.

Writing was a long process and a massive struggle for me. Support from the Institute for Advanced Study (IAS) at Princeton afforded me the time to complete this monograph. There was no better place for me to write this book than IAS, where scholars like Albert Einstein, Robert Oppenheimer, John von Neumann, and Hideki Yukawa lived and researched. Their engagements with nuclear things have forever changed the global nuclear landscape. Writing this book, I modeled my work after Yukawa's goal of using his knowledge of atomic science to disseminate "its knowledge to the general public to urge them to experience this tiny, invisible reality [the existence of the atom] as familiar as possible" ([1943] 1976, 148–49). I hope my book has lived up to his aspiration.

I thank the faculty of the School of Social Science, Didier Fassin and Alondra Nelson, for hosting me in 2020–21. I could not have asked for a better role model for how to navigate through academia's sticky politics and gridlock than Alondra. I appreciate the IAS staff for working tirelessly to make it possible for us to continue pursuing scholarly exchanges during the early COVID-19 chaos. I also want to acknowledge the members of the Science and the State seminar, especially faculty leader Charis Thompson and fellow member Sarah Vaughn. Sarah in particular pushed me to consider the broader implications of 3.11 on the environment and climate change and adaptation.

I had numerous opportunities to workshop chapters and ideas. I thank Susan Lindee and Elly Truitt at the Department of History of Science and Sociology of Knowledge and Frederick Dickinson at the Department of East Asian Languages and Civilizations, both at the University of Pennsylvania. I also benefited from conversations with others: Ken Buesseler at the Woods Hole Oceanographic Institution;

Katheryn Goldfarb and Timothy Oaks at the University of Colorado, Boulder; Rihan Yeh and Michael Berman at the Linguistic Anthropology Workshop at the University of San Diego; Yukiko Koga at the Columbia Modern Japan Seminar; and Christopher Nelson, Anne Allison, Valerie Lambert, Michael Lambert, and others who participated in my talk at the Carolina Asia Center. I have been fortunate to be a member of the oikography workshop at the Society for Advanced Research organized by Federico Neiburg and João Biehl. I thank them and the participants for their generous feedback: Susan Ellison, Carlos Fausto, Catherine Fennel, Angela Garcia, Ann Kelly, Javier Lezaun, Alexandra Middleton, Eugênia Motta, Sebastián Ramírez, Thiago da Costa Oliveira, and Kaya Williams.

Many other individuals and scholars in the West helped me with the process of writing this book. I thank Vickey Bestor, Dmitri Brown, Flint Espil, Michael Fischer, Kim Fortun, Flavia Fulco, Tristan Grunow, Salvatore Guisto, Hugh Gusterson, Jeanne Haffner, Stefan Helmreich, Jyunko Habu, Susanna Hoffman, Arnold Howitt, Suma Ikeuchi, Miyako Inoue, William Johnston, Graham Jones, Nina Kammerer, Norman Kleiman, Hiroko Kumaki, Sam Levine, Andrew Littlejohn, Joseph Masco, Allison McFarlane, Hirokazu Miyazaki, David Odo, Heather Paxson, Adriana Petryna, Annemarie Samuels, Donna Semel, Hillary Semel, Sarah Semel, Scott Semel, Alpen Sheth, Mrinalini Tankha, Taro Tsuda, Eve Tuck, and Christine Yano.

For those in Japan, I thank Norio Akasaka, Arinobu Hori, Makoto Iokibe, Yoshiteru Iwamoto, Hiroshi Kainuma, Fumitoshi Kanazawa, Emi Kaneko, Kumi Kato, Wataru Kumagai, Ryosuke Maeda, Mori Masako, Yukihiko Mori, Koichi Nemoto, Shunsuke Nozawa, Hiroshi Okumura, Yukiko Okimoto, Kazuko Sasaki, Keiichi Sato, Yoshiaki Seto, David Slater, Kenji Tamura, Masararu Tsubokura, Hidenori Watanave, Toshitaka Yamao, Shingo Yano, Hirokazu Yoshie, and Shunya Yoshimi. I also would like to extend my gratitude to Hideaki Yanami at Kahoku Shimpō, Masahisa Tashiro at Fukushima Minpō, Thomas Hinton at Fukushima University, Kouhei Hori and Fumihiko Futakami at Minamisōma City Museum, and Akio Koike, Takafumi Masaki, and others at Tokyo Electric Power Company, as well as various organizations and communities such as Date City, Fukushima Central Television, Fukushima Prefectural Museum, the Japan Atomic Energy Agency, the Japanese Red Cross Society, Kashima Rekishi Aikoukai, the National Diet Library, Date city, the National Institute of Technology (KOSEN) Fukushima College, Shichigō Community Center, Sōma city, and Tagajyo High School.

I want to thank University of California Press, in particular Kate Marshall, for having faith in my work and patiently waiting as I slowly prepared the manuscript. After the submission of the initial draft, it felt like everything happened so quickly. I am amazed by her competence and efficiency. I would also like to thank the editor of the California Series in Public Anthropology, Ieva Jusionyte, who found my ethnography worthwhile and has provided both academic and moral support. I thank the five readers, including Ieva, whose valuable comments and suggestions helped strengthen the manuscript.

As a nonnative speaker of English, I could not have written this book without the tireless and meticulous work of Hollianna Bryan. As my personal editor, Holli not only helped me put together this book with accessible language but also somehow magically learned to understand what I was trying to say in broken English and suggested edits. At the same time, Holli preserved my voice in the process. Holli's support was indispensable in every step, from organizing the book to writing and editing. Holli is one of the dialogic partners of this book, whose perspective enabled me to learn how best to tell the story about coastal Fukushima to an English-speaking audience. Thank you, Holli, for being a true professional and for never admitting you were tired of reading my writing over and over, even when I was tired of it. Sharron Wood provided an additional pair of eyes to make the manuscript more readable. Any errors in the book are my own.

Since leaving Japan to pursue my education in the United States, I have been fortunate to weave *en* with those who made me feel like the United States is my home. I want to acknowledge the Heberton family and my honorary uncle and friend, Marian Brezina. We have survived many bumpy roads together as a team. I am in utter awe of your willingness to treat me as a member of your family even after I left Denver. I hope this book serves as proof that I have not forgotten your family and your continuous support. I am not good at verbalizing it, but I am always in the presence of the loving spirit, and I trust you are as well.

Above all others, Beth Semel made it possible for me to complete this manuscript. Beth, you have witnessed my ups and downs and listened with pure patience to my often-nonsensical musings. Without your unconditional care and support, I would not have been able to come this far. I deeply respect you as a person and scholar. Together we grow and will continue to do so. Thank you, Beth, for being there and making it possible for me to experience new things and positive feelings in this world. You never fail to inspire me. And to my adopted cat, Millie Semel: I am not sure if you understand human language. Still, your inquisitive look as you stare at my laptop has led me to believe in your miraculous ability to sense something valuable in my writing.

Finally, I would like to acknowledge readers like you who have found this book out of millions of others. Thank you. It must be because of some mysterious working of *en*.

# NOTES

## INTRODUCTION

1. See, for example, Greenpeace (2021), Koide (2012, 2021), and Norimatsu and Busby (2011).

2. See also Tierney (2008) on elite panic.

3. Here I follow the Japanese convention by using Chornobyl, the Ukrainian spelling of Chernobyl.

4. For a more comprehensive account of the accident, see Asia Pacific Initiative (2021) and NHK Meltdown Syuzaihan (2021) in Japanese. In English, see Funabashi (2021) and Vetter (2020).

5. For examples, see Kasai (2020), Ozaki (2013), and Yoshida (2020). A 2017 documentary film by Chiaki Kasai titled *Life: Ikiteyuku* features Takayuki Ueno, a Minamisōma resident who lost his family members. The search for the missing continues in 2022.

6. Ozawa-De Silva (2021) also writes about her experience as a Japanese person living outside the country and witnessing 3.11 from afar.

7. On the subway attack, see Lifton (2000) and Murakami (2001).

8. In one interview about his writing on the Tokyo subway gas attack, Murakami describes an earthquake as a metaphor for the collective sense of insecurity: "No ground is solid" (Ellis, Hirabayashi, and Murakami 2005, 558). Elsewhere (Morimoto 2012), I have written about the effects of disaster as "shaking grounds." Retrospectively speaking, I might not have come to the United States if it were not for my experience of living through 1995's dark time as a teenager.

9. Davis and Hayes-Conroy (2018) write about the evacuees' reported experience of radiation revealing "who we are as people." In this sense, my fieldwork was about reflecting critically on my own positionality and assumptions about radiation exposure.

10. See Numazaki (2012, 27) for the general postdisaster sentiment of reporting that other people are suffering more.

11. Lambek (1996, 246) discusses the act of elaboration in memories by stating that "the smoother the story, the more evident that it is the product of secondary reworking."

12. On the discussion on the roles and ethics of anthropologists studying 3.11, see Gill (2014), Kimura (2014), Miyazawa (2018), and Numazaki (2012).

13. The size of Fukushima Prefecture is 5,322 square miles, which is comparable to the size of Connecticut, which is 5,543 square miles.

14. In June 2022 the Japanese supreme court ruled against the state's responsibility for the accident and its failure to supervise TEPCO's disaster preparedness and dismissed its duty to compensate Fukushima evacuees.

15. See https://web.archive.org/web/20220614181631/https://www.youtube.com/watch?v=Hwov3yaAipI&feature=youtu.be.

16. Sakurai and Kainuma (2012, 139); see also Yamaoka (2012).

17. Between July 2013 and 2014 I lived in Kashima district in Minamisōma. Since then I have made more than twenty-five follow-up visits, each over two weeks between 2014 and 2019, and lived in Kashima and Odaka. I was able to visit Japan frequently between 2014 and 2018 as a manager of the Japan Disasters Digital Archive (https://jdarchive.org), which is managed by the Reischauer Institute of Japanese Studies at Harvard University.

18. Eguchi et al. (2021) and Hirosaki et al. (2018) explore the relationship between the frequency of laughter and people's lifestyle change resulting from 3.11.

19. See Aritsuka and Sudo (2016), Harigane et al. (2021), Hori et al. (2017), Iwasaki (2021), Longmuir and Agyapong (2021), Orui et al. (2020), Nomura, Oikawa, and Tsubokura (2019), and Tsubokura (2018).

20. For detailed information about thyroid cancers, see https://web.archive.org/web/20211020223319/https://www.pref.fukushima.lg.jp/site/portal-english/en03-03.html. As of June 2022 there are 287 confirmed cases in Fukushima. For more on this, see the discussion in chapter 5.

21. See, for example, Inaizumi (2021) and Katayama (2020). Hiraoka, Tateishi, and Mori (2015) and Yasui (2018) discuss cleanup workers' health.

22. See, for example, DiNitto (2019), Morris-Suzuki (2014), Petryna and Cole (2011), Phillips (2013), Plokhy (2022), and von Hippel (2011).

23. The former Nuclear and Industrial Safety Agency at the Ministry of Economy, Trade, and Industry (now the Nuclear Regulation Authority at the Ministry of the Environment) initially designated the accident as level four on March 12, 2011, and then changed it to level five on March 18, 2011.

24. Peta stands for $10^{15} = 1,000,000,000,000,000$.

25. For more information, see https://web.archive.org/web/20220619114558/https://www.reconstruction.go.jp/topics/main-cat2/sub-cat2-1/hinanshasuu.html.

26. See https://web.archive.org/web/20210306182039/https://www.mri.co.jp/knowledge/column/dia6ou000002kkbn-att/MTR_Fukushima_2012.pdf.

27. See https://web.archive.org/web/20211221192226/https://www.pref.fukushima.lg.jp/uploaded/attachment/461556.pdf.

28. See also Joy Parr (2010) on radiation perception and safety practices among nuclear power plant workers in Canada.

29. Yuri Lotman (1990) also discusses a periphery as the location of experiments, where new things and ideas emerge out of uncertainties.

30. See also Goldfarb (2016).

31. Even though the film director and I never saw eye to eye and fought all the time, I acknowledge that he first introduced me to Minamisōma.

32. See Allison (2013, 8) for the use of *muen* to express the lack of social relations, and see also Rowe (2011, ch. 2).

33. Kudō (2016) reported that a few taxi drivers in the tsunami-stricken Ishino-maki city, Miyagi Prefecture, which had lost over 3,500 people, shared the experience of giving rides to a ghost. The story goes something like this: A taxi driver picks up a customer who is wearing off-season winter clothes. As the driver arrives at the requested destination or on the way to the destination, the customer vanishes. Similarly, in *The Legends of Tōno,* Japanese ethnologist Kunio Yanagita (1875–1962) recorded a story he had heard of a widower's unexpected encounter with his wife, who died from the 1896 Meiji-Sanriku tsunami in Iwate Prefecture. In the story, the wife was with another man. Disheartened, the husband asked about their son, who was also swept away by the tsunami. The wife ran away crying. For other ghost-related stories and spiritual experiences from the survivors of the 2011 tsunami, see Kanebishi (2018), Ishido (2017), Okuno (2017), Takahashi and Horie (2021), and Utagawa (2016).

34. Lifton and Falk (1982) discuss the psychosocial impacts of atomic weapons in a manner uncannily similar to Masco's later conceptualization of it.

35. Of the 2.7 Peta Bq of cesium-137 deposited onto the terrestrial environment, 67 percent fell out onto the forests, compared to 5 percent onto urban areas (Onda et al. 2020).

36. If it were not for my long-term fieldwork, my monograph would have told a story about how the evil state and electric company made innocent people's lives miserable. Although this is undeniably true, and Hatsumi did narrate her experience to the film director in this way in 2012, my fieldwork taught me that it is only half of the story. I concur with Numazaki (2012, 34) that ethnography is fundamentally a "slow science," and that it takes years to learn about the history of a place and discover and cultivate *en* with a group of people under investigation. Anthropology is intrinsically "slow" due to its methodological commitment to ethnography, unlike natural science, which Isabelle Stengers (2018) believes should "slow down" to disengage from the fast-paced knowledge economy. In this sense, the untimeliness of ethnography—the time it requires for an ethnographer to understand local concerns and hear how they are narrated—affords a temporal distance that enables a more objective stance (Rabinow et al. 2008) toward political and economic interests or leads to an unexpected discovery. An ethnographer is able to observe not only what changes but also what remains unchanged over time. One lesson for an ethnographic study of disaster is that although a sense of urgency is critical (Slater 2013), it could make researchers more vulnerable to an extractive mode of inquiry.

37. To avoid any potential confusion of the two terms, newspapers and other publications do not use the Chinese character for *baku* and instead use the neutral, Japanese kana spelling (ばく).

38. There are other incidences of radiation exposure, such as the nuclear fallout exposure from the Bravo Testing in 1954 by the crews of the Lucky Dragon No. 5 and the Tōkaimura Criticality accident in 1999.

39. In her book *Precarious Japan,* Allison (2013) mentions her experience volunteering in Minamisōma. McNeill (2013) discusses Minamisōma as one of the contentious sites of the national and international media's early reporting, or lack thereof, on the TEPCO accident. Loh and Amir (2019) discuss the role played by medical professionals in coastal Fukushima.

40. The specific hue of gray Kurokawa had in mind was *rikyū nezu,* the RGB color model numbers 136, 142, and 126.

41. See also Figal (1999, 52–73) on Minakata.

42. See Kyoko Matsunaga (2014) for an in-depth analysis of Silko and her critique of American nuclearism. Lou Cornum's "The Irradiated International" (2018) attempts to make broader connections between different countries and groups mediated by nuclear things. Similarly, Jacobs, in *Nuclear Bodies: The Global Hibakusha* (2022), argues for the need for solidarity among the victims of fallout exposure across geographical sites.

43. I drew this framing from Gabrielle Hecht's concept of nuclearity. While Hecht elaborates upon many definitions of the term, I use the most general one: "*Nuclearity* is a *technopolitical* phenomenon that emerges from political and cultural configurations of technical and scientific things, from the social relations where knowledge is produced" (2012a, 15; emphasis in original).

44. Contrary to what is argued elsewhere (Polleri 2022), suspending radiological damage is different from moving toward "post-victimization." Here, I use the word *suspension* to signal the act of decentering our focus from radiological damage to allow for the exploration of other types of harm caused by 3.11. In other words, I argue that suspension is one analytical stance to stay with and get hold of the spatiotemporally dislocating effects of slow violence (Nixon 2011).

45. Some examples include Caldicott (2014), Kimura (2016), Lochbaum, Lyman, and Stranahan (2014), Morris-Suzuki (2014, 2015), Polleri (2019, 2021), Slater, Morioka, and Danzuka (2014), and Sternsdorff-Cisterna (2020).

46. Biehl and Neiburg (2021) suggest engaging in "oikography," an ethnographic approach to studying ways people make their home, to explore the significance of the home as the ground of the political.

47. I discuss the concept elsewhere (Morimoto 2021a, 2021b, and 2022).

48. See, for example, Fukushima Minpō Shimbun (2013, 2014, 2015), Fukushima Minyū Shimbun (2022), Higuchi (2021), Imai (2016), Kanno (2020), Manabiai Minamisōma (2018), Meguro (2016), Ōnuma and Yoshihara (2018), Ōwada (2016), Ōwada and Kitazawa (2013), Sato (2020), Shibui et al. (2014), Sugano (2022), Thanks & Dream (2017), and Voice of Fukushima (2015). Also, Fukushima Medical

University has a list of selected publications by its associated scholars on disaster and radiation. See www.fmu.ac.jp/univ/en/papers/index.html.

49. Stawkowski (2016) also writes against atomic victimhood in her ethnography of the Koyan residents near the Polygon site in Kazakhstan.

50. Here my use of the term *atomic* was inspired by historian of science Angela Creager's seminal book *Life Atomic: A History of Radioisotopes in Science and Medicine* (2013a).

51. See Ralph (2020) and Garcia (2017) on nontraditional archives.

52. Whenever possible, I have also used the Internet Archive's Wayback Machine (https://web.archive.org/) to preserve the URLs of the born-digital materials I used as references (web articles, documents, videos, etc.). Unlike printed media, born-digital materials can have a short half-life. Paying attention to the rapid cycle of born-digital materials is one critical intervention in studying disasters in the twenty-first century.

53. See the World Nuclear Association's report in May 2022, https://web .archive.org/web/20220607104528/https://www.world-nuclear.org/information- library/current-and-future-generation/nuclear-power-in-the-world-today.aspx.

## CHAPTER I. NAMING THE NUCLEAR GHOSTS

1. Lafcadio Hearn ([1904] 1971) translated a series of Japanese ghost stories and published them as *Kwaidan: Stories and Studies of Strange Things.*

2. See https://web.archive.org/web/20210102102921/https://news.yahoo.co.jp /byline/onomasahiro/20150103-00041982/.

3. 「今の日本には白昼堂々おばけが歩き回っている。放射能おばけというおばけが。おばけは人々に恐怖を吹き込み、恐怖は毒となって社会の全身を巡り、放射線問題の解決を困難にするばかりか、民主政治を麻痺させている。」

4. On the history and debates about permissible dosages, see Caufield (1990), Cram (2016a), Iguchi (2019), Imanaka (2012), Lindee (1994), Maruyama (2017), Murakami et al. (2019), Nakagawa (2011), Natori (2013), Onaga (2018), Shimazono (2019), Sutou (2016), Walker (2000), and Watanabe, Endo, and Yamada (2016).

5. Interestingly, unlike during the COVID-19 pandemic, when the phrase "with Corona/ウィズ コロナ" became popular in Japanese, there was never a discussion of life "with radiation" in postfallout Japan.

6. On the feminist counterperspective about the irrationality and baselessness of the fear of radiation exposure in fallout, see Aoki (2018), Kimura (2016), and Morioka (2014).

7. See, for example, Anzai (2021), Caldicott (2014), Koide (2021), Koide and Field (2019), Muto (2021), Yamaguchi and Muto (2012), and Ian Thomas Ash's 2013 documentary film, *A2-B-C.*

8. Although it is not much discussed, the TEPCO accident produced radioactive waste outside Fukushima Prefecture in places such as Miyagi, Tochigi, Ibaraki, Gunma, and Chiba.

9. For Japanese anthropological scholarship on coastal Fukushima, see Sekiya and Takakura (2019), Takagi, Sato, and Kanai (2021), Takahashi (2018), Takezawa (2022), Tsujiuchi and Gill (2022), Tsujiuchi and Masuda (2019), and Yoshihara (2013, 2016, 2021). Also, some exciting new ethnographic work in English about Fukushima is coming out by scholars like Hiroko Kumaki, who has conducted extensive fieldwork in coastal Fukushima. See Kumaki (2022).

10. Riken Komatsu (2021) also writes about driving on Route 6. Komatsu lives in Iwaki, the southernmost part of coastal Fukushima, and thus his driving direction is opposite from mine but complementary.

11. Nagatoshi Shimoeda, a Tōhoku resident, hosts a collection of pictures of tsunami-affected Minamisōma on his personal website, https://web.archive.org/web/20211115171404/http://shi.na.coocan.jp/tohokukantodaijisin-4.html.

12. In 2021, with the completion of Tōhoku Chūō Expressway/Sōma-Fukushima Road (about forty-five kilometers), access between central and coastal Fukushima became easier.

13. Most of the Jyoban line was devastated by the tsunami, and it took six years for it to reopen. Prior to that, access by train was cut between Sendai and Minamisōma and southward to Tokyo.

14. A similar phenomenon was observed in Miyagi Prefecture in the north, where the Sendai-Tōbu Road functioned as a tsunami wall and a temporary evacuation shelter.

15. Local resident and surfer Kōji Suzuki used to own a surf shop, Sun Marin, near the Migita beach. Although Suzuki lost his shop, he reopened the shop inland near Rokkoku in 2012. He has been working on surf tourism to bring people to Minamisōma. See https://web.archive.org/web/20220626192749/https://www.tfm.co.jp/hand/index.php?catid=3728&itemid=170491.

16. A surviving pine tree in Rikuzentakata in Iwate Prefecture (Wakui 2012) is more famous than the one in Kashima.

17. Before the tree died, the group worked with scientists to extract its DNA, and now clones of the tree are growing.

18. For more on the unusual black spots on digital radiographs, see Kashimura and Chida (2015).

19. Scientists are divided about the health effects of low-dose radiation, and the question is highly politicized. See Nakagawa (2011) for the history of radiation exposure. For technoscientific debates about Fukushima, see Hida (2012), Imanaka (2012), Kodama, Shimizu, and Noguchi (2014), and Murakami et al. (2014).

20. Fabian defines a denial of coevalness in anthropological thinking as *"a persistent and systematic tendency to place the referent(s) of anthropology in a Time other than the present of the producer of anthropological discourse"* (1983, 31; emphasis in original).

21. By 2013, one of the most prominent contaminants from the TEPCO accident, cesium-134 (with a half-life of about two years), reached its first half-life. Thus, the radiation level in coastal Fukushima was objectively less than before.

22. Although *kanrenshi* is not clearly defined, the state is usually more likely to recognize any death contiguous with an evacuation, such as moving to a different hospital, decreased quality of care, separation from family, and so on, within a year of the disaster. The government provides a fixed onetime payment of 5,000,000 yen (approximately $40,000) for each recognized *kanrenshi* case. For more information in Japanese, see https://web.archive.org/web/20220615155029/https://www.bousai.go.jp/taisaku/hisaisyagyousei/pdf/siryo2.pdf. For the number of *kanrenshi*, see the Ministry of Reconstruction site, https://web.archive.org/web/20220604061456/https://www.reconstruction.go.jp/topics/main-cat2/sub-cat2-6/20140526131634.html.

23. On July 21, 2011, the state named fifty-seven locations (fifty-nine households) within Minamisōma as designated areas. See map 3 in chapter 3.

24. The twenty-kilometer zone is about 107 square kilometers, and the thirty-kilometer zone is about 181 square kilometers.

25. *Nomakake* is one of the 318 Important Intangible Folk Cultural Properties of Japan.

26. On April 22, 2011, the state set the initial evacuation zones in the following way: restricted zones (within twenty kilometers), deliberate evacuation zones (areas with potential high exposure), and evacuation preparation zones (within thirty kilometers). This zoning underwent further modification in November 2011, and then again in March 2012. See more details in Morimoto (2015) and Fukushima Prefecture's "Transition of Evacuation Designated Zones," https://web.archive.org/web/20220602082836/http://www.pref.fukushima.lg.jp:80/site/portal-english/en03-08.html.

27. For the cabinet office's 2021 white paper on the rate of society's aging, see https://web.archive.org/web/20220604092018/https://www8.cao.go.jp/kourei/whitepaper/w-2020/html/zenbun/s1_1_1.html.

28. For Minamisōma's demographic data, see MachiDsu, https://web.archive.org/web/20220614210041/https://www.city.minamisoma.lg.jp/material/files/group/6/machidesu_2021.pdf, and also Morita et al. (2016, 2018).

29. The so-called *kariage* or *minashi kasetsu* is a type of housing for which the state paid for rent on behalf of the evacuees. There were 3,835 *kariage* units in multiple cities occupied by 9,706 Minamisōma residents at its peak.

30. Environmental historian Sara Pritchard (2012) uses the term "envirotechnical" to capture 3.11's complex entanglement of natural, technological, and political factors.

31. See Kanebishi and Tōhoku Gakuin Daigaku Shinsai no Kiroku Project (2020) for the multiplicity of losses caused by 3.11.

32. In Fukushima Prefecture as a whole, there were a total of 16,800 *kasetsu* units and 33,016 individuals living in *kasetsu* at its peak. In Minamisōma, there were 2,783 units with 4,741 individuals at its peak. As of March 2022, there were only 168 temporary housing units left. Of these, only three units are inhabited, by a total of four individuals. See https://web.archive.org/web/20220614211031/https://www.pref.fukushima.lg.jp/uploaded/life/622869_1734684_misc.pdf.

33. At its peak, thirty-four locations of temporary housing units were spread throughout Minamisōma.

The epigraph is my translation. An alternative translation is available in Wakamatsu and Binard (2016). The original Japanese text is 神隠しされた街 私たちの神隠しはきょうかもしれない/うしろで子どもの声がした気がする/ふりむいてもだれもいない/なにかが背筋をぞくっと襲う/広場にひとり立ちつくす。

1. His exact words were "Some may have concerns about Fukushima. Let me assure you, the situation is under control."

2. Although the total number of MPs is constantly changing, as of 2021 Iwaki city had the most MPs (458) in the prefecture. Following Iwaki were Fukushima city (369), Koriyama city (360), Minamisōma (257), Aizu-Wakamatsu (152), and Tamura city (132). An archived site of real-time monitoring results can be accessed at https://web.archive.org/web/20201029035734/https://radioactivity.nsr.go.jp/map/ja/#. The cost of maintaining the MPs is estimated at around $6 million per year. There have been several issues with and concerns about MPs. For example, they are usually installed after the surrounding area has been cleaned up, leaving people to wonder about the accuracy and generalizability of a reading from an MP to a nearby region with different environmental and geological conditions. In fact, in 2012 the 675 portable MPs installed in Fukushima Prefecture and elsewhere turned out to be incorrectly assembled. As a result, they had been displaying inaccurate numbers, 10 percent less than the correct readings. Moreover, the thirty-four MPs installed in March 2015 malfunctioned; one MP in Kashima displayed 54 μSv/h, more than a thousand times the average reading of the neighboring MPs (*Fukushima Minpo,* https://web.archive.org/web/20150705202759/http://www.minpo.jp/news/detail/2015040822037). In 2018 the Nuclear Regulation Authority proposed removing around 2,400 MPs located outside coastal Fukushima Prefecture by the end of the year, but the residents of Fukushima rejected the proposal.

3. Jennifer Robertson (2012) used a similar term, "inter-disaster," to discuss the prolonged temporality of 3.11. Nihei, however, focuses on the experience of being in between space, like how Kurokawa (1977) conceptualizes the Japanese aesthetic longing for an in-between, ambiguous, ambivalent space, or *en.*

4. These numbers include a few Haramachi residents whose residences were within the twenty-kilometer zone.

5. For a more comprehensive account of nuclear accident compensation, see *Hinanhakusho,* the Nuclear Accident Evacuation White Paper, by Kansai Gakuin University, JCN, and SAFLAN (2015).

6. For the Act on Nuclear Damage Compensation, see https://web.archive.org/web/20140303182714/http://www.tepco.co.jp/en/press/corp-com/release/betu11_e/images/110830e19.pdf. See also Endo (2013).

7. See the June 14, 2011, cabinet decision, "Framework of Government Support to the Tokyo Electric Power Company (TEPCO) to Compensate for Nuclear Damage Caused by the Accident at Fukushima Nuclear Power Plant (Provisional Translation)," https://web.archive.org/web/20220120070644/https://www.meti.go.jp/english/earthquake/nuclear/pdf/20110513_nuclear_damages.pdf.

8. For the original version, see https://web.archive.org/web/20140303182714/https://www4.tepco.co.jp/en/press/corp-com/release/betu11_e/images/110830e19.pdf, and see an example of a claim at https://web.archive.org/web/20200111081616/https://www.tepco.co.jp/cc/press/betu11_j/images/111011b.pdf.

9. Various guidelines and their suggested changes are available at https://web.archive.org/web/20220308072852/http://www.mext.go.jp/a_menu/genshi_baisho/jiko_baisho/index.htm.

10. See Buesseler (2020) for more on the contaminated water.

11. It includes the following categories: evacuation expenses, homecoming expenses, expenses for temporarily returning home, life and bodily damage, damages for incapacity to work, psychological damage by evacuation, medical examination expenses, expenses for examination of property, loss or diminishment of property value, and business damage for a corporation, agriculture, or fishery. In addition, sociologist Yukio Yotsumoto and political scientist Shunichi Takekawa (2016) suggest eleven dimensions of damage by the TEPCO accident.

12. Sato et al. (2020) note that in 2016 the average compensation for individual residents (excluding the loss of property) was around $100,000 for those in the emergency preparation and deliberate evacuation zones. For the exclusion zone, it was around $150,000.

13. "Outline of the Nuclear Damage Compensation Facilitation Corporation Act," August 2011, Cabinet Secretariat, https://web.archive.org/web/20220503203052/https://www.meti.go.jp/english/earthquake/nuclear/roadmap/pdf/20111012_nuclear_damages_2.pdf. According to the Board of Audit of Japan, the total spending for the TEPCO accident and its aftermath (compensation, decontamination, and decommissioning) is around twelve trillion yen, or eighty-five billion dollars, as of November 2022. They estimate that it might take forty-three more years (until 2065) for TEPCO to be able to return the money.

## CHAPTER 3. KALEIDOSCOPIC HARM

1. Here I focus on the evacuation orders associated only with 1F and not with the other nuclear power plant, Fukushima Dai-ni (2F), located between Tomioka and Naraha town.

2. See Shimazaki et al. (2015) for research on *kasetsu* residents in Minamisōma and their quality of life.

3. Wenger et al. (1975) discuss how human and material convergence upon a disaster site is a typical issue during the postdisaster response.

4. See Aoki (2019), Matsui (2021), Sawano et al. (2018), Seki (2015), and Son et al. (2015) for examples of residents' and evacuees' experiences of discrimination.

5. Although the state officially reopened Odaka in July 2016, many evacuees could not go back right away for various reasons, such as that the state had not completed decontaminating their residences or their houses were demolished as part of the decontamination efforts and still needed to be rebuilt.

6. After the accident the state incorporated the ICRP's 2007 recommendation for public annual effective dose limits of 1 mSv/y, which later posed many political and practical issues for decontaminating parts of coastal Fukushima. During the emergency phase, the ICRP recommendation suggested setting the dose limits in the 20–100 mSv/y range. See the dose limits comparison by the Ministry of the Environment at https://web.archive.org/web/20201027210458/https://www.env.go.jp/en/chemi/rhm/basic-info/1st/pdf/basic-1st-04–02.pdf.

7. See, for example, Sternsdorff-Cisterna (2020) on a zero becquerel movement in Tokyo.

8. The value for drinking water was provisionally set to 200 Bq/kg for cesium and then in April 2012 it was switched to 10 Bq/kg. See also Hamada et al. (2012), Iwaoka (2016), and Yamaguchi et al. (2021).

9. For the U.S. standard, see U.S. Food and Drug Administration CPG Sec. 555.880 Guidance Levels for Radionuclides in Domestic and Imported Foods, https://web.archive.org/web/20220331085459/http://www.fda.gov/media/72014/download.

10. A radioisotope releases radiation when it goes through a decay process. *Radioactivity* refers to the ability of the radioisotope to emit radiation. One popular explanation of the relationship between radiation and radioactivity is the following: Radiation is to radioactivity as light is to its source's ability to illuminate. A source of light determines its intensity. A lightbulb emits weaker light than the sun.

11. Almost every bag of rice commercially produced in Fukushima since 2012 is traceable at https://fukumegu.org/ok/contentsV2/kome_summary.html. Since 2013, the amount of cesium in agricultural products has been decreasing for a few reasons, including decontamination. Also, the half-life of cesium-134 (about two years) has been reached, and farmers have been learning how to reduce the transfer of cesium to foodstuffs. The above site shows the data for all bags of rice inspected between 2012 and 2021.

12. For more detailed data, see the Fukushima Prefectural Government's report, https://web.archive.org/web/20220418113721/https://www.pref.fukushima.lg.jp/uploaded/attachment/120037.pdf. The Ministry of Agriculture and Fishery's 2020 report suggests further continuation of the consumer avoidance of Fukushima foodstuffs. See https://web.archive.org/web/20220616162003/https://www.maff.go.jp/j/shokusan/ryutu/attach/pdf/R1kekka-12.pdf.

13. For an English-language translation of the guidelines, see Nuclear Energy Agency 2012, https://web.archive.org/web/20220616162645/https://inis.iaea.org/collection/NCLCollectionStore/_Public/44/052/44052662.pdf?r=1.

14. The Buffett Institute for Global Affairs at Northwestern University recently published a call to action to reconsider the structure of nuclear compensation. See "Nuclear Compensation: Lessons from Fukushima," https://web.archive.org/web/20220614221458/https://nuclear-compensation.northwestern.pub/. Osaka (2019) provides a critique of the institutional failure. For the legal history of nuclear compensation in Japan, see Feldman (2014).

1. For the individual variation in radiation sensitivity, see Fukunaga and Yokoya (2016).

2. For more on suicides in Fukushima and their characteristics, see Kuroda, Orui, and Hori (2021) and Takebayashi et al. (2020).

3. See Andrews (2016), Avenell (2012), Brown (2019), Geilhorn and Iwata-Weickgennant (2017), Manabe (2015), Novak (2017), and Wiemann (2018).

4. See the Nuclear Regulatory Authority, November 20, 2013, "The Basic Thinking on Safety and Security Measures for Reopening the Zones," https://web.archive.org/web/20220308084212/https://www.nsr.go.jp/data/000254661.pdf. See the Ministry of the Environment's comparison chart of annual radiation exposure. https://web.archive.org/web/20220916135003/https://www.env.go.jp/chemi/rhm/h29kisoshiryo/h29kiso-02-05-03.html.

5. Many issues concerning the opening of previously alienated areas in coastal Fukushima come from the logic that equates the opening with the end of harm, whereby the state decides to reopen a region when the majority of the town's population is concentrated in a low-radiation area, while the others who live in still highly contaminated regions are forced to go back or lose their compensation.

6. Torpey (2001) discusses this act of coming to terms with the past through economic compensation as "reparation politics," which, as he notes, seems to have become common since the Holocaust.

7. See also Yamashita, Ichimura, and Sato (2013) on the impact of the compensation policy on local communities.

8. Ralph (2020) discusses the role of open secrets in reproducing police violence.

9. For individual cases settled through the ADR in Minamisōma, see https://web.archive.org/web/20181122124012/http://www.city.minamisoma.lg.jp/index.cfm/10,15968,88,html.

10. Formally known as the Japan Atomic Energy Research Institute (JAERI), JAEA was established in 2005 through the consolidation of JAERI and the Power Reactor and Nuclear Fuel Development Corporation (PNC), later the Japan Nuclear Cycle Development Institute (JNC).

11. PRR-1 was a boiling water reactor (BWR), which uses demineralized water as both a coolant and neutron moderator. The heat produced by nuclear fission in the reactor core vaporizes the coolant, and the steam is used to drive a turbine to generate electricity. The steam is then cooled in a condenser to be reused as the coolant. The Idaho National Laboratory and General Electronics (currently GE Hitachi Nuclear Energy) in the mid-1950s developed this reactor model.

12. JRR-1 first reached criticality in August 1957 and was retired in March 1970.

13. Two reactors, JAEA's FBR and JAPC's BWR (110k kw), are currently under maintenance, and GCR (16.6k kw) has been in the process of decommissioning since 1998.

14. See Saito (2002) and Tatewaki, Kumazawa, and Ariga (2009) for more information about the accident.

15. In September 1990, at a JCO-operated uranium-processing facility (a subsidiary of Sumitomo Metal Mining Co.), three workers were processing a batch of fuel for the FBR "the quick way," by adding an aqueous uranyl nitrate solution to the tank of around 18.8 percent enriched uranium with fissile radioisotope U-235 in a bucket larger than suggested in the safety standards. As the bucket took about seven times more solution (16.6 kg) than was standard (2.4 kg), the tank reached the critical mass of uranium, leading to criticality and a self-sustained nuclear chain reaction. Strong gamma radiation from this event killed two of the three workers. The village independently released an emergency evacuation order before JAEA or the national government released an order (Nanasawa [2005] 2011). As a result of this accident, more than 663 people, including the village residents, were exposed to a high dose of radiation, and the damages (including victims' compensation) amounted to over 150,000,000 yen. There is no known long-term study of the exposure residents were subjected to and no information about its potential health effects. This accident's severity was ranked at four on the seven-point International Nuclear and Radiological Event Scale.

## CHAPTER 5. RADIOACTIVE MOSQUITOS AND THE SCIENCE OF HALF-LIVES

1. For the English version of the city's first revitalization plan, see https://web.archive.org/web/20190803020106/https://www.city.minamisoma.lg.jp/material/files/group/7/english.pdf.

2. Dr. Tsubokura was affiliated with the Institute of Medical Science at the University of Tokyo. He spent a few days a week at Minamisōma's city hospital to examine and research the residents' internal contamination using a whole body counter, a device that is used to identify and measure the radioactive material in the bodies of the humans and animals.

3. Once inside the body, radioactive substances are metabolized, speeding up their natural half-life processes. For example, the biological half-life of cesium is about one hundred days for adults and thirty days for children due to the difference in their rates of metabolism.

4. Some radiation, such as gamma waves and X-rays, go through the body. The X-ray is an example of radiation that permeates through some parts of the body without being absorbed or scattered, thereby leaving marks only for impermeable areas, like bones, which are more likely to absorb the rays.

5. Although sometimes radiocontrast agents, such as iodine and barium, are used to improve the visibility of a CT scan, they are considered a relatively safe medical technology with only a few common side effects. It is important to note that many medical procedures, like CT scans and mammography, involve low-dose internal and external radiation exposure.

6. Dr. Tsubokura has contributed to a local newspaper, *Fukushima Minyū,* writing a column called "Dr. Tsubokura's Radiations Seminars" since January 2015, https://web.archive.org/web/20220531131732/https://www.minyu-net.com/kenkou/housyasen/. In his December 26, 2020, column, Dr. Tsubokura points out that Japan is one of the most medically irradiated countries in the world. According to him, on average, Japanese individuals receive about 3.9 mSv per year from medical procedures, six times more than the world average. See https://web.archive.org/web/20210301101546/https://www.minyu-net.com/kenkou/housyasen/FM20201226-571361.php.

7. See also the CDC fact sheet for acute radiation syndrome, https://web.archive.org/web/20220520094819/https://www.cdc.gov/nceh/radiation/emergencies/arsphysicianfactsheet.htm.

8. Arguably, the most heated controversy is the increased risk of thyroid cancer among Fukushima's children. The Chornobyl accident in 1987 taught the international medical community that a nuclear accident could lead to increased rates of thyroid cancer among youth (Cardis et al. 2006). Like cesium, iodine-131 is common fallout debris that is known to cause thyroid cancers, but iodine-131 has a much shorter physical half-life, around eight days. For example, after the TEPCO accident, iodine-131 that was suspended in the atmosphere and fell on the terrestrial environment disappeared (was no longer detectable) in about eighty days. Compared to the Chornobyl accident, the TEPCO accident released about ten times less iodine-131 (Steinhauser, Brandl, and Johnson 2014). See also Toki et al. (2020).

9. Since October 2011 the local government has been closely monitoring more than three hundred thousand children who resided in Fukushima in March 2011 and were under the age of eighteen at the time of the accident. Under the Fukushima Health Management Survey (Kenmin Kenkō Chōsa), the local government offers free screening to those born between April 1992 and April 2012 to monitor their thyroid glands for abnormalities. As of 2021 there had been five rounds of testing, and each round had fewer participants than the one before. Toru was tested once in 2011 and again in 2012, but since then he has opted out of the screening program, claiming, "I do not want to know if there is some problem with me." According to Fukushima Prefecture, 219 individuals have had their thyroid surgically operated on.

10. Fukushima Medical University lists a set of research articles on thyroid issues at https://web.archive.org/web/20220126140126/http://kenko-kanri.jp/publications/thyroid-examination/. See also Midorikawa and Ohtsuru (2022) and Ohtsuru and Midorikara (2021) on the importance of voluntary participation to avoid the risk of overdiagnosis. Ochi et al. (2016) and Suzuki (2021) discuss scientists' responsibility to not stir fear among the public and to prevent the stigmatization of children. The U.S. Preventive Services Task Force's recommendation for screening for thyroid cancer can be found at https://web.archive.org/web/20220305104334/https://jamanetwork.com/journals/jama/fullarticle/2625325. For a more critical perspective on the screening process, see Sakakibara (2021).

11. See the 2017 UNSCEAR report, "Evaluation of Data on Thyroid Cancer in Regions Affected by the Chernobyl Accident," https://web.archive.org

/web/20220422214702/https://www.unscear.org/docs/publications/2017/Cherno-byl_WP_2017.pdf.

12. See the 2020/2021 report by the United Nations Scientific Committee on the Effects of Atomic Radiation, "Scientific Annex B: Levels and Effects of Radiation Exposure Due to the Accident at the Fukushima Daiichi Nuclear Power Station: Implications of Information Published since the UNSCEAR 2013 Report," https://web.archive.org/web/20220521154755/https://www.unscear.org/unscear/uploads/documents/unscear-reports/UNSCEAR_2020_21_Report_Vol.II.pdf.

13. See the U.S. Nuclear Regulatory Commission's "33.15 Requirements for the Issuance of a Type-C Specific License of Broad Scope," https://web.archive.org/web/20201026160819/https://www.nrc.gov/reading-rm/doc-collections/cfr/part033/part033-0015.html.

14. Yasui et al. (2017) discuss the post–nuclear disaster medical curriculum in Japan.

15. Observing the state-sponsored radiation-information centers in Fukushima and textbooks after the TEPCO accident, Maxime Polleri (2021) describes the state's efforts to normalize radiation exposure as "radioactive performance."

16. See Shannon Cram's discussion of the problems of the construction of the "Reference Man" (2015) for setting a standard for radiation exposure.

17. Elizabeth Roberts (2017) describes the shifting idea of "exposure" that developed in relation to the idea of the environment.

18. *Housha nou*/放射脳 is a play on words. In Japanese, radio-activity/放射-能, or the ability of an isotope to radiate, is a homophone for brain, or *nou*, so *housha nou* literally means "radioactive brain."

19. "Poor Risk Communication in Japan Is Making the Risk Much Worse," https://web.archive.org/web/20220308143000/https://blogs.scientificamerican.com/guest-blog/poor-risk-communication-in-japan-is-making-the-risk-much-worse/.

20. An edited volume by Shimazono, Goto, and Sugita (2016) explores lessons for scientists from the TEPCO accident. See also Shimazono (2019) and Shimura et al. (2015).

21. Denise Normile (2021) wrote a brief biography of Dr. Tsubokura's last decade of work in Fukushima in *Science*. My account here draws from both my personal knowledge of Dr. Tsubokura and Normile's article.

22. Dr. Tsubokura wrote about his experience in the May 2021 special issue of *Life*, https://web.archive.org/web/20210509112741/https://facta.co.jp/article/202105025.html.

23. According to the ICRP's official 1974 definition, the "Reference man is defined as being between 20–30 years of age, weighing 70 kg, is 170 cm in height, and lives in a climate with an average temperature of from 10°C to 20°C. He is a Caucasian and is a Western European or North American in habitat and custom" (International Commission on Radiological Protection 1975, 4).

24. The exact quantity of radioisotopes released from the TEPCO accident is unknown. Some estimates claim that the total release of cesium-137 to be as much as 8.8 Peta Bq (Terada et al. 2020).

25. Measuring for other radioisotopes takes much longer than measuring for the most prominent contaminant, cesium, so the presence of other isotopes is calculated based on their ratio to cesium.

26. There are different interpretations of this. See, for example, Jacobs (2021).

27. Cesium is by far the most critical radioisotope released by the TEPCO accident for a couple of reasons. First, the half-life of cesium is longer than that of other radioisotopes, such as iodine-131 and xenon-133. Although the half-life of strontium-90 is about as long as that for cesium-137 (about thirty years), less of it was released into the air compared to cesium-137 (a ratio of 1:1,000). Moreover, unlike cesium, strontium tends to bind to calcium, such as bones, which are less likely to be consumed. Cesium is more likely to bind to muscle cells, soil, and other bodies.

28. Between 2011 and 2021 an average of around six thousand people were tested for internal exposure annually. The data for WBC results in Minamisōma is accessible here: https://web.archive.org/web/20220503233005/https://www.city.minamisoma .lg.jp/portal/health/hoshasen_hibaku/3/4/index.html.

29. See also Murakami et al. (2020) for the positive correlation between the residents' subjective sense of their well-being and their decision to return.

30. Almost all cattle within the twenty-kilometer zone were killed rather than circulated in the market. Also, milk from dairy cattle was thrown away until the protocol for a contamination measurement was set.

31. In May 2011 the state ordered farmers inside the twenty-kilometer zone to euthanize their surviving cows. About 1,700 cows were killed as a result. A few dairy farmers in coastal Fukushima refused to kill their cows and took care of them by commuting inside the evacuation zone or living at the edge of the zone (Hasegawa and Hasegawa 2014; Shinnami 2015).

32. Another tablet like the one in Odaka is located in Tomioka town, Futaba district, Fukushima.

33. An image of the "babyscan" WBC can be seen at https://web.archive.org /web/20220624210626/https://www.city.minamisoma.lg.jp/portal/sections/14 /1440/14405/9/3/7815.html. More than 4,300 children under the age of six underwent the scan between July 2014 and March 2020, and the results can be accessed at https:// web.archive.org/web/20220624211526/https://www.city.minamisoma.lg.jp/portal /sections/14/1440/14405/9/4/1/index.html.

34. Katsuma Yagasaki, an emeritus professor of physics at Ryukyu University, has criticized Dr. Tsubokura as a safety myth believer and promoter. See https:// web.archive.org/web/20220616194107/https://www.sting-wl.com/category/%E6% 95%99%E3%81%88%E3%81%A6%EF%BC%81%E7%9F%A2%E3%83%B6%E5%B4 %8E%E5%85%8B%E9%A6%AC%E6%95%99%E6%8E%88.

35. Tsubokura reiterates this point in a recent interview with journalist Misaki Hattori (2021).

36. YomiDr, March 13, 2017, "Masaharu Tsubokura Part 2. How to Communicate the Radiation Information: You Need More Than the Right Information to Persuade People," https://web.archive.org/web/20220616194312/https://yomidr .yomiuri.co.jp/article/20170308-OYTET50014/.

37. See Loh and Amir's discussion (2019) of a new type of expertise that emerged in coastal Fukushima. Abeysinghe et al. (2022) also explores diverse experiences among local health professionals.

38. On the importance of the reconstruction not only of infrastructures but also of lives, see Seki (2015).

39. See, for example, Hisamura et al. (2017), Hori et al. (2017), Murakami et al. (2018), Nomura et al. (2016), Tsubokura (2018), and Tsubokura et al. (2017).

## CHAPTER 6. BETWEEN *FŪHYŌ* AND *FŪKA*

1. Since March 2018, anyone has been able to take a virtual tour inside 1F at https://www.tepco.co.jp/en/insidefukushimadaiichi/index-e.html.

2. This has been a controversial proposal. There has also been a debate about approaching Fukushima as a place for both "hope tourism" and "dark tourism" (see, e.g., Azuma 2013; Gerster, Boret, and Shibayama 2021; Ide 2018; Jang, Sakamoto, and Funck 2021; Stone 2012). For Fukushima Prefecture's definition of hope tourism, see https://web.archive.org/web/20220207183702/https://www.hopetourism .jp/en/concept.html. Netflix's 2018 show *Dark Tourist* had an episode on Fukushima, https://web.archive.org/web/20210612104449/https://real-fukushima.com /netflixs-dark-tourist/. The Fukushima prefectural government's Real Fukushima project expressed deep concern about the program.

3. Since the summer of 2020, the COVID-19 pandemic has reduced the number of visitors.

4. See https://web.archive.org/web/20220616202416/https://laurakkerr.com /wp-content/uploads/2022/01/LKKerr_Synchronicity.pdf.

5. I borrow the expression "phantom" from Osseo-Asare's work on Ghana's history of atomic power development (2019). In Ghana, the absence of an NPP and the rumors associated with the plant still produced tangible effects on the local development and national politics surrounding atomic energy.

6. Three laws related to power generation were passed in 1974: the Electric Power Development Taxation Law, the Special Budget Law for the Development of Electric Power, and the Law for the Adjustment of Areas Adjacent to Power Generating Facilities. For example, the city of Minamisōma received about $500,000 a year as a neighboring community of 1F. Minamisōma has been using the fund to pay for nursery school teachers. See https://web.archive.org/web/20190802194439 /https://www.city.minamisoma.lg.jp/portal/sections/12/1210/12101/8/2487.html.

7. The information in Japanese on how different municipalities in costal Fukushima used the nuclear energy–related subsidies is available at https://web.archive .org/web/20220607113628/https://www.pref.fukushima.lg.jp/site/portal/ps-gen-shindengen.html.

8. For a detailed account of the Odaka-Namie nuclear power plant and antinuclear activism in coastal Fukushima, see Katsunobu Onda (2011). Shibata (2018) discusses key activists in coastal Fukushima.

9. For more information about Minamisōma's declaration as an anti–nuclear energy city (*datsu-genpatsu toshi sengen*), see https://web.archive.org/web/20190803032036 /https://www.city.minamisoma.lg.jp/portal/sections/13/1320/13204/1/3108.html.

10. On March 31, 2019, most of Namie, except the highly contaminated western area, became open to all.

11. In 2018 TEPCO reopened its public relations center in Tomioka Town as the TEPCO Decommissioning Archive Center, which is now the orientation place for the 1F tour. Information about the center is available in the digital leaflet at https:// web.archive.org/web/20220420045343/https://www4.tepco.co.jp/fukushima_hq /decommissioning_ac/pdf/leaflet-e.pdf.

12. TEPCO's "Fukushima Daiichi Timeline after March 11, 2011" offers detailed information about the progress of decontamination at 1F; see https://web.archive .org/web/20220420045522/https://www4.tepco.co.jp/en/decommissiontraject /index-e.html. See also Kainuma (2016) for detailed information about 1F's decommissioning.

13. The state lifted Naraha town's evacuation order on September 1, 2015.

14. See https://web.archive.org/web/20220420045403/https://www.tepco .co.jp/en/hd/responsibility/index-e.html.

15. For more information about frozen soil walls, see https://web.archive.org /web/20220401061919/https://www.tepco.co.jp/en/decommision/planaction /landwardwall/index-e.html.

16. For more information about the fuel removal operation, see https://web .archive.org/web/20210909232530/https://www.tepco.co.jp/en/decommision /planaction/removal-e.html.

17. Noma (2018) writes about her experience of the site tour and discusses the TEPCO guide's emphasis on radiation exposure. In 2018, the site of exposure shifted to the path between Reactor Units 2 and 3. Kingston (2022a) also writes about his tour experience.

18. For TEPCO's photo and video archive of the accident and its mitigation processes, see https://www.tepco.co.jp/en/hd/library/index-e.html.

19. For detailed accounts of 1F laborers, see Fuse (2012), Happy (2013), Inaizumi (2021), Katayama (2020), and Tatsuta (2017).

20. For residents' various accounts of their experience of discrimination, see Ikeda et al. (2018) and also chapter 3.

21. Lisa Onaga and Harry Yi-Jui Wu (2018) discuss the term *genba* in detail. While they use *genba* in the context of chemical and radiological exposure and contamination, the more general definition of *genba* in everyday Japanese use (and as it is used by Jyunko here) is "a site of action where something significant has occurred or has been produced" (198).

22. See Fuji and Seki (2015) on the residents' attachment to the land.

23. For similar preservation activities by Tomioka town, see Abe (2013) and Tomioka town and Fukushima University (2017).

24. The survey results can be accessed at https://web.archive.org/web /20220418115005/https://www.town.fukushima-futaba.lg.jp/9269.htm.

25. Many pieces of the "disaster heritage" (Good 2016; Littlejohn 2020, 2021) are being archived using 3-D modeling technology. See Kanō (2017) for more information. Kingston (2020b) discusses different narratives of the accident by comparing the two coastal Fukushima disaster museums, the TEPCO Decommissioning Archive Center and Fukushima Prefecture's Great East Japan Earthquake, Tsunami, and Nuclear Disaster Memorial Museum. See Gerster and Maly (2022) for a more general discussion of 3.11 disaster museums.

26. Many other books touch on the history of the development of nuclear energy in Japan. See, for example, see Akimoto (2014), Atomic Energy Commission (1986), Genshiryoku Gijyutsushi Kenkyukai (2015), Fuse (2011), Inoue (2014), Kajita (2014), Kamata (2006), Leatherbarrow (2022), Low (2020), Nakajima and Kihara (2019), Takebayashi (2001), and Yoshioka (2011).

27. Kazutaka Kigawada, then vice president of TEPCO, made the executive decision to name this the "nuclear power station" department to signify the hope of rapidly transforming nuclear technology into a peaceful force and undoing its negative image from the war. Around the same time, TEPCO sent its employees to places such as the Argonne National Laboratory in the United States; the Shippingport Atomic Power Station, Calder Hall Nuclear Power Station, and Atomic Energy Research Establishment in the United Kingdom; and the Chalk River Nuclear Laboratories in Canada for technical internships (Tokyo Electric Power Company 1983, 350). Kigawada was born in Yanagawa town in Fukushima prefecture, and his personal tie to Fukushima helped TEPCO's negotiations with the prefecture and hosting town.

28. The Chōjyahara district of Ōkuma town was a Japanese army air force base during World War II and was turned into a salt manufacturing field after the war. The National Developer Company (currently Kokudo), Seibu's parent company, owned some of these fields, which made it easier for TEPCO to purchase the land.

29. For the land purchase, TEPCO was only responsible for dealing with a development company, Kokudo, the owner of the former military base in the region. Therefore, the Fukushima prefectural developer and the town office of Ōkuma dealt with negotiating the purchase of land owned by individual families. Ultimately, they negotiated with around 290 families; for the NPP site, 320,000 square meters (3,444,451 square feet) of residential land were purchased for approximately 50,000 yen. See also Yoshikawa (2021).

30. Additionally, it was agreed that there would be additional compensation for the occupation loss that farmers would incur; all land property (including the land TEPCO was acquiring from the National Developer Company) would be valued equally; the town mayor would do his best to sell the national forest to compensate for the loss of woods, a prime source of fuel that the forest generated; and, finally, the town would seek a special tax categorization (Ōkuma Town Editorial Association 1985, 834–35).

31. A fishing right is equivalent to land for fishers. Each fisher has to buy the right to be allowed to fish in a specific area, but usually a fishing association owns a large area in which its members can fish.

32. The compensation was given to nine fishing associations of the neighboring regions, including the northern end of Shinchi town to the southern end of Yotsukura in Iwaki city.

33. See, for example, Kingston (2014), Kurokawa and Ninomiya (2018), Nakajima (2014), Nöggerath, Geller, and Gusiakov (2011), and Saito (2021).

34. *Kizuna* (bonds) became a buzzword after 3.11 (see, e.g., Bestor 2013, Gerster 2019, and Sternsdorff-Cisterna 2015). In Japanese, *kizuna* connotes an unbreakable, obligatory relation, or encumbrance, and thus it is a much stronger form of relationality than *en*.

35. The model for the development of underdevelopment was established through the nuclearization of Tōkai village in the early 1950s. A local newspaper, *Ibaraki Shimbun* (Ibaraki Shimbun Company 2003, 85–86), reflected on the radioactive development accordingly: "The ideal model of the nuclear center was a place where nuclear facilities were surrounded by park and forestation, areas that were themselves surrounded by nonresidential factory areas."

36. In *Chugoku Shimbun,* a Hiroshima local discussed what the postwar nuclear development meant for Hiroshima on May 26, 1957: "Although we, the residents of Hiroshima, always had a strong angst about and fear of nuclear power [*genshiryoku*], we need to accept the fact that the rest of the world has been gathering their best knowledge and talents for achieving the peaceful use of atoms—the most substantial energy that the human civilization has ever attained. We need to suspend our subjective feelings and emotions and observe the peaceful use of atoms objectively and scientifically in order to assess its true value."

CHAPTER 7. *FRECON BAGGU* AND THE ARCHIVE OF
(HALF-)LIVES

1. See Parmentier (1985) and Sahlins (1981, 68) for a discussion of "sedimentation."

2. A preliminary English translation of the act is available at https://web.archive.org/web/20220308172949/http://josen.env.go.jp/en/framework/pdf/special_act.pdf?20130118. I use this English version.

3. In 2014 Takashi Murakami, an artist and associate professor at Miyagi University of Education, tried to exhibit his artwork made with one of these bags and a solar panel at the art exhibition *Power to the People* at Tohoku Electric Green Plaza in Sendai. The owner of the plaza where the exhibit was taking place, Tohoku Electric Company, identified his artwork as a suspicious object and asked Murakami to remove it. The organizer had to postpone the entire exhibition for three days. After a long negotiation, the artist and Tohoku Electric came to an agreement to exhibit the artwork in a separate room, away from the main exhibition space.

4. See Stoler (2008).

5. On the social, cultural, and political constructions of the idea of home, see, for example, Bourdieu (1970), Kondo (1990), Morimoto (2021b), and Robertson (1988). For the locals' attachment to their hometown, see Koyama (2022), and for a

similar psychological attachment to home among the residents of Chornobyl, see Alexievich ([1997] 2019, 215–16).

6. In 2013 the International Atomic Energy Agency suggested against adhering strictly to the 1 mSv/y standard and instead encouraged spending time and money on building consensus across local communities. The IAEA final report, "The Follow-up IAEA International Mission on Remediation of Large Contaminated Areas Off-Site the Fukushima Daiichi Nuclear Power Plant," can be found at https://web .archive.org/web/20220524230356/https://www.iaea.org/sites/default/files/final_ report230114_0.pdf.

7. When I talked to the former minister of reconstruction, Takumi Nemoto, in May 2015 at Harvard University, he implied that ministries other than the Ministry of the Environment should have been responsible for the cleanup. Moreover, he mentioned that the Ministry of the Environment did not know much about nuclear science and had to learn more about it after it was assigned the role, and thus it took them some time to act.

8. See the Ministry of the Environment's website, https://web.archive.org /web/20211021210615/https://josen.env.go.jp/en/decontamination/. For discussions of decontamination, see Kawasaki (2018a, 2018b), Kirby (2019), Kogure (2013), Kurokawa (2017, 2020), Morimoto (2020), and Nakanishi (2021).

9. See Samuels (2013) and Aldrich (2014) for a discussion of the resiliency of many predisaster policies such as national security, energy, and local governance in post-3.11 Japan. Gabrielle Hecht (2012b) discusses the invisibility of subcontracted workers in the nuclear industry.

10. Kurokawa (2017, 2020) explores the issue of decontamination in Date city, which is outside the SDA. Date city approached decontamination as a solution to its residents' psychological issues, or *kokoro no jyosen.*

11. Peter Galison (2011) calls such wastescape "waste-wilderness." His and Robb Moss's 2015 documentary film *Containment* features a Minamisōma resident who used to live near Sanrokusen and also Namie's antinuclear "cowboy."

12. For detailed information about the Interim Storage Facility and the transportation of decontaminated waste within Fukushima, see https://web.archive.org /web/20211226080520/http://josen.env.go.jp/en/storage/.

13. As of August 2021 the state was exploring negotiations with international law organizations to allow the disposal of domestic nuclear waste outside Japan. However, according to the current framework, nuclear waste does not include decontaminated waste from the TEPCO accident.

14. For more detailed discussions on *satoyama,* see Knight (2010), Ishizawa (2018), and Satsuka (2014).


CHAPTER 8. IN SEARCH OF THE INVISIBLE

1. See https://web.archive.org/web/20220407012626/http://aboutjapan.japan-society.org/are_the_japanese_people_religious.

2. I had the greatest pleasure to support Nishiyama and the co-organizer Wataru Kumagai for this event. I thank the Toyota Foundation for helping with it.

3. The Odaka district produced a booklet of the monster serpent legend in 1988 and reprinted it in 1997 and 2013 using a state subsidiary from the Three Electric Laws.

4. During the Meiji period (1868–1912), the state created a series of laws to separate Buddhism and Shintoism. Previously, the functional divisions between the two in people's everyday lives were less clear (*shinbutsu-shūgō*).

5. As the domain's foundational spiritual venue, the Hiwashi shrine has significance beyond protecting its neighboring communities. For example, the shrine serves a critical function for the Nomaoi festival. The holy water (*omitarashi*) preserved at Hiwashi used to be used for the final-day Nomakake ritual to wish for the prosperity of the domain and its people.

6. The Sōma family is related to the Chiba family. Sōma was given the land in coastal Fukushima.

7. See Takase, Yoshida, and Kumagai (2012).

8. Here, the connection between how adherents of Shintoism and many Indigenous communities throughout the world relate to their land is evident. See, for example, LaDuke (1999), Kimmerer (2015), and Kirsch (2001).

9. Sato (2013) describes in detail how and why the state failed to communicate with the residents using its emergency system, the System for Prediction of Environmental Emergency Dose Information (SPEEDI).

10. For more information about Namie town, see https://web.archive.org /web/20220617014224/https://www.town.namie.fukushima.jp/uploaded/attachment/14466.pdf.

11. In April 2011 Namie town moved its operation to an office in Nihonmatsu city, Fukushima. Various parts of Namie reopened in April 2017, and the town hall resumed operation. After 2021 one could visit and explore most parts of Namie without a permit. As of May 2022, about 1,878 residents are living in Namie.

12. Burglary became an issue in the exclusion zones, and some towns installed security cameras in 2014.

13. A full version of the Ministry of the Environment's report "The Review of the Response to Disaster-Affected Animals" is available in Japanese at https://web .archive.org/web/20210317063502/https://www.env.go.jp/nature/dobutsu/aigo/2_ data/pamph/h2508c/full.pdf. See also Itoh (2018).

14. In 2013, at least 1 kilogram of a sample was necessary for maximum possible accuracy in testing. Later, the amount was reduced to 500 grams. In the process, staff would cut up foodstuffs to fit them in a special container. Now, a more advanced instrument makes it possible not to destroy the item being measured.

15. There were 1,120 Bq/kg of cesium-137 and 504 Bq/kg of cesium-134. The total amount of cesium was 1,620 Bq. It was tested on November 5, 2013, at the Kashima Continuing Education Center with a NaI(TI) scintillation detector with the detection limits of cesium-137 = 13.0 and of cesium-134 = 20.3.

16. The Tenmei famine also impacted other parts of the country; Tōhoku and North Kantō were the hardest-hit regions. The Sōma domain learned from the

experience and adapted the teachings of Sontoku (Kinjirō) Ninomiya. Although Ninomiya never visited the domain, his pupil Koukei Tomita implemented his teachings. As a result, the domain survived the Tenpō famine between 1833 and 1839 without too much damage.

17. For a detailed history of the religion-led migration in Japan, see Drixler (2016). For the history of immigration by Kaga residents, see Ikebata (1995). Satoshi Aohara's documentary film *Dotoku Ryūri* covers this topic by focusing on Shin Buddhism. Since 3.11 the city of Minamisōma and Nanto city in Toyama Prefecture have been recultivating their migration-based *en*. According to the record, around forty-seven families (231 people) from Nanto migrated to the Sōma region (Chiaki 2009).

18. In addition to coming from Kaga and Etchū, migrants came from Echigo (Niigata Prefecture), Inaba (Tottori Prefecture), Satsuma (Kagoshima Prefecture), and Hyūga (Miyazaki Prefecture). Sōma used the folk song "Sōma nihen gaeshi" as a promotional song to attract migrants to the region (Ikebata 1995).

19. "The financing of immigrants to Sōma also used an unusual formula which avoided any direct funding from the domain. Private investors (*kinshu*) would assist settlers with the costs of housing, farm tools, seed, and food. The land they resettled would be tax free for 15 years" (Drixler 2016, 25).

20. One characteristic of Shin Buddhism is the keeping of a large and ornamental Buddhist altar. One could identify Shin Buddhists by looking at their family altar.

21. On the history of *benkei*, see Kawasaki (2014).

22. See, for example, Shōfuku Ji's website, https://web.archive.org /web/20070825010112/http://www.newcs.futaba.fukushima.jp/shofukuji/history .html.

23. See Fukumoto (2020) for the effects of low-dose radiation on animals.

24. For the movement of radiocesium by decontamination into the downstream river, see Feng et al. (2022).

### CHAPTER 9. A WILD BOAR CHASE

A version of this chapter was published as "A Wild Boar Chase: Ecology of Harm and Half-Life Politics in Coastal Fukushima," in *Cultural Anthropology* 37, no. 1: 69–98.

1. See https://city-ms-chozyu.maps.arcgis.com/apps/webappviewer/index.html ?id=a3dfb082665847e38c9e39cfe44042b6.

2. See the information about TEPCO's initiative at https://web.archive.org /web/20200826050304/https://www.tepco.co.jp/fukushima_hq/decontamination /archive/2018/20181031_01-j.html. Since February 2020 TEPCO has been using drones at night to map out boars' habitats and estimate the headcount in Tomioka. See https://web.archive.org/web/20220127095411/https://www.tepco.co.jp /fukushima_hq/decontamination/archive/2020/20200527_01-j.html.

3. Craft goods made with boar skin by the company Ino DATE can be seen at https://web.archive.org/web/20220217031301/http://www.ino-date.com/.

4. The original name of the book in Japanese is *Hitsuji wo Meguru Bouken,* which can be translated literally as "An Adventure Surrounding Sheep." Anne McClintock at Princeton University once suggested that the English title might have come from the expression "wild-goose chase," or a futile search, and that the title reminded her of Shakespeare's *Romeo and Juliet,* act 2, scene 4, in which Mercutio says, "Nay, if our wits run the wild-goose chase, I am done; for thou hast more of the wild goose in one of thy wits than, I am sure, I have in my whole five."

5. Sato et al. (2020) analyze the 2017 Futaba district evacuee-survey data while paying attention to different types of networks, including weak connection (*yowai tsunagari*).

6. Fukushima Prefecture's radiological-monitoring results of wildlife can be accessed at https://web.archive.org/web/20220120062325/https://www.pref .fukushima.lg.jp/site/portal/wildlife-radiationmonitoring1.html.

7. NPR's 2020 Special Series on Recovering Fukushima included a story about wild monkeys in Minamisōma. See https://web.archive.org/web/20220618000937/ https://www.npr.org/2020/09/10/904356338/in-rural-fukushima-the-border-between-monkeys-and-humans-has-blurred%20.

8. See Saito et al. (2019) for data on wild boars' internal exposure in Sōma.

9. A recent study by Masilkova et al. (2021) suggests that boars are prosocial creatures that engage in rescue behavior when a trap captures other boars in a herd.

10. See https://web.archive.org/web/20220310022229/https://www.city .minamisoma.lg.jp/portal/shi_joho/shinsaikanrenjouhou/houshasenmonitarin-gukekka/kankyo_monitoring/10098.html.

11. The ICRP largely ignored nonhuman organisms as the targets of protection until the publication of "ICRP Publication 108" in 2008. See the discussion on "Reference Animals and Plants," https://web.archive.org/web/20220525000044 /https://www.icrp.org/publication.asp?id=icrp%20publication%20108.

12. For the report in Japanese, see https://web.archive.org/web/20210613212357 /https://news.tiiki.jp/articles/4619.


EPILOGUE

1. See *Minamisōma City Magazine,* May 2022, p. 2, https://web.archive.org /web/20220615193942/https://www.city.minamisoma.lg.jp/material/files/group/3 /koho_040501.pdf.

2. Acemoglu and Restrepo (2022) observe that having an aging population is related closely to enthusiasm for developing automation technologies.

3. On October 12, 2019, massive Typhoon No. 19 (Hagibis) landed on mainland Japan. Torrential rain and the resulting floods caused devastating damage to a large part of eastern Japan, including Fukushima. Typhoon No. 19 killed sixty-four people, thirty of whom lived in Fukushima Prefecture.

4. For the changes in the number of workers at 1F, see TEPCO's decommissioning magazine, *Hairomichi,* no. 32, p. 1, https://web.archive.org/web/20220611214350 /https://www.tepco.co.jp/decommission/visual/magazine/pdf/hairomichi_032 .pdf.

5. On June 12, 2022, a small section of the former exclusion zone (0.37 square miles of 6.17 square miles) in Katsurao village reopened. Ōkuma town opened 3.32 square miles of its land (11 percent) on June 30, 2022. Futaba town will open 2.14 square miles of its land (9 percent) on August 30, 2022. Both towns are in the process of determining additional portions of land to be decontaminated starting in 2024.

6. The abridged version of the survey result in Japanese can be accessed at https://web.archive.org/web/20220608234031/https://www.env.go.jp/chemi/rhm /portal/communicate/result/.

7. The idea of human and more-than-human kinship is also evident among many Indigenous communities. See, for example, LaDuke (1999), TallBear (2019), and Yazzie and Baldly (2018).

# BIBLIOGRAPHY

Abe, Koichi. 2013. *Fukushima saisei to rekishi/bunkaisan*. Tokyo: Yamakawa Shuppan.

Abe, Tamaki. 2014. *Nomaoi wo ikiru Minamisōma wo ikiru*. Tokyo: East Press.

Abeysinghe, Sudeepa, Claire Leppold, Akihiko Ozaki, and Alison Lloyd William, eds. 2022. *Health, Wellbeing and Community Recovery in Fukushima*. London: Routledge.

Acemoglu, Daron, and Pascual Restrepo. 2022. "Demographics and Automation." *Review of Economic Studies* 89, no. 1: 1–44. https://doi.org/10.1093/restud/rdab031.

Akasaka, Norio. 2007. *Houhoutoshiteno Tōhoku*. Tokyo: Kashiwa Shobo.

———. 2014. *Shinsaikou 2011.3–2014.2*. Tokyo: Fujiwara Shoten.

Akasaka, Norio, and Eiji Oguma, eds. 2012. *Henkyokara hajimaru: Tokyo/Tohoku Ron*. Tokyo: Akashi Shoten.

Akimoto, Kenji. 2014. *Genshiryokusuishin no gendaishi: Genshiryoku soumeiki kara Fukushima genpatsu jiko made*. Tokyo: Gendai Shokan.

Akiyama J., S. Kato, M. Tsubokura, J. Mori, T. Tanimoto, K. Abe, et al. 2015. "Minimal Internal Radiation Exposure in Residents Living South of the Fukushima Dai-ichi Nuclear Power Plant Disaster." *PLoS ONE* 10, no. 10: e0140482. https://doi:10.1371/journal.pone.0140482.

Aldrich, Daniel P. 2005. "Japan's Nuclear Power Plant Siting: Quelling Resistance." *Asia-Pacific Journal: Japan Focus* 3, no. 6. https://apjjf.org/-Daniel-P-Aldrich/2047/article.pdf.

———. 2008. *Site Fights: Divisive Facilities and Civil Society in Japan and the West*. Ithaca, New York: Cornell University Press.

———. 2014. "Revisiting the Limits of Flexible and Adaptive Institutions: The Japanese Government's Role in Nuclear Power Plant Siting over the Postwar Period." In *Critical Issues in Contemporary Japan*, edited by Jeff Kingston, 79–91. New York: Routledge.

———. 2019. *Black Wave: How Networks and Governance Shaped Japan's 3/11 Disasters*. Chicago: University of Chicago Press.

Alexievich, Svetlana. (1997) 2019. *Voices from Chernobyl: The Oral History of a Nuclear Disaster.* Dublin: Dalkey Archive Press.

Allison, Anne. 2013. *Precarious Japan.* Durham, NC: Duke University Press.

Andrews, William. 2016. *Dissenting Japan: A History of Japanese Radicalism and Counterculture from 1945 to Fukushima.* London: Hurst & Company.

Andrews-Speed, Philip. 2020. "Governing Nuclear Safety in Japan after the Fukushima Nuclear Accident: Incremental or Radical Change?" *Journal of Energy & Natural Resources Law* 38, no. 2: 161–81. https://doi.org/10.1080/02646811.2020.1741990.

Anzai, Ikerou. 2021. *Watashino hangenpatsu jinsei to "Fukushima Project" no ashiato.* Kyoto: Kamogawa Press.

Aoki, Miki. 2018. *Chizukara kesareru machi. 3.11 gono "ittehaikenai shinjitsu."* Tokyo: Toppan.

Aritsuka, Ryoji, and Yasuhiro Sudo. 2016. *3.11 to kokoronosaigai: Fukushima nimiru sutoresu shoukougun.* Tokyo: Otsuki Shoten.

Asahi News Book. 1957. *Toukai Mura: Genshiro no hiha moeru.* Tokyo: Asahi Shimbun.

Asahi Shimbun Investigation Team. 2014. *Soredemo nihonjin ha genpatsu wo eranda: Tōkai mura to genshiryoku mura no hanseiki.* Tokyo: Asahi Shimbun.

Asia Pacific Initiative. 2021. *Fukushima genpatsu jiko 10 nen kenshouiinkai: Minkainjiko saishūhokokusho.* Tokyo: Discover Twenty-one.

Atomic Energy Commission. 1986. *Genshiryoku kaihatsu sanjyunenshi.* Tokyo: Japan Atomic Energy Relations Organization.

Auyero, Javier, and Debora Swistun. 2008. "The Social Production of Toxic Uncertainty." *American Sociological Review* 73, no. 3: 357–79. https://doi.org/10.1177/000312240807300301.

Avenell, Simon. 2012. "From Fearsome Pollution to Fukushima: Environmental Activism and the Nuclear Blind Spot in Contemporary Japan." *Environmental History* 17, no. 2: 244–76.

Awaji, Takehisa, Ryoichi Yoshimura, and Masafumi Yokemoto. 2015. *Fukushima genpatsu jiko baishō no kenkyu.* Tokyo: Nippon Hyoronsha.

Azuma, Hiroki. 2014. *Yowai tsunagari: Kensaku keyword wo sagasutabi.* Tokyo: Gentosha.

———, ed. 2013. *Fukushima dai-ichi genpatsu kankouka keikaku: Shisouchizu vol. 4-2.* Tokyo: Genron.

Baba, Makoto. 2021. *Naishin hibaku: Fukushima haranomachi no 10 nenn.* Tokyo: Ushio Press.

Bankoff, Greg. 2008. "The Historical Geography of Disaster: 'Vulnerability' and 'Local Knowledge' in Western Discourse." In *Mapping Vulnerability: Disasters, Development & People,* edited by Greg Bankoff, Georg Frerks, and Dorothea Hilhorst, 25–36. New York: Routledge.

Bannerman, Ty 2018. "The First." *American Literature Review,* Spring 2015. https://web.archive.org/web/20220129020731/https://americanliteraryreview.com/2018/12/05/ty-bannerman-the-first/.

Barrios, Roberto E. 2017. *Governing Affect: Neoliberalism and Disaster Reconstruction*. Lincoln: University of Nebraska Press.

Barthes, Roland. 1972. *Mythologies*. New York: Hill and Wang.

———. 1981. *Camera Lucida: Reflections on Photography*. New York: Hill and Wang.

Basso, Keith H. 1996. *Wisdom Sits in Places: Landscape and Language among the Western Apache*. Albuquerque: University of New Mexico Press.

Baudrillard, Jean. (1981) 1994. *Simulacra and Simulation*. Michigan: University of Michigan Press.

Beck, Ulrich. 1987. "The Anthropological Shock: Chernobyl and the Contours of the Risk Society." *Berkeley Journal of Social Sociology* 32: 153–65.

———. 1992. *Risk Society: Towards a New Modernity*. London: SAGE.

———. 1999. *World Risk Society*. Cambridge: Polity.

Bestor, Theodore C. 2013. "Disasters, Natural and Unnatural: Reflections on March 11, 2011, and Its Aftermath." *Journal of Asian Studies* 72, no. 4: 763–82. https:// doi.org/10.1017/S0021911813001770.

Biehl, João, and Federico Neiburg. 2021. "Oikography: Ethnographies of House-ing in Critical Times." *Cultural Anthropology* 36, no. 4: 539–47. https://doi.org/10 .14506/ca36.4.01.

Bird, Isabella L. 1880. *Unbeaten Tracks in Japan*. London: J. Murray.

Boudia, Soraya, Angela N. H. Creager, Scott Frickel, Emmanuel Henry, Nathalie Jas, Carsten Reinhardt, and Jody A. Roberts. 2018. "Residues: Rethinking Chemical Environments." *Engaging Science, Technology, and Society* 4: 165–78. https:// doi.org/10.17351/ests2018.245.

Bourdieu, Pierre. 1970. "The Berber House or the World Reversed." *Social Science Information* 9, no. 2: 151–70. https://doi.org/10.1177/053901847000900213.

Brown, Alexander. 2019. *Anti-Nuclear Protest in Post-Fukushima Tokyo: Power Struggles*. London: Routledge.

Brown, Azby. 2021. "Ushering in the New Normal: Viability and Informal Community Leadership in Fukushima Ten Years after 3.11." *Asia-Pacific Journal: Japan Focus* 19, issue 17, no. 8. https://apjjf.org/2021/17/brown.html.

Brugge, Doug, Timothy Benally, and Esther Yazzie-Lewis, eds. 2006. *The Navajo People and Uranium Mining*. Albuquerque: University of New Mexico Press.

Buesseler, Ken O. 2020. "Opening the Floodgates at Fukushima: Tritium Is Not the Only Radioisotope of Concern for Stored Contaminated Water." *Science* 369, no. 6504: 621–22.

Button, Gregory. 2010. *Disaster Culture: Knowledge and Uncertainty in the Wake of Human and Environmental Catastrophe*. Walnut Creek, CA: Left Coast.

Caldicott, Helen, ed. 2014. *Crisis without End: The Medical and Ecological Consequences of the Fukushima Nuclear Catastrophe*. New York: New Press.

Campbell, Kym Rouse, and Todd S. Campbell. 2001. "The Accumulation and Effects of Environmental Contaminants on Snakes: A Review." *Environmental Monitoring and Assessment* 70, no. 3: 253–301. https://doi.org/10.1023/A:1010731409732.

Cardis, Elisabeth, Geoffrey Howe, Elaine Ron, Vladimir Bebeshko, Tetyana Bogdanova, Andre Bouville, Zhanat Carr, et al. 2006. "Cancer Consequences of the Chernobyl Accident: 20 Years On." *Journal of Radiological Protection* 26, no. 2: 127–40. https://doi.org/10.1088/0952-4746/26/2/001.

Carroll, Patrick. 2006. *Science, Culture, and Modern State Formation*. Berkley: University of California Press.

Carver, Steve. 2019. "Rewilding through Land Abandonment." In *Rewilding*, edited by Nathalie Pettorelli, Sarah M. Durant, and Johan T. du Toit, 99–122. Cambridge: Cambridge University Press.

Caufield, Catherine. 1990. *Multiple Exposures: Chronicles of the Radiation Age*. Chicago: University of Chicago Press.

Chiaki, Kenji. 2009. "Tonamiimin no Sōma-Nakamurahan heno imin." *Tonami Sansonchiiki Kenkyu Kiyou*, 26: 1–17.

Churchill, Ward, and Winona LaDuke. 1986. "Native America: The Political Economy of Radioactive Colonialism." *Critical Sociology* 13, no. 3: 51–78.

Clarke, Lee, and Caron Chess. 2008. "Elites and Panic: More to Fear Than Fear Itself." *Social Forces* 87, no. 2: 993–1014. https://doi.org/10.1353/sof.0.0155.

Cleveland, Kyle. 2014. "'Significant Breaking Worse': The Fukushima Nuclear Crisis as a Moral Panic." *Critical Asian Studies* 46, no. 3: 509–39.

Cornum, Lou. 2018. "The Irradiated International." Future Project, Data and Society Research Institute. https://datasociety.net/wp-content/uploads/2018/06/ii-web.pdf.

Cox, Aimee Meredith. 2015. *Shapeshifters: Black Girls and the Choreography of Citizenship*. Durham, NC: Duke University Press.

Cram, Shannon. 2015. "Becoming Jane: The Making and Unmaking of Hanford's Nuclear Body." *Environment and Planning D: Society and Space* 33, no. 5: 796–812. https://doi.org/10.1177/0263775815599317.

———. 2016a. "Living in Dose: Nuclear Work and the Politics of Permissible Exposure." *Public Culture* 28, no. 3: 519–39. https://doi.org/10.1215/08992363-3511526.

———. 2016b. "Wild and Scenic Wasteland: Conservation Politics in the Nuclear Wilderness." *Environmental Humanities* 7, no. 1: 89–105. https://doi.org/10.1215/22011919-3616344.

Creager, Angela N. H. 2013a. *Life Atomic: A History of Radioisotopes in Science and Medicine*. Chicago: University of Chicago Press.

———. 2013b. "Timescapes of Radioactive Tracers in Biochemistry and Ecology." *History and Philosophy of the Life Sciences* 35, no. 1: 83–89.

Cunningham, Kelly, Thomas G. Hinton, Jared J. Luxton, Aryn Bordman, Kei Okuda, Lynn E. Taylor, Josh Hayes, et al. 2021. "Evaluation of DNA Damage and Stress in Wildlife Chronically Exposed to Low-Dose, Low-Dose Rate Radiation from the Fukushima Dai-Ichi Nuclear Power Plant Accident." *Environment International* 155 (October): 106675. https://doi.org/10.1016/j.envint.2021.106675.

Daniel, E. Valentine. 1984. *Fluid Signs: Being a Person the Tamil Way*. Berkley: University of California Press.

Das, Veena. 2000. "Suffering, Legitimacy, and Healing: The Bhopal Case." In *Illness and the Environment: A Reader in Contested Medicine,* edited by J. Stephen Kroll-Smith, Phil Brown, and V. J. Gunter, 270–87. New York: New York University Press.

Davis, Sasha, and Jessica Hayes-Conroy. 2018. "Invisible Radiation Reveals Who We Are as People: Environmental Complexity, Gendered Risk, and Biopolitics after the Fukushima Nuclear Disaster." *Social & Cultural Geography* 19, no 6: 720–40. https://doi.org/10.1080/14649365.2017.1304566.

De Togni, Giulia. 2022. *Fall-out from Fukushima: Nuclear Evacuees Seeking Compensation and Legal Protection after the Triple Meltdown.* London: Routledge.

DiNitto, Rachel. 2019. *Fukushima Fiction: The Literary Landscape of Japan's Triple Disaster.* Honolulu: University of Hawai'i Press.

Douglas, Mary. (1966) 2002. *Purity and Danger: An Analysis of the Concepts of Pollution and Taboo.* London: Routledge.

Drixler, Fabian. 2016. "The Politics of Migration in Tokugawa Japan: The Eastward Expansion of Shin Buddhism." *Journal of Japanese Studies* 42, no. 1: 1–28. https://doi.org/10.1353/jjs.2016.0024.

Edelstein, Michael R. 2004. *Contaminated Communities: Coping with Residential Toxic Exposure.* 2nd ed. Boulder, CO: Westview.

Eguchi, Eri, Tetsuya Ohira, Hironori Nakano, Fumikazu Hayashi, Kanako Okazaki, Mayumi Harigane, Narumi Funakubo, et al. 2021. "Association between Laughter and Lifestyle Diseases after the Great East Japan Earthquake: The Fukushima Health Management Survey." *International Journal of Environmental Research and Public Health* 18, no. 23: 12699. https://doi.org/10.3390/ijerph182312699.

Ellis, Jonathan, Mitoko Hirabayashi, and Haruki Murakami. 2005. "'In Dreams Begins Responsibility': An Interview with Haruki Murakami." *Georgia Review* 59. no. 3: 548–67.

Endo, Kiyoji. 2015. "Kasetsude shinunda shouga naina." In *Fukkō nante shiteimasen,* edited by Tetsuya Shibui, Nagaoko Yoshiyuki, and Shin Watanabe, 94–110. Tokyo: Daisan Shokan.

Endo, Noriko. 2013. *Genshiryoku songaibaishōseido no kenkyu: Tokyo denryoku Fukushima genpatsu jiko karano kousatsu.* Tokyo: Iwanami Shoten.

Erickson, Jon D., and Duane Chapman. 1993. "Sovereignty for Sale: Nuclear Waste in Indian Country." *Native Americas: Akwe:kon's Journal of Indigenous Issues* (Fall): 3–10.

Erikson, Kai. 1991. "Radiation's Lingering Dread." *Bulletin of the Atomic Scientists* 47, no. 2: 34–39.

———. 1994. *A New Species of Trouble: The Human Experience of Modern Disasters.* New York: W. W. Norton.

Fabian, Johannes. 1983. *Time and the Other: How Anthropology Makes Its Object.* New York: Columbia University Press.

Farmer, Paul. 1996. "On Suffering and Structural Violence: A View from Below." *Daedalus* 125, no. 1: 261–83.

Feldman, Eric A. 2014. "Fukushima: Catastrophe, Compensation, and Justice in Japan." *DePaul Law Review* 62, no. 2: 335–56.

Feng Bin, Yuichi Onda, Yoshifumi Wakiyama, Keisuke Taniguchi, Asahi Hashimoto, and Yupan Zhang. 2022. "Persistent Impact of Fukushima Decontamination on Soil Erosion and Suspended Sediment." *Nature Sustainability* 5: 879–89. https://doi.org/10.1038/s41893-022-00924-6.

Figal, Gerald. 1999. *Civilization and Monsters: Spirits of Modernity in Meiji Japan.* Durham, NC: Duke University Press.

Fortun, Kim. 2001. *Advocacy after Bhopal: Environmentalism, Disaster, New Global Orders.* Chicago: University of Chicago Press.

————. 2012. "Ethnography in Late Industrialism." *Cultural Anthropology* 27, no. 3: 446–64. https://doi.org/10.1111/j.1548-1360.2012.01153.x.

Freudenburg, William R. 1997. "Contamination, Corrosion and the Social Order: An Overview." *Current Sociology* 45, no. 3: 19–39.

Frohmberg, Eric, Robert Goble, Virginia Sanchez, and Dianne Quigley. 2000. "The Assessment of Radiation Exposures in Native American Communities from Nuclear Weapons Testing in Nevada." *Risk Analysis* 20, no. 1: 101–12. https://doi.org/10.1111/0272-4332.00010.

Fujigaki, Yuko. 2015. "The Processes through Which Nuclear Power Plants Are Embedded in Political, Economic, and Social Contexts in Japan." In *Lessons from Fukushima: Japanese Case Studies on Science Technology and Society,* edited by Yuko Fujigaki, 7–26. Cham, Switzerland: Springer.

Fujii, Kensei, and Reiko Seki. 2015. "Gosenzosama ga nemuru machi: Joudoshinshu no chi kara." In *"Ikiru" jikan no paradigm: Hisai genchi kara egaku gempatsu jiko no sekai,* edited by Reiko Seki, 146–63. Tokyo: Nippon Hyoronsha.

Fujiki, Kai. 2016. *Minamisōma ni yakudousuru kodai no gunyakusho: Izumikanga iseki.* Tokyo: Shinsensha.

Fukumoto, Manabu, ed. 2020. *Low-Dose Radiation Effects on Animals and Ecosystems: Long-Term Study of the Fukushima Nuclear Accident.* Springer Open. https://doi.org/10.1007/978-981-13-8218-5.

Fukunaga, Hisanori, and Akinari Yokoya. 2016. "Low-Dose Radiation Risk and Individual Variation in Radiation Sensitivity in Fukushima." *Journal of Radiation Research* 57, no. 1: 98–100. https://doi.org/10.1093/jrr/rrv053.

Fukushima Minpō Shimbun. 2013. *Fukushima to genpatsu vol. 1: Yuuchi kara daishinsai heno gojyunen.* Tokyo: Waseda University Press.

————. 2014. *Fukushima to genpatsu vol. 2: Houshasen tono tatakai + sen nichi no kioku.* Tokyo: Waseda University Press.

————. 2015. *Fukushima to genpatsu vol. 3: Genpatsujiko kanrenshi.* Tokyo: Waseda University Press.

Fukushima Minyū Shimbun. 2022. *Higashinihon daishinsai: Shoge anotoki.* Fukushima: Fukushima Minyū Shimbun.

Funabashi, Yoichi. 2021. *Meltdown: Inside the Fukushima Nuclear Crisis.* Washington, DC: Brookings Institution.

Fuse, Tetsuya. 2011. *Fukushima genpatsu no machi to mura.* Tokyo: Nanatsumori Shokan.

Fuse, Yuuji. 2012. *Rupo ichiefu: Fukushima daiichi genpatsu level 7 no genba.* Tokyo: Iwanami.

Galison, Peter. 2011. "Waste-Wilderness: A Conversation with Peter L. Galison." *FOP.* https://web.archive.org/web/20220121114326/https://fopnews.wordpress.com/2011/03/31/galison/.

Galtung, Johan. 1969. "Violence, Peace, and Peace Research." *Journal of Peace Research* 6, no. 3: 167–91.

Garcia, Angela. 2017. "The Ambivalent Archive." In *Crumpled Paper Boat: Experiments in Ethnographic Writing,* edited by Anand Pandian and Stuart McLean, 29–44. Durham, NC: Duke University Press.

Gardner, James B. 2011. "September 11: Museums, Spontaneous Memorials, and History." In *Grassroots Memorials: The Politics of Memorializing Traumatic Death,* edited by Peter Jan Margry and Cristina Sánchez-Carretero, 285–303. New York: Berghahn.

Geertz, Clifford. 1992. "'Local Knowledge' and Its Limits: Some *Ober Dicta.*" *Yale Journal of Criticism* 5, no. 2: 129–35.

Geilhorn, Barbara, and Kristina Iwata-Weickgenannt, eds. 2017. *Fukushima and the Arts: Negotiating Nuclear Disaster.* London: Routledge.

Genshiryoku Gijyutsushi Kenkyukai, ed. 2015. *Fukushima jiko ni itaru genshiryoku kaihatsushi.* Tokyo: Chuo University Press.

George, Timothy S. 2001. *Minamata: Pollution and the Struggle for Democracy in Postwar Japan.* Cambridge, MA: Harvard University Press.

Gerke, Hannah C., Thomas G. Hinton, and James C. Beasley. 2021. "Movement Behavior and Habitat Selection of Rat Snakes (*Elaphe* spp.) in the Fukushima Exclusion Zone." *Ichthyology & Herpetology* 109, no. 2: 545–56. https://doi.org/10.1643/h2019282.

Gerster, Julia. 2019. "Hierarchies of Affectedness: Kizuna, Perceptions of Loss, and Social Dynamics in Post-3.11 Japan." *International Journal of Disaster Risk Reduction* 41: 101304. https://doi.org/10.1016/j.ijdrr.2019.101304.

Gerster, Julia, and Elizabeth Maly. 2022. "Japan's Disaster Memorial Museums and Framing 3.11: Othering the Fukushima Daiichi Nuclear Disaster in Cultural Memory." *Contemporary Japan,* 34, no. 2: 187–209. https://doi.org/10.1080/1869 2729.2022.2112479.

Gerster, Julia, Sebastien Penmellen Boret, and Akihiro Shibayama. 2021. "Out of the Dark: The Challenges of Branding Post-Disaster Tourism Ten Years after the Great East Japan Earthquake." *EATSJ-Euro-Asia Tourism Studies Journal,* vol. 2. https://web.archive.org/web/20221106155609/https://www.eatsa-researches.org/journal/out-of-the-dark-the-challenges-of-branding-post-disaster-tourism-ten-years-after-the-great-east-japan-earthquake/?pdf=1349.

Gill, Tom. 2013. "This Spoiled Soil: Place, People and Community in an Irradiated Village in Fukushima Prefecture." In *Japan Copes with Calamity: Ethnographies of the Earthquake, Tsunami and Nuclear Disasters of March 2011,* edited by Tom Gill, Brigitte Steger, and David H. Slater, 201–33. Bern: Peter Lang.

———. 2014. "Radiation and Responsibility: What Is the Right Thing for an Anthropologist to Do in Fukushima?" *Japanese Review of Cultural Anthropology* 15: 151–63. https://doi.org/10.14890/jrca.15.0_151.

Goffman, Erving. 1971. "The Territories of the Self." In *Relations in Public: Microstudies of the Public Order*, 28–61. New York: Basic Books.

———. 1989. "On Fieldwork." *Journal of Contemporary Ethnography* 18, no. 2: 123–32.

Goldfarb, Kathryn E. 2016. "'Coming to Look Alike': Materializing Affinity in Japanese Foster and Adoptive Care." *Social Analysis* 60, no. 2: 47–64. https://doi.org/10.3167/sa.2016.600204.

Good, Megan. 2016. "Shaping Japan's Disaster Heritage." In *Reconsidering Cultural Heritage in East Asia*, edited by Akira Matsuda and Luisa E. Mengoni, 139–61. London: Ubiquity.

Gordon, Avery F. 1997. *Ghostly Matters: Haunting and the Sociological Imagination*. Minneapolis: University of Minnesota Press.

Greenpeace Japan. 2021. "Fukushima Daiichi 2011–2021: The Decontamination Myth and A Decade of Human Rights Violations." Greenpeace. https://web.archive.org/web/20210313153241/https://www.greenpeace.org/static/planet4-japan-stateless/2021/03/ff71ab0b-finalfukushima2011-2020_web.pdf.

Gusterson, Huge. 2008. "Ethnographic Research." In *Qualitative Methods in International Relations: A Pluralist Guide*, edited by Audie Klotz and Deepa Prakash, 93–113. New York: Palgrave Macmillan.

Hall, Eric J., and Amato J. Giaccia. 2019. *Radiobiology for the Radiologist*. 8th ed. Philadelphia, PA: Lippincott, Williams, and Wilkins.

Hamada, Nobuyuki, Haruyuki Ogino, and Yuki Fujimichi. 2012. "Safety Regulations of Food and Water Implemented in the First Year Following the Fukushima Nuclear Accident." *Journal of Radiation Research* 53, no. 5: 641–71. https://doi.org/10.1093/jrr/rrs032.

Hannigan, John. 2012. *Disasters without Borders: The International Politics of Natural Disasters*. Hoboken, NJ: Wiley.

Happy. 2013. *Fukushima dai-ichi genpatsu shūsoku sagyō nikki: 3.11 karano 700 nichikan*. Tokyo: Kawade Shobo.

Harada, Kouji H., Tamon Niisoe, Mie Imanaka, Tomoyuki Takahashi, Katsumi Amako, Yukiko Fujii, Masatoshi Kanameishi, et al. 2014. "Radiation Dose Rates Now and in the Future for Residents Neighboring Restricted Areas of the Fukushima Daiichi Nuclear Power Plant." *Proceedings of the National Academy of Sciences* 111, no. 10: E914–E923. https://doi.org/10.1073/pnas.1315684111.

Harigane, Mayumi, Yoshitake Takebayashi, Michio Murakami, Masaharu Maeda, Rie Mizuki, Yuichi Oikawa, Saori Goto, et al. 2021. "Higher Psychological Distress Experienced by Evacuees Relocating outside Fukushima after the Nuclear Accident: The Fukushima Health Management Survey." *International Journal of Disaster Risk Reduction* 52 (January): 101962. https://doi.org/10.1016/j.ijdrr.2020.101962.

Hasegawa, Kenichi, and Hanako Hasegawa. 2014. *Rakunouka Hasegawa Kenichi ga kataru madeinamura Iitate.* Tokyo: Nanatsumori Shokan.

Hasegawa, Shin, Teppei Suzuki, Ayako Yagahara, Reiko Kanda, Tatsuo Aono, Kazuaki Yajima, and Katsuhiko Ogasawara. 2020. "Changing Emotions about Fukushima Related to the Fukushima Nuclear Power Station Accident—How Rumors Determined People's Attitudes: Social Media Sentiment Analysis." *Journal of Medical Internet Research* 22, no. 9: e18662. https://doi.org/10.2196/18662.

Hashimoto, Kazutaka. 2013. *En no shakaigaku: Fukushi shakaigaku no shitenkara.* Tokyo: Harvestsha.

Hashimoto, Shigeatsu, Masato Nagai, Tetsuya Ohira, Shingo Fukuma, Mitsuaki Hosoya, Seiji Yasumura, Hiroaki Satoh, et al. 2020. "Influence of Post-Disaster Evacuation on Incidence of Hyperuricemia in Residents of Fukushima Prefecture: The Fukushima Health Management Survey." *Clinical and Experimental Nephrology* 24, no. 11: 1025–32. https://doi.org/10.1007/s10157-020-01924-6.

Hastrup, Frida. 2011. *Weathering the World: Recovery in the Wake of the Tsunami in a Tamil Fishing Village.* New York: Berghahn.

Hattori, Misaki. 2021. *Tokyo denryoku Fukushima dai-ichi genpatsu jiko kara 10 nen no chiken: Fukkō suru Fukushima no kagaku to rinri.* Tokyo: Maruzen Shuppan.

Hayakawa Tadao, and Houzen Matsuno. 2012. "Tsunamini nagasare Manokoudo wo nanpashita waga gyosen Inarimaru." *Kashima Rekishi Aikoukaishi,* 5, no. 27: 159–78.

Hearn, Lafcadio. (1904) 1971. *Kwaidan: Stories and Studies of Strange Things.* Tokyo: Charles E. Tuttle.

Hecht, Gabrielle. 2012a. *Being Nuclear: Africans and the Global Uranium Trade.* Cambridge, MA: MIT Press.

———. 2012b. "Nuclear Nomads: A Look at the Subcontracted Heroes." *Bulletin of the Atomic Scientists.* https://web.archive.org/web/20160912103104/http://thebulletin.org/nuclear-nomads-look-subcontracted-heroes.

Hida, Syuntarō. 2012. *Naibu hibaku.* Tokyo: Fusosha.

Higuchi, Kenji. 2021. *Fukushima genpatsu kimin rekishi no shōnin: Owarinaki genpatsu jiko.* Tokyo: Hachigatsu Shokan.

Hinton, T. G., J. S. Bedford, J. C. Congdon, and F. W. Whicker. 2004. "Effects of Radiation on the Environment: A Need to Question Old Paradigms and Enhance Collaboration among Radiation Biologists and Radiation Ecologists." *Radiation Research* 162, no. 3: 332–38. https://doi.org/10.1667/RR3222.

Hiraoka, Koh, Seiichiro Tateishi, and Koji Mori. 2015. "Review of Health Issues of Workers Engaged in Operations Related to the Accident at the Fukushima Dai-ichi Nuclear Power Plant." *Journal of Occupational Health* 57, no. 6: 497–512. https://doi.org/10.1539/joh.15-0084-RA.

Hirosaki, Mayumi, Tetsuya Ohira, Seiji Yasumura, Masaharu Maeda, Hirooki Yabe, Mayumi Harigane, Hideto Takahashi, Michio Murakami, Yuriko Suzuki, Hironori Nakano, Wen Zhang, Mayu Uemura, Masafumi Abe, and Kenji Kamiya. 2018. "Lifestyle Factors and Social Ties Associated with the Frequency of Laughter after the Great East Japan Earthquake: Fukushima Health Management Survey." *Quality of Life Research* 27, no. 3: 639–50.

Hisamura, Masaki, Arinobu Hori, Akira Wada, Itaru Miura, Hiroshi Hoshino, Shuntaro Itagaki, Yasuto Kunii, et al. 2017. "Newly Admitted Psychiatric Inpatients after the 3.11 Disaster in Fukushima, Japan." *Open Journal of Psychiatry* 7, no. 3: 131–46. https://doi.org/10.4236/ojpsych.2017.73013.

Hobsbawm, Eric J., and Terrence O. Ranger. (1983) 2009. *The Invention of Tradition*. Cambridge: Cambridge University Press.

Hoffman, Susanna. 2003. "The Hidden Victims of Disaster." *Environmental Hazards* 5, no. 2: 67–70.

Hopson, Nathan. 2017. *Ennobling Japan's Savage Northeast: Tōhoku as Japanese Postwar Thought, 1945–2011*. Illustrated edition. Cambridge, MA: Harvard University Press.

Hori, Arinobu, Hiroshi Hoshino, Itaru Miura, Masaki Hisamura, Akira Wada, Shuntaro Itagaki, Yasuto Kunii, et al. 2017. "Psychiatric Outpatients after the 3.11 Complex Disaster in Fukushima, Japan." *Annals of Global Health* 82, no. 5: 798–805. https://doi.org/10.1016/j.aogh.2016.09.010.

Ibaraki Shimbun Company. 2003. *Genshiryoku mura*. Ibaraki: Nakashobou.

Ide, Akira. 2018. *Dark Tourism: Kanashimi no kioku wo meguru tabi*. Tokyo: Gentosha.

Igarashi, Yasumasa. 2018. *Genpatsu jiko to "shoku."* Tokyo: Chuokōron.

Iguchi, Satoshi. 2019. *Post-3.11 no risk shakaigaku: Genpatsu jiko to houshasen risk ha donoyouni katararetanoka*. Kyoto: Nakanishiya Shuppan.

Iimura, Satoshi. 2005. *Ritsuryo kokka no tai enishi seisaku: Sōma no seitetsu isekigun*. Tokyo: Shinsensha.

Ikebata, Daiji. 1995. *Kitano mumeihi: Kagaimin no ashiato wo tadoru*. Kanazawa: Hokuriku Shimbun.

Ikeda, Kayoko, Shuji Shimizu, Hiroshi Kainuma, Kunikazu Noguchi, Kazuya Kodama, Haruno Matsumoto, Ikurou Anzai, Masaki Ichinose, Makoto Omori, Sae Ochi, Hideo Konami, Ryugo Hayano, Sachiko Banba, and Masaharu Maeda. 2018. *Shiawaseni narutameno "Fukushima sabetsu" ron*. Kyoto: Kamogawa Press.

Imai, Akira. 2016. *Fukushima Inside Story: Yakuba shokuin ga mita genpatsu hinan to shinsai fukko*. Tokyo: Koujin no Tomo.

Imanaka, Tetsuji. 2012. *Teisenryo houshasenhibaku: Chernobyl kara Fukushima he*. Tokyo: Iwanami Shoten.

Inaizumi, Ren. 2021. *Hairo: "Haiboku no genba" de hataraku hokori*. Tokyo: Shinchōsha.

Independent Investigation Commission on the Fukushima Nuclear Accident. 2014. *The Fukushima Daiichi Nuclear Power Station Disaster: Investigating the Myth and Reality*. New York: Routledge.

Inose, Kohei. 2014. "'Living with Uncertainty': Public Anthropology and Radioactive Contamination." *Japanese Review of Cultural Anthropology* 15: 141–50.

Inoue, Takeshi. 2014. *Genshiryoku to chiiki seisaku: "Kokusaku he no kyōryoku" to "jichi no jissen" no tenkai*. Kyoto: Koyo Shobō.

International Commission on Radiological Protection. 1975. "Report of the Task Group on Reference Man ICRP Publication 23." https://web.archive.org /web/20210925090758/https://journals.sagepub.com/pb-assets/cmscontent /ANI/P_023_1975_Report_on_the_Task_Group_on_Reference_Man_revo.pdf.

Ishido, Satoru. 2017. *Risk to ikiru shisha to ikiru*. Tokyo: Akishobō.

———. 2021. *Mienai sen wo aruku*. Tokyo: Kōsansha.

Ishizawa, Maya. 2018. "Cultural Landscapes Link to Nature: Learning from *Satoyama* and *Satoumi*." *Built Heritage* 2, no. 4: 7–19. https://doi.org/10.1186 /BF03545680.

Itoh, Mayumi. 2018. *Animals and the Fukushima Nuclear Disaster*. Cham, Switzerland: Palgrave Macmillan.

Iwamoto, Yoshiteru. 1994. *Tōhoku kaihatsu 120 nen*. Tokyo: Tosui Shobō.

———. 2000. *Rekishi toshiteno Sōma: Hana ha Sōma ni mi ha Date ni*. Tokyo: Tosui Shobō.

———, ed. 2013a. *Rekishi toshiteno higashi nihon dai shinsai*. Tokyo: Tosui Shobō.

———. 2013b. "The Immigration of Jōdō Shinshu-sect Buddhists into the Mutsu-Nakamura Domain during the Tokugawa Period: The Actual Situation as Seen from the Memoranda of Kowata Hikobei." [Japanese]. *Sonraku Shakai Kenkyu Journal* 17, no. 2: 19–29. https://doi.org/10.9747/jars.17.2_18.

Iwasaki, Keiko. 2021. *Fukushima genpatsu jiko to kokoro no kenkou: Jisshō keizaigaku desaguru gensai fukkō no kagi*. Tokyo: Nippon Hyoronsha.

Iwasaki, Toshio. 1970. *Ninomiya Sontoku no Sōma shihō*. Tokyo: Kinseisha.

Jacobs, Robert A. 2014. "The Radiation That Makes People Invisible: A Global Hibakusha Perspective." *Asia-Pacific Journal: Japan Focus* 12, issue 31, no. 1. https://apjjf.org/2014/12/31/Robert-Jacobs/4157/article.html.

———. 2021. "Fukushima Radiation Inside Out." In *Legacies of Fukushima: 3.11 in Context,* edited by Kyle Cleveland, Scott Gabriel Knowles, and Ryuma Shineha, 50–62. Philadelphia: University of Pennsylvania Press.

———. 2022. *Nuclear Bodies: The Global Hibakusha*. New Haven, CT: Yale University Press.

Jang, Kyungjae, Kengo Sakamoto, and Carolin Funck. 2021. "Dark Tourism as Educational Tourism: The Case of 'Hope Tourism' in Fukushima, Japan." *Journal of Heritage Tourism* 16, no. 4: 481–92. https://doi.org/10.1080/17438 73X.2020.1858088.

Jasanoff, Sheila. 1999. "The Songlines of Risk." *Environmental Values* 8, no. 2: 135–52.

Jasanoff, Sheila, and Sang-Hyun Kim. 2009. "Containing the Atom: Sociotechnical Imaginaries and Nuclear Power in the United States and South Korea." *Minerva* 47, no. 2: 119–46.

Jensen, Casper Bruun, Miho Ishii, and Philip Swift. 2016. "Attuning to the Webs of *En:* Ontography, Japanese Spirit Worlds, and the 'Tact' of Minakata Kumagusu." *HAU: Journal of Ethnographic Theory* 6, no. 2: 149–72. https://doi.org/10.14318/hau6.2.012.

Jobin, Paul. 2020. "The Fukushima Nuclear Disaster and Civil Actions as a Social Movement." *Asia-Pacific Journal: Japan Focus* 18, issue 9, no. 1. https://apjjf .org/2020/9/Jobin.html.

Jorgenson, Timothy J. 2016. *Strange Glow: The Story of Radiation*. Princeton, NJ: Princeton University Press.

Jung, Carl G. 1984. *Dream Analysis*. Princeton, NJ: Princeton University Press.

Kainuma, Hiroshi. 2011. *Fukushima ron: Genshiryokumura ha nazeumaretaka*. Tokyo: Sekidōsha.

———. 2012. *Fukushima no seigi: "Nihon no kawaranasa" tono tatakai*. Tokyo: Tougensha.

———. 2015. *Hajimeteno Fukushima-gaku*. Tokyo: East Press.

———, ed. 2016. *Encyclopedia of the "1F": A Guide to the Decommissioning of the Fukushima Daiichi Nuclear Power Station*. Tokyo: Ota Shuppan.

Kajita, Makoto. 2014. "Establishment of Nuclear Power Stations and Reconfiguration of Local Society and Economy: A Case Study of Tomioka, Fukushima Prefecture." [Japanese]. *Geographical Review of Japan Series A* 87, no. 2: 108–27.

Kamata, Satoshi. 2006. *Nihon no genpatsu chitai*. Tokyo: Shinchōsha.

Kanbe, Hidehiko. 2021. *Fukushima dai-ichi genpatsu jikogo no minji soshō: Genpatsu no saikadou sashidome to genpatsu higai no baishō/genjyō kaifuku*. Kyoto: Horitsu Bunkasha.

Kanebishi, Kiyoshi, ed. 2018. *Watashi no yume made aini kitekureta*. Tokyo: Asahi Shimbun Press.

Kanebishi, Kiyoshi, and Tohoku Gakuin Daigaku Shinsai no Kiroku Project. 2020. *Shinsai to yukuefumei: Aimaina soushitsu to jyuyō no monogatari*. Tokyo: Shinyousha.

Kanno, Norio. 2018. *"Madei no mura" ni kaerou: Iitate sonchō kunou to ketsudan to kansha no 7 nen*. Tokyo: Wani Books.

Kanno, Tetsu. 2020. *"Zenson hinan" wo ikiru: Seizon/Seikatsuken wo hakaishita Fukushima dai-ichi genpatsu "kakoku" jiko*. Tokyo: Gensousha.

Kanō, Harunao. 2017. "Higashinihon daishinsai ikō tou no 3D digital archive sakusei niokeru keisoku shuhō nitsuite." *Sentan Sokuryo Gijyutsu* 110: 32–42.

Kansai Gakuin University, JCN, and SAFLAN. 2015. *Genpatsu hinan hakusho*. Tokyo: Jinbun Shoin.

Kasai, Chiaki. 2020. *Kazoku shashin: 3.11 genpatsu jiko to wasurerareta tsunami*. Tokyo: Shōgakkan.

Kashimura, Yasuhiro, and Koichi Chida. 2015. "Nuclear Reactor Accident Fallout Artifacts: Unusual Black Spots on Digital Radiographs." *American Journal of Roentgenology* 205, no. 6: 1240–43. https://doi.org/10.2214/AJR.15.14557.

Katayama, Izumi. 2015. *Fukushima no kome ha anzendesuga tabetekurenakutemo kekkoudesu*. Kyoto: Kamogawa Shuppan.

Katayama, Natsuko. 2020. *Fukushima genpatsu sagyoin nisshi: Ichiefu no shinjitsu 9 nenkan no kiroku*. Tokyo: Asahi Shimbunsha.

Kawasaki, Kota. 2018a. *Kankyo fukkō: Higashinihon daishinsai/Fukushima genpatsu jiko no hisaichikara*. Tokyo: Hassakusha.

———. 2018b. *Fukushima no jyosen to fukkō*. Tokyo: Maruzen Shuppan.

Kawasaki, Yu. 2014. "Benkei" no kita michi: Fukushima-ken Minamisōma-shi Kaibama no kyōdoryōri wo megutte." *Fukushima no Minzoku* 42: 93–107.

Kikuchi, Masaya. 2013. "The Extent of Damage from the Nuclear Accident on Agriculture Produce from Fukushima Prefecture: A View from the Wholesale Market Trends." [Japanese]. *Agriculture Economics Research* 85, no. 3: 140–50.

Kimmerer, Robin Wall. 2015. *Braiding Sweetgrass: Indigenous Wisdom, Scientific Knowledge and the Teachings of Plants*. Minneapolis, MN: Milkweed Editions.

Kimura, Aya H. 2015. "Risk Communication Programs after the Fukushima Nuclear Accident: A Comparison of Epistemic Cultures." *United Nation University Fukushima Global Communication Programme Working Paper Series,* 1–7. https://web.archive.org/web/20220121045326/https://i.unu.edu/media/fgc.unu.edu-en/page/922/FGC-WP-13-FINAL.pdf.

———. 2016. *Radiation Brain Moms and Citizen Scientists*. Durham, NC: Duke University Press.

———. 2020. "The Potentials and Challenges of Citizen Science: 9 Years of Experience from Post-Fukushima Japan." *Somatosphere.* https://web.archive.org/web/20210411012522/http://somatosphere.net/2020/citizen-science-fukushima-japan.html/.

Kimura, Shūhei. 2014. "Visualizing with 'Soft Light': A Reflection on Public Anthropology and 3/11." *Japanese Review of Cultural Anthropology* 15: 127–40. https://doi.org/10.14890/jrca.15.0_127.

Kingston, Jeff. 2014. "Japan's Nuclear Village: Power and Resilience." In *Critical Issues in Contemporary Japan,* edited by Jeff Kingston, 107–19. New York: Routledge.

———. 2022a. "Contesting Fukushima." *The Asia-Pacific Journal: Japan Focus* 20, issue 12, no. 1. https://apjjf.org/2022/12/Kingston.html.

———. 2022b. "Fukushima's Dueling Museums." *The Asia-Pacific Journal: Japan Focus* 20, issue 12, no. 2. https://apjjf.org/2022/12/Kingston2.html.

Kirby, Peter Wynn. 2019. "Slow Burn: Dirt, Radiation, and Power in Fukushima." *Asia-Pacific Journal: Japan Focus* 17, issue 19, no. 3. https://apjjf.org/2019/19/Kirby.html.

Kirsch, Stuart. 2001. "Lost Worlds: Environmental Disaster, 'Culture Loss,' and the Law." *Current Anthropology* 42, no. 2: 167–98. https://doi.org/10.1086/320006.

Klein, Naomi. 2008. *The Shock Doctrine: The Rise of Disaster Capitalism.* 1st ed. New York: Picador.

Klose, Alexander. 2015. *The Container Principle: How a Box Changes the Way We Think.* Translated by Charles Marcrum II. Cambridge, MA: MIT Press.

Knight, Catherine. 2010. "The Discourse of 'Encultured Nature' in Japan: The Concept of *Satoyama* and Its Role in 21st-Century Nature Conservation." *Asian Studies Review* 34, no. 4: 421–41. https://doi.org/10.1080/10357823.2010.527920.

Knight, John. 2006. *Waiting for Wolves in Japan: An Anthropological Study of People-Wildlife Relations.* Honolulu: University of Hawai'i Press.

Kobashi, Yurie, Tomohiro Morita, Akihiko Ozaki, Toyoaki Sawano, Nobuaki Moriyama, Naomi Ito, and Masaharu Tsubokura. 2021. "Long-Term Care Utilization Discrepancy among the Elderly in Former Evacuation Areas, Fukushima."

*Disaster Medicine and Public Health Preparedness* (March): 1–3. https://doi
.org/10.1017/dmp.2020.481.

Kodama, Kazuya, Shuji Shimizu, and Kunikazu Noguchi. 2014. *Houshasen hibaku
no rika/shakai: Yonenme no "Fukushima no shinjitsu."* Kyoto: Kamogawa
Shuppan.

Kogure, Keiji. 2013. *Houshanō jyosen to hakibutsu shori.* Tokyo: Gihoudō Shuppan.

Koide, Hiroaki. 2012. "Japanese Radiation Expert Koide on Fukushima Dangers."
*Asia-Pacific Journal: Japan Focus* 10, issue 54, no. 83. https://apjjf.org/-Koide-
Hiroaki/4690/article.html.

———. 2021. *Genpatsu jiko ha owatteinai.* Tokyo: Mainichi Shimbunsha.

Koide, Hiroaki, and Norma Field. 2019. "The Fukushima Nuclear Disaster and the
Tokyo Olympics." *Asia-Pacific Journal: Japan Focus* 17, issue 5, no. 3. https://apjjf
.org/2019/05/Koide-Field.html.

Komatsu, Riken. 2018. *Shin fukkō ron.* Tokyo: Genron.

———. 2021. *Shin fukkō ron zouho edition.* Tokyo: Genron.

Kondo, Dorinne K. 1990. *Crafting Selves: Power, Gender, and Discourses of Identity
in a Japanese Workplace.* Chicago: University of Chicago Press.

Kowata, Jin, and Masumi Kowata. 2012. *Genpatsu ricchi: Ōkuma choumin ha
uttaeru.* Tokyo: Takushoku Shobō Shinsha.

Kowata, Shinsuke, Tsuguo Idogawa, and Kazuhiko Monnma. 1996. "Hiwashi Jinjya." *Odaka Rekishi* 6: 1–34.

Koyama, Yohei. 2022. "Living with Contradiction—Risk Perception, Self-Determi-
nation, and Life after Fukushima." *Environmental Advances* 7: 100181.

Kuchinskaya, Olga. 2014. *The Politics of Invisibility: Public Knowledge about Radiation Health Effects after Chernobyl.* Cambridge, MA: MIT Press.

———. 2019. "Citizen Science and the Politics of Environmental Data." *Science,
Technology, & Human Values* 44, no. 5: 871–80.

Kudō, Yuka. 2016. "Shishaga kayou machi." In *Yobisamasareru reisei no shinsaigaku,*
edited by Kanabishi Kiyoshi, 1–23. Tokyo: Shinyousha.

Kuletz, Valerie. 1998. *The Tainted Desert: Environmental and Social Ruin in the
American West.* New York: Routledge.

Kumaki, Hiroko. 2021. "Living in Paradox: Technopolitics of Health and Well-
Being in Fukushima." Paper presented at the A Decade of Fukushima: Socio-
Technical Perspectives on Surviving the Nuclear Age Conference, University of
Colorado, Boulder.

———. 2022. "Suspending Nuclearity: Ecologies of Planting Seeds after the Nuclear
Fallout in Fukushima, Japan." *Cultural Anthropology* 37, no. 4: 707–37. https://
doi.org/10.14506/ca37.4.05.

Kunii, Nobuaki, Maya Sophia Fujimura, Yukako Komasa, Akiko Kitamura,
Hitoshi Sato, Toshihiro Takatsuji, Masamine Jimba, and Shinzo Kimura. 2018.
"The Knowledge and Awareness for Radiocesium Food Monitoring after the
Fukushima Daiichi Nuclear Accident in Nihonmatsu City, Fukushima Prefec-
ture." *International Journal of Environmental Research and Public Health* 15, no.
10: 2289. https://doi.org/10.3390/ijerph15102289.

Kurgan, Laura. 2013. *Close Up at a Distance: Mapping, Technology, and Politics.* New York: Zone Books.

Kuroda, Yujiro, Masatsugu Orui, and Arinobu Hori. 2021. "Trends in Suicide Mortality in 10 Years around the Great East Japan Earthquake: Analysis of Evacuation and Non-Evacuation Areas in Fukushima Prefecture." *International Journal of Environmental Research and Public Health* 18, no. 11: 6005. https://doi.org/10.3390/ijerph18116005.

Kuroda, Yujiro, Masaharu Tsubokura, Kiyoshi Sasaki, Takashi Hara, Atsushi Chiba, Keishin Mashiko, and Thierry Schneider. 2020. "Development of Radiation Education in Schools after the Fukushima Daiichi Nuclear Power Plant Accident—A Study from the Perspectives of Regionality, Multidisciplinarity and Continuity." *Radioprotection* 55, no. 4: 317–24. https://doi.org/10.1051/radiopro/2020078.

Kurokawa, Kishō. 1977. *Gurei no bunka: Nihonteki kūkan toshiteno "en."* Tokyo: Souseiki.

Kurokawa, Kiyoshi, and Andrea Ryoko Ninomiya. 2018. "Examining Regulatory Capture: Looking Back at the Fukushima Nuclear Power Plant Disaster, Seven Years Later." *University of Pennsylvania Asian Law Review* 13, no. 2: 47–71.

Kurokawa, Shoko. 2017. *"Kokoro no jyosen" toiu kyokou: Jyosen senshintoshi ha naze jyosen wo yametaka.* Tokyo: Shūeisha International.

———. 2020. *Kokoro no jyosen: Genpatsu suishinha no jikkentoshi Fukushima-ken Date-shi.* Tokyo: Shūeisha.

LaDuke, Winona. 1999. *All Our Relations: Native Struggles for Land and Life.* Cambridge, MA: South End.

Lambek, Michael. 1996. "The Past Imperfect: Remembering as Moral Practice." In *Tense Past: Cultural Essays in Trauma and Memory,* edited by Paul Antze and Michael Lambek, 235–54. New York: Routledge.

Leatherbarrow Andrew 2022. *Melting Sun: The History of Nuclear Power in Japan and the Disaster at Fukushima Daiichi.* New Haven, CT: Andrew Leatherbarrow.

Leppold, Claire, Tetsuya Tanimoto, and Masaharu Tsubokura. 2016. "Public Health after a Nuclear Disaster: Beyond Radiation Risks." *Bulletin of the World Health Organization* 94, no. 11: 859–60.

Lesbirel, S. Hayden. 1998. NIMBY *Politics in Japan: Energy Siting and the Management of Environmental Conflict.* Ithaca, NY: Cornell University Press.

Lifton, Robert Jay. 2000. *Destroying the World to Save It: Aum Shinrikyo, Apocalyptic Violence, and the New Global Terrorism.* New York: Henry Holt and Company.

Lifton, Robert Jay, and Richard A. Falk. 1982. *Indefensible Weapons: The Political and Psychological Case against Nuclearism.* Toronto: Canadian Broadcasting Corporation.

Lindee, Susan. 1994. *Suffering Made Real: American Science and the Survivors at Hiroshima.* Chicago: University of Chicago Press.

———. 2016. "Survivors and Scientists: Hiroshima, Fukushima, and the Radiation Effects Research Foundation, 1975–2014." *Social Studies of Science* 46, no. 2: 184–209. https://doi.org/10.1177/0306312716632933.

Littlejohn, Andrew. 2020. "Museums of Themselves: Disaster, Heritage, and Disaster Heritage in Tohoku." *Japan Forum* 33, no. 4: 476–96. https://doi.org/10.1080/09555803.2020.1758751.

———. 2021. "Ruins for the Future: Critical Allegory and Disaster Governance in Post-Tsunami Japan." *American Ethnologist* 48, no. 1: 7–21.

Liu, Jing, and Michael Faure. 2016. "Compensation for Nuclear Damage: A Comparison among the International Regime, Japan and China." *International Environmental Agreements: Politics, Law and Economics* 16, no. 2: 165–87. https://doi.org/10.1007/s10784-014-9252-7.

Lochbaum, David, Edwin Lyman, and Susan Q. Stranahan. 2014. *Fukushima: The Story of a Nuclear Disaster.* New York: New Press.

Loh, Shi Lin, and Sulfikar Amir. 2019. "Healing Fukushima: Radiation Hazards and Disaster Medicine in Post-3.11 Japan." *Social Studies of Science* 49, no. 3: 333–54. https://doi.org/10.1177/0306312719854540.

Longmuir, Caley, and Vincent I. O. Agyapong. 2021. "Social and Mental Health Impact of Nuclear Disaster in Survivors: A Narrative Review." *Behavioral Sciences* 11, no. 113: 1–23. https://doi.org/10.3390/bs11080113.

Lotman, Yuri M. 1990. *Universe of the Mind: A Semiotic Theory of Culture.* Translated by Ann Shukman. Bloomington: Indiana University Press.

Low, Morris. 2020. *Visualizing Nuclear Power in Japan: A Trip to the Reactor.* Cham, Switzerland: Palgrave Macmillan.

Lyons, Phillip C., Kei Okuda, Matthew T. Hamilton, Thomas G. Hinton, and James C. Beasley. 2020. "Rewilding of Fukushima's Human Evacuation Zone." *Frontiers in Ecology and the Environment* 18, no. 3: 127–34. https://doi.org/10.1002/fee.2149.

Manabe, Noriko. 2015. *The Revolution Will Not Be Televised: Protest Music after Fukushima.* New York: Oxford University Press.

Manabiai Minamisoma. 2018. *Kataritsugu furusato Minamisōma: Ikita akashi to ikiteikuomoi to.* Fukushima: Manabiai Minamisōma.

Maruyama, Teruhisa. 2017. *Fukushima genpatsu jiko no houteki sekinin ron 2: Teisenryo hibaku to kenkou higai no ingakankei wo tou.* Tokyo: Akashishoten.

Masco, Joseph. 2006. *The Nuclear Borderlands: The Manhattan Project in Post–Cold War New Mexico.* Princeton, NJ: Princeton University Press.

———. 2015. "The Age of Fallout." *History of the Present* 5, no. 2: 137–68. https://doi.org/10.5406/historypresent.5.2.0137.

Masilkova, Michaela, Miloš Ježek, Václav Silovský, Monika Faltusová, Jan Rohla, Tomáš Kušta, and Hynek Burda. 2021. "Observation of Rescue Behaviour in Wild Boar (*Sus scrofa*)." *Scientific Reports* 11, no. 1: 16217. https://doi.org/10.1038/s41598-021-95682-4.

Matsui, Katsuhiro. 2021. *Genpatsu hinan to saisei heno mosaku: "Jibungoto" toshite kangaeru.* Tokyo: Toushindo.

Matsunaga, Kyoko. 2014. "Leslie Marmon Silko and Nuclear Dissent in the American Southwest." *Japanese Journal of American Studies* 25: 67–87.

McNeill, David. 2013. "Them versus Us: Japanese and International Reporting of the Fukushima Nuclear Crisis." In *Japan Copes with Calamity: Ethnographies of the Earthquake, Tsunami and Nuclear Disasters of March 2011*, edited by Tom Gill, Brigitte Steger, and David H. Slater, 127–50. Bern: Peter Lang.

Meguro, Tomiko. 2016. *Watashitachino shōgenshū: Futaba-machi wo osotta houshanō kara nogarete*. Fukushima: Shousyukai.

Midorikawa, Sanae, and Akira Ohtsuru. 2022. "Young People's Perspectives of Thyroid Cancer Screening and Its Harms after the Nuclear Accident in Fukushima Prefecture: A Questionnaire Survey Indicating Opt-Out Screening Strategy of the Thyroid Examination as an Ethical Issue." *BMC Cancer* 22, no. 1: 235. https://doi.org/10.1186/s12885-022-09341-6.

Minakata, Kumagusu, and Shinichi Nakazawa. 2015. *Minakata mandala*. Tokyo: Kawade Shobō Shinsha.

Miura, Hideyuki. 2020. *Shiroi tochi: Rupo Fukushima "kitakukonnankuiki" to sono syūhen*. Tokyo: Shūeisha.

Miyazawa, Kaoru. 2018. "Becoming an Insider and an Outsider in Post-Disaster Fukushima." *Harvard Educational Review* 88, no. 3: 334–54. https://doi.org/10.17763/1943-5045-88.3.334.

Morioka, Rika. 2014. "Gender Difference in the Health Risk Perception of Radiation from Fukushima in Japan: The Role of Hegemonic Masculinity." *Social Science & Medicine* 107 (April): 105–12. https://doi.org/10.1016/j.socscimed.2014.02.014.

Morimoto, Ryo. 2012. "Shaking Grounds, Unearthing Palimpsests: Semiotic Anthropology of Disaster." *Semiotica* 192: 263–74.

———. 2015. "Interpretative Frameworks of Disaster in Society Close-Up." In *Natural Hazards, Risks, and Disasters in Society: A Cross-Disciplinary Overview*, edited by John F. Shroder, Andrew E. Collins, Samantha Jones, Bernard Manyena, and Janaka Jayawickrama, 323–51. Amsterdam: Elsevier.

———. 2020. "From Nuclear Things to Things Nuclear: Minding the Gap at the Knowledge-Policy-Practice Nexus in Post-Fallout Fukushima." In *Disaster Upon Disaster: Exploring the Gap between Knowledge, Policy and Practice*, edited by Susanna Hoffman and Roberto Barrios, 218–40. NY: Berghahn Books.

———. 2021a. Commentary—Ethnographic Lettering: "Pursed Lips: A Call to Suspend Damage in the Age of Decommissioning." *Critical Asian Studies*, March 22, 2021.

———. 2021b. "Home Otherwise: Living Archives and Half-Life Politics in Coastal Fukushima." *Cultural Anthropology* 36, no. 4: 573–79. https://doi.org/10.14506/ca36.4.05.

———. 2022. "A Wild Boar Chase: Ecology of Harm and Half-Life Politics in Coastal Fukushima." *Cultural Anthropology* 37, no 1: 69–98. https://doi.org/10.14506/ca37.1.08.

Morita, Tomohiro, Claire Leppold, Masaharu Tsubokura, Tsuyoshi Nemoto, and Yukio Kanazawa. 2016. "The Increase in Long-Term Care Public Expenditure following the 2011 Fukushima Nuclear Disaster." *Journal of Epidemiology and Community Health* 70, no. 7: 738. https://doi.org/10.1136/jech-2015-206983.

Morita, Tomohiro, Shuhei Nomura, Tomoyuki Furutani, Claire Leppold, Masaharu Tsubokura, Akihiko Ozaki, Sae Ochi, Masahiro Kami, Shigeaki Kato, and Tomoyoshi Oikawa. 2018. "Demographic Transition and Factors Associated with Remaining in Place after the 2011 Fukushima Nuclear Disaster and Related Evacuation Orders." *PLOS ONE* 13, no. 3: e0194134. https://doi.org/10.1371/journal.pone.0194134.

Morris-Suzuki, Tessa. 2014. "Touching the Grass: Science, Uncertainty and Everyday Life from Chernobyl to Fukushima." *Science, Technology and Society* 19, no. 3: 331–62. https://doi.org/10.1177/0971721814548115.

———. 2015. "Re-animating a Radioactive Landscape: Informal Life Politics in the Wake of the Fukushima Nuclear Disaster." *Japan Forum* 27, no. 2: 167–88. https://doi.org/10.1080/09555803.2015.1040817.

Murakami, Haruki. 1989. *A Wild Sheep Chase.* Translated by Alfred Birnbaum. New York: Kodansha International.

———. 2000. *Norwegian Wood.* Translated by Jay Rubin. New York: Vintage.

———. 2001. *Underground: The Tokyo Gas Attack and the Japanese Psyche.* Translated by Alfred Birnbaum and Philip Gabriel. Illustrated edition. New York: Vintage.

———. 2005. *Kafka on the Shore.* Translated by Philip Gabriel. New York: Knopf.

Murakami, Kazuo, Takashi Nagai, Kyoko Ono, and Atsuo Kishimoto. 2014. *Kijyunchi no karakuri: Anzen ha koushite suujini natta.* Tokyo: Koudansha.

Murakami, Michio, Mayumi Hirosaki, Yuriko Suzuki, Masaharu Maeda, Hirooki Yabe, Seiji Yasumura, and Tetsuya Ohira. 2018. "Reduction of Radiation-Related Anxiety Promoted Wellbeing after the 2011 Disaster: 'Fukushima Health Management Survey.'" *Journal of Radiological Protection* 38, no. 4: 1428–40. https://doi.org/10.1088/1361-6498/aae65d.

Murakami, Michio, Atsushi Kumagai, Aleksandr N. Stojarov, and Masaharu Tsubokura. 2019. "Radiation Is Not a Political Tool." *Science* 366, no. 6465: 581–82. https://doi.org/10.1126/science.aaz3408.

Murakami, Michio, Yoshitake Takebayashi, Kyoko Ono, Aya Kubota, and Masaharu Tsubokura. 2020. "The Decision to Return Home and Wellbeing after the Fukushima Disaster." *International Journal of Disaster Risk Reduction* 47 (August): 101538. https://doi.org/10.1016/j.ijdrr.2020.101538.

Murillo, Luis Felipe R., and Sean Bonner. 2021. "Building a Community-Based Platform for Radiation Monitoring after 3.11." In *Legacies of Fukushima: 3.11 in Context,* edited by Kyle Cleveland, Scott Gabriel Knowles, and Ryuma Shineha, 183–96. Philadelphia: University of Pennsylvania Press.

Murphy, Michelle. 2017. "Alterlife and Decolonial Chemical Relations." *Cultural Anthropology* 32, no. 4: 494–503. https://doi.org/10.14506/ca32.4.02.

Muto, Ruiko. 2021. *10 nego no Fukushima kara anata he.* Tokyo: Otsuki Shoten.

Nabeshima, Sou. 2020. *Darega inochi wo sukuunoka: Genpatsu jiko to tatakatta ishitachi no kiroku.* Tokyo: Ronsousha.

Nakagawa, Yasuo. 2011. *Houshasen hibaku no rekishi: America genbaku kaihatsu kara Fukushima jiko made.* Tokyo: Akashi Shoten.

Nakajima, Hajime. 2013. *Genpatsu baishō chūkanshishin no kangaekata.* Tokyo: Shojihomu.

Nakajima, Hisato. 2014. *Sengoshi no nakano Fukushima genpatsu: Kaihatsu seisaku to chiiki shakai.* Tokyo: Otsuki Shoten.

Nakajima, Tokunosuke, and Masao Kihara. 1979. *Genshiryoku sangyoukai.* Tokyo: Kyoikusha.

Nakanishi, Tomoko. 2021. *Fukushima dojyō osen no 10 nen: Houshasei cesium ha dokohe ittanoka.* Tokyo: NHK Shuppan.

Nakasuji, Jyun. 2016. *Kasabuta: Fukushima the Silent Views.* Tokyo: Toua Shuppan.

———. 2021. *Konsento no mukougawa.* Tokyo: Shougakkan.

Nanasawa, Kiyoshi. (2005) 2011. *Tōkai mura rinkai jiko heno michi.* Tokyo: Iwanami Shoten.

Natori, Haruhiko. 2013. *Houshasen ha naze wakarinikuinoka: Houshasen no kenkou heno eikyou wakatteirukoto wakaranaikoto.* Tokyo: Apple Shuppansha.

Nemoto, Keisuke. 2017. *Genpatsu jiko to Fukushima no nougyō.* Tokyo: Tokyo Daigaku Shuppankai.

NHK Meltdown Syuzaihan. 2021. *Fukushima dai-ichi genpatsu jiko no "shinjitsu."* Tokyo: Koudansha.

Nihei, Norihiro. 2012. "'Saikan' no shikou: Kurikaesu hizukeno tameni." In *Henkyo kara hajimaru: Tokyo/Tohoku hen,* 122–58. Tokyo: Akashi Shoten.

Niitsu, Takeshi. 2011. *Inoshishi no bunkashi kouko hen: Hakkutsu shiryo nadokaramita inoshishi no sugata.* Tokyo: Yūzankaku.

Nixon, Rob. 2011. *Slow Violence and Environmentalism of the Poor.* Cambridge, MA: Harvard University Press.

Nöggerath, Johannis, Robert J. Geller, and Viacheslav K. Gusiakov. 2011. "Fukushima: The Myth of Safety, the Reality of Geoscience." *Bulletin of the Atomic Scientists* 67, no. 5: 37–46. https://doi.org/10.1177/0096340211421607.

Noma, Michiko. 2018. "Fukushima dai-ichi genpatsu wo shisatsushite." *Journal of Japan Atomic Agency* 60, no. 12: 57–59.

Nomura, Shuhei, Marta Blangiardo, Masaharu Tsubokura, Akihiko Ozaki, Tomohiro Morita, and Susan Hodgson. 2016. "Postnuclear Disaster Evacuation and Chronic Health in Adults in Fukushima, Japan: A Long-Term Retrospective Analysis." *BMJ Open* 6, no. 2: e010080. https://doi.org/10.1136/bmjopen-2015-010080.

Nomura, Shuhei, Michio Murakami, Wataru Naito, Tetsuo Yasutaka, Toyoaki Sawano, and Masaharu Tsubokura. 2020. "Low Dose of External Exposure among Returnees to Former Evacuation Areas: A Cross-Sectional All-Municipality Joint Study Following the 2011 Fukushima Daiichi Nuclear Power Plant Incident." *Journal of Radiological Protection* 40, no. 1: 1–18. https://doi.org/10.1088/1361-6498/ab49ba.

Nomura, Shuhei, Tomoyoshi Oikawa, and Masaharu Tsubokura. 2019. "Low Dose from External Radiation among Returning Residents to the Former Evacuation

Zone in Minamisoma City, Fukushima Prefecture." *Journal of Radiological Protection* 39, no. 2: 548–63. https://doi.org/10.1088/1361-6498/ab0f87.

Norimatsu, Satoko Oka, and Chris Busby. 2011. "Fukushima Is Worse than Chernobyl–on Global Communication." *Asia-Pacific Journal: Japan Focus* 9, issue 29, no. 1. https://apjjf.org/2011/9/29/Satoko-Norimatsu/3563/article.html.

Normile, Denise. 2021. "This Physician Has Studied the Fukushima Disaster for a Decade—and Found a Surprising Health Threat." *Science*. https://web.archive.org/web/20220518165614/https://www.science.org/content/article/physician-has-studied-fukushima-disaster-decade-and-found-surprising-health-threat.

Novak, David. 2017. "Project Fukushima! Performativity and the Politics of Festival in Post-3/11 Japan." *Anthropological Quarterly* 90, no. 1: 225–53. https://doi.org/10.1353/anq.2017.0008.

Nozawa, Shunsuke. 2015. "Phatic Traces: Sociality in Contemporary Japan." *Anthropological Quarterly* 88, no. 2: 373–400. https://doi.10.1353/anq.2015.0014.

Nuclear Energy Agency, ed. 2012. *Japan's Compensation System for Nuclear Damage: As Related to the TEPCO Fukushima Daiichi Nuclear Accident.* Issy-les-Moulineaux: OECD, Nuclear Energy Agency.

Numazaki, Ichiro. 2012. "Too Wide, Too Big, Too Complicated to Comprehend: A Personal Reflection on the Disaster That Started on March 11, 2011." *Asian Anthropology* 11, no. 1: 27–38.

Ochi, Sae. 2021. "'Life Communication' after the 2011 Fukushima Nuclear Disaster: What Experts Need to Learn from Residential Non-Scientific Rationality." Supplement, *Journal of Radiation Research* 62: 188–94. https://doi.org/10.1093/jrr/rraa135.

Ochi, Sae, Shigeaki Kato, Masaharu Tsubokura, Claire Leppold, Masahiro Kami, and Kenji Shibuya. 2016. "Voice from Fukushima: Responsibility of Epidemiologists to Avoid Irrational Stigmatization of Children in Fukushima." *Thyroid* 26, no. 9: 1332–33. https://doi.org/10.1089/thy.2016.0120.

Oguma, Eiji. 2013. "Nobody Dies in a Ghost Town: Path Dependence in Japan's 3.11 Disaster and Reconstruction." Translated by Shin-Lin Loh. *Asia-Pacific Journal: Japan Focus* 11, issue 44, no. 1. https://apjjf.org/2013/11/44/Oguma-Eiji/4024/article.html.

Ohashi, Haruka, Masae Saito, Reiko Horie, Hiroshi Tsunoda, Hiromu Noba, Haruka Ishii, Takashi Kuwabara, et al. 2013. "Differences in the Activity Pattern of the Wild Boar *Sus scrofa* Related to Human Disturbance." *European Journal of Wildlife Research* 59, no. 2: 167–77. https://doi.org/10.1007/s10344-012-0661-z.

Ohtsuru, Akira, and Sanae Midorikawa. 2021. "Lessons Learned from Conducting Disease Monitoring in Low-Dose Exposure Conditions as a Counter-Measure after a Nuclear Disaster." Supplement, *Journal of Radiation Research* 62: 164–70. https://doi.org/10.1093/jrr/rraa105.

Ōkubo, Takashi. 2019. "The Status Quo of Those Who Have Lost Their Home: Restoration Progress of Iitate Village." [Japanese]. *Annals of Sociological Research* 26: 19–28.

Ōkuma Town History Editorial Association. 1985. *Ōkuma-chou shi dai ikkan tuushi hen*. Fukushima: Ōkuma Town.

Okuno, Shuji. 2017. *Tamashiidemoiikara sobaniite: 3.11 gono reitaiken wo kiku*. Tokyo: Shinchosha.

Oliver-Smith, Anthony, and Susanna M. Hoffman. 2002. "Why Anthropologists Should Study Disasters." In *Catastrophe & Culture: The Anthropology of Disaster*, edited by Susanna M. Hoffman and Anthony Oliver-Smith, 3–22. Santa Fe, NM: School of American Research Press.

Onaga, Lisa. 2018. "Measuring the Particular: The Meanings of Low-Dose Radiation Experiments in Post–1954 Japan." *Positions* 26, no. 2: 265–304.

Onaga, Lisa, and Harry Yi-Jui Wu. 2018. "Articulating Genba: Particularities of Exposure and Its Study in Asia." *Positions* 26, no. 2: 197–212.

Onda, Katsunobu. 2011. *Genpatsuni shison no inochiha urenai: Genpatsuga dekinakatta Fukushima Namie-machi*. New edition. Tokyo: Nanatsumori Sokan.

Onda, Yuichi, Keisuke Taniguchi, Kazuya Yoshimura, Hiroaki Kato, Junko Takahashi, Yoshifumi Wakiyama, Frederic Coppin, and Hugh Smith. 2020. "Radionuclides from the Fukushima Daiichi Nuclear Power Plant in Terrestrial Systems." *Nature Reviews Earth & Environment* 1, no. 12: 644–60. https://doi.org/10.1038/s43017-020-0099-x.

O'Neill, Daniel. 2019. "Rewilding Futures: Japan's Nuclear Exclusion Zone and Post 3.11 Eco-Cinema." *Journal of Japanese and Korean Cinema* 11, no. 1: 85–100. https://doi.org/10.1080/17564905.2019.1600697.

Onitsuka, Hiroshi. 2011. "Hooked on Nuclear Power: Japanese State-Local Relations and the Vicious Cycle of Nuclear Dependence." *Asia-Pacific Journal: Japan Focus* 10, issue 3, no. 1. https://apjjf.org/2012/10/3/Hiroshi-Onitsuka/3677/article.html.

Onuma, Yyunichi, and Yoshihara Naoki. 2018. *Fukko? Kizuna? Fukushima no Ima*. Tokyo: Kaihou Shuppan.

Orui, Masatsugu, Chihiro Nakayama, Nobuaki Moriyama, Masaharu Tsubokura, Kiyotaka Watanabe, Takeo Nakayama, Minoru Sugita, and Seiji Yasumura. 2020. "Current Psychological Distress, Post-Traumatic Stress, and Radiation Health Anxiety Remain High for Those Who Have Rebuilt Permanent Homes Following the Fukushima Nuclear Disaster." *International Journal of Environmental Research and Public Health* 17, no. 24: 9532. https://doi.org/10.3390/ijerph17249532.

Ortiz, Alfonzo. 1977. "Some Concerns Central to the Writing of 'Indian' History." *The Indian Historian* 10, no. 1: 17-22.

Osaka, Eri. 2019. "Current Status and Challenges in the Fukushima Nuclear Disaster Compensation Scheme: An Example of Institutional Failure?" *SSRN Electronic Journal*. https://doi.org/10.2139/ssrn.3318877.

———. 2020. "Fukushima genpatsu jiko baishō no keika to ronten: Genpatsu ADR oyobi syudansosho wo chushin ni." *Fukkō* 8, no. 5: 27–32.

Ōseko, Noriyuki. 2013. *Sōma Futaba no genfūkei: Fukushima hamadōri no rekishi to minzoku*. Tokyo: Iwata Shoin.

Osseo-Asare, Abena Dove. 2019. *Atomic Junction: Nuclear Power in Africa after Independence.* Cambridge: Cambridge University Press.

Ota, Keisuke. 2011. *Minamisōma toukakan no kyumeiiryo: Tsunami, genpatsu saigai to tatakatta ishi no kiroku.* Tokyo: Jiji Tsushin Shuppan Kyoku.

Owada, Arata. 2016. *Owada note.* Fukushima: Fukushima Minpō.

Owada, Takeshi, and Takuya Kitazawa. 2013. *Genpatsu hinanmin dokoku no note.* Tokyo: Akashi Shoten.

Owatari, Misaki. 2016. *Soredemo Iitate-mura ha sokoniaru: Mura kisha ga mitsumeta kokyo no gonen.* Tokyo. Sankei Shimbun.

Ozaki, Takashi. 2013. *Yuna wo sagashite: Genpatsu no machi Ōkuma no 3.11.* Kyoto: Kamogawa Shuppan.

Ozawa, Shoji. 2012. *Iitate mura: Rokusennin ga utsukushii mura wo owareta.* Tokyo: Nanatsumori Shokan.

Ozawa-De Silva, Chikako. 2021. *The Anatomy of Loneliness: Suicide, Social Connection, and the Search for Relational Meaning in Contemporary Japan.* Berkeley: University of California Press.

Parkes, C. Murray. 1967. "Psycho-social Transitions: A Field for Study." *Social Science & Medicine* 5, no. 2: 101–15.

Parmentier, Richard J. 1985. "Semiotic Mediation: Ancestral Genealogy and Final Interpretant." In *Semiotic Mediation: Sociocultural and Psychological Perspectives,* edited by Elizabeth Mertz and Richard J. Parmentier, 359–85. Orlando, FL: Academic Press.

———. 1994. *Signs in Society: Studies in Semiotic Anthropology.* Bloomington: Indiana University Press.

———. 2012. "The World Has Changed Forever: Methodological Reflections on the Experience of Sudden Change." *Semiotica* 192: 235–42.

Parr, Joy. 2010. *Sensing Changes: Technologies, Environments, and the Everyday, 1953–2003.* Vancouver: UBC Press.

Parry, Richard Lloyd. 2017. *Ghosts of the Tsunami: Death and Life in Japan's Disaster Zone.* New York: MCD/Farrar, Strauss and Giroux.

Pascale, Celine-Marie. 2017. "Vernacular Epistemologies of Risk: The Crisis in Fukushima." *Current Sociology* 65, no. 1: 3–20. https://doi.org/10.1177/0011392115627284.

Peirce, Charles S. 1960. *Collected Papers.* 8 vols. Edited by Charles Hartshorne and Paul Weiss. Cambridge, MA: Harvard University Press.

Perko, Tanja. 2016. "Risk Communication in the Case of the Fukushima Accident: Impact of Communication and Lessons to Be Learned: Lessons from Fukushima Risk Communication." *Integrated Environmental Assessment and Management* 12, no. 4: 683–86. https://doi.org/10.1002/ieam.1832.

Petryna, Adriana. 2013. *Life Exposed: Biological Citizens after Chernobyl.* Princeton, NJ: Princeton University Press.

Petryna, Adriana, and Blake Cole. 2011. "Lessons from Chernobyl: Adriana Petryna Says the Decades-Old Disaster May Prove Invaluable in Responding to Japan's Fukushima Nuclear Power Plant Accident." *Omnia.* Penn Arts & Sciences.

https://web.archive.org/web/20210301014705/https://omnia.sas.upenn.edu
/story/lessons-chernobyl.

Phillips, Sarah. 2013. "Fukushima Is Not Chenorbyl? Don't Be So Sure." *Somatosphere*,
March 11, 2013. https://web.archive.org/web/20220117192152/http://somatosphere
.net/2013/fukushima-is-not-chernobyl-dont-be-so-sure.html/.

Plokhy, Serhii. 2022. *Atoms and Ashes: A Global History of Nuclear Disasters*. New
York: W. W. Norton.

Polleri, Maxime. 2019. "Conflictual Collaboration: Citizen Science and the Governance of Radioactive Contamination after the Fukushima Nuclear Disaster."
*American Ethnologist* 46, no. 2: 214–26. https://doi.org/10.1111/amet.12763.

———. 2021. "Radioactive Performances: Teaching about Radiation after the Fukushima Nuclear Disaster." *Anthropological Quarterly* 94, no. 1: 93–123. https://doi
.org/10.1353/anq.2021.0015.

———. 2022. "Ethnographies of Nuclear Life: From Victimhood to Post-Victimization." *Platypus: The CASTAC Blog.* https://web.archive.org/web
/20220622053309/https://blog.castac.org/2022/03/ethnographies-of-nuclear-
life-from-victimhood-to-post-victimization/.

Pritchard, Sara B. 2012. "An Envirotechnical Disaster: Nature, Technology, and
Politics at Fukushima." *Environmental History* 17, no. 2: 219–43. https://doi
.org/10.1093/envhis/ems021.

Rabinow, Paul, George E. Marcus, James D. Faubion, and Tobias Rees. 2008.
*Designs for an Anthropology of the Contemporary*. Durham, NC: Duke University
Press.

Ralph, John C., and Bruce D. Maxwell. 1984. "Relative Effects of Human and Feral
Hog Disturbance on a Wet Forest in Hawaii." *Biological Conservation* 30: 291–303.

Ralph, Laurence. 2020. *The Torture Letters: Reckoning with Police Violence*. Chicago: University of Chicago Press.

Robbins, Joel. 2013. "Beyond the Suffering Subject: Toward an Anthropology of the
Good." *Journal of the Royal Anthropological Institute* 19, no. 3: 447–62. https://
doi.org/10.1111/1467-9655.12044.

Roberts, Elizabeth F. S. 2017. "What Gets Inside: Violent Entanglements and Toxic
Boundaries in Mexico City." *Cultural Anthropology* 32, no. 4: 592–619. https://
doi.org/10.14506/ca32.4.07.

Robertson, Jennifer. 1988. "Furusato Japan: The Culture and Politics of Nostalgia."
*International Journal of Politics, Culture, and Society* 1, no. 4: 494–518.

———. 2012. "From Uniqlo to NGOs: The Problematic 'Culture of Giving' in
Inter-Disaster Japan." *Asia-Pacific Journal* 10, issue 18, no. 2. https://apjjf.org
/2012/10/18/Jennifer-Robertson/3747/article.html.

Rowe, Mark Michael. 2011. *Bonds of the Dead: Temples, Burial, and the Transformation of Contemporary Japanese Buddhism*. Chicago: University of Chicago Press.

Sahlins, Marshall. 1981. *Historical Metaphors and Mythical Realities: Structure in the
Early History of the Sandwich Island Kingdom*. Ann Arbor: University of Michigan Press.

Saito, Hiro. 2021. "The Sacred and Profane of Japan's Nuclear Safety Myth: On the Cultural Logic of Framing and Overflowing." *Cultural Sociology* 15, no. 4: 486–508. https://doi.org/10.1177/17499755211001046.

Saito, Kimiaki, Satoshi Mikami, Masaki Andoh, Norihiro Matsuda, Sakae Kinase, Shuichi Tsuda, Tadayoshi Yoshida, et al. 2019. "Summary of Temporal Changes in Air Dose Rates and Radionuclide Deposition Densities in the 80 Km Zone over Five Years after the Fukushima Nuclear Power Plant Accident." *Journal of Environmental Radioactivity* 210 (December): 105878. https://doi.org/10.1016/j.jenvrad.2018.12.020.

Saito, Mitsuhiro. 2002. *Genshiryoku jiko to Tōkai mura no hitobito: Genshiryoku shisetsu to machizukuri.* Ibaraki: Nakashobō.

Saito, Rie, Hitoshi Ohmachi, Yui Nemoto, and Masahiro Osako. 2019. "Estimation of the Total Amount of the Radiocaesium in the Wild Boar in Their Body—Each Organs Survey and Incineration Residue Survey." *Journal of the Society for Remediation of Radioactive Contamination of the Environment* 7, no. 3: 165–73.

Saito, Toshi. 2018. *Hiroshima no hibaku to Fukushima no hibaku.* Kyoto: Kamogawa Shuppan.

Sakakibara, Takahiro. 2021. *Fukushima ga chinmoku shita hi: Genpatsu jiko to koujyōsen hibaku.* Tokyo: Shūeisha.

Sakurai, Katsunobu, and Hiroshi Kainuma. 2012. *Tatakau shichō: Hisaichikara mieta kono kunino shinjitsu.* Tokyo: Tokuma Shoten.

Samuels, Richard J. 2013. *3.11: Disaster and Change in Japan.* Ithaca, NY: Cornell University Press.

Sanpei, Houji, and Eiji Futakami. 2018. *Sanpei chougi funtousu: Namie-machi Tsushima no kiroku.* Fukushima: Dourinsha.

Sato, Keiichi, Fuminori Tanba, Naoya Sekiya, and Masafumi Yokemoto. 2020. "Fukushima genshiryoku hatsudensho jiko go no futabagun jyumin no seishinteki kenkoujyōtai no shakaiteki youin: 2017 nen dai 2 kai futabagun jyumin jittai chōsa no bunseki." *Chikianzengakkai ronbunshu* 37: 97–107.

Sato, Masao. 2020. *Fukushima genpatsu jiko 10 nengo no yukueto aratana kadai: Fukkō wo ikinuku Fukushima.* Tokyo: Godō Forest.

Sato, Yasuo. 2013. *Houshanō kakusan yosoku system SPEEDI: Naze ikasarenakattaka.* Tokyo: Tokyo Shoten.

Satsuka, Shiho. 2014. "The Satoyama Movement: Envisioning Multispecies Commons in Postindustrial Japan." *RCC Perspective, Asian Environments: Connections across Borders, Landscapes, and Times* 3: 87–94.

Sawano, Toyoaki, Toshiyuki Kambe, Yuki Seno, Ran Konoe, Yoshitaka Nishikawa, Akihiko Ozaki, Yuki Shimada, Tomohiro Morita, Hiroaki Saito, and Masaharu Tsubokura. 2019. "High Internal Radiation Exposure Associated with Low Socio-Economic Status Six Years after the Fukushima Nuclear Disaster: A Case Report." *Medicine* 98, no. 47: e17989. https://doi.org/10.1097/MD.0000000000017989.

Sawano, Toyoaki, Michio Murakami, Akihiko Ozaki, Yoshitaka Nishikawa, Aoi Fukuda, Tomoyoshi Oikawa, and Masaharu Tsubokura. 2021. "Prevalence of Non-Communicable Diseases among Healthy Male Decontamination Workers after the Fukushima Nuclear Disaster in Japan: An Observational Study." *Scientific Reports* 11, no. 1: 21980. https://doi.org/10.1038/s41598-021-01244-z.

Sawano, Toyoaki, Yoshitaka Nishikawa, Akihiko Ozaki, Claire Leppold, and Masaharu Tsubokura. 2018. "The Fukushima Daiichi Nuclear Power Plant Accident and School Bullying of Affected Children and Adolescents: The Need for Continuous Radiation Education." *Journal of Radiation Research* 59, no. 3: 381–84. https://doi.org/10.1093/jrr/rry025.

Sawano, Toyoaki, Akihiko Ozaki, and Masaharu Tsubokura. 2020. "Review of Health Risks among Decontamination Workers after the Fukushima Daiichi Nuclear Power Plant Accident." *Radioprotection* 55, no. 4: 277–82. https://doi.org/10.1051/radiopro/2020080.

Sawano, Toyoaki, Masaharu Tsubokura, Akihiko Ozaki, Claire Leppold, Shuhei Nomura, Yuki Shimada, Sae Ochi, et al. 2016. "Non-Communicable Diseases in Decontamination Workers in Areas Affected by the Fukushima Nuclear Disaster: A Retrospective Observational Study." *BMJ Open* 6, no. 12: e013885. https://doi.org/10.1136/bmjopen-2016-013885.

Schneider, Thierry, Mélanie Maître, Jacques Lochard, Sylvie Charron, Jean-François Lecomte, Ryoko Ando, Yumiko Kanai, et al. 2019. "The Role of Radiological Protection Experts in Stakeholder Involvement in the Recovery Phase of Post-Nuclear Accident Situations: Some Lessons from the Fukushima-Dai-ichi NPP Accident." *Radioprotection* 54, no. 4: 259–70. https://doi.org/10.1051/radiopro/2019038.

Schwab, Gabriele. 2020. *Radioactive Ghosts*. Minneapolis: University of Minnesota Press.

Sebeok, Thomas Albert. 1981. *The Play of Musement*. Bloomington: Indiana University Press.

Seki, Reiko. 2015. "Kyōsei sareta hinan, kyōyō sareru kikan: 'Kouzousai' karano ridatsu to seikatsu no fukkō." In *"Ikiru" jikan no paradigm: Hisai genba kara egaku genpatsu jiko no sekai,* edited by Reiko Seki, 120–40. Tokyo: Nippon Hyoronsha.

Sekiya, Naoya. 2003. "Social Psychology of 'Image Contamination'—The Fact of 'Image Contamination' and Its Mechanism." [Japanese]. *Saigai jyouhou* 1: 78–89.

———. 2011. *Fūhyō higai: Sono mechanism wo kangaeru.* Tokyo: Koubunsha.

———. 2016. "Research Survey of Consumer Psychology about Radioactive Contamination after the Accident at TEPCO's Fukushima Daiichi Nuclear Power Stations: Factor Analyses for Agriculture Revitalization of Fukushima and for Countermeasure against Economic Damage by Harmful Rumor." [Japanese]. *Journal of Regional Safety* 29: 143–53.

Sekiya, Yuichi, and Hiroki Takakura. 2019. *Shinsai fukkō no koukyo jinruigaku: Fukushima genpatsu jiko hisaisha tono kyodou.* Tokyo: Tokyodaigaku Shuppankai.

Shibata, Tetsuo. 2018. *Fukushima Teikoushatachino kingendaishi: Hirata Ryobei, Iwamoto Tadashi, Hangai Seiji, Suzuki Yasuzo.* Tokyo: Sairyusha.

Shibui, Tetsuya, Kazumi Murakami, Makoto Watabe, and Nobuyuki Ota. 2014. *Shinsai ikou: Higashinihon daishinsai report.* Tokyo: San-ichi Shobo.

Shimazaki, Hiroko, Hideki Miyaguchi, Chinami Ishizuki, Syuichi Onoda, Keitaro Harasawa, Tomoyosi Oikawa, and Yukio Kanazawa. 2015. "The Characteristic of Life of the Residents in Temporary Housing after Disaster of Fukushima Dai-Ichi Nuclear Power Station in Minamisoma, Fukushima." [Japanese]. *Journal of Japan Primary Care* 38, no. 1: 9–17.

Shimazono, Susumu. 2019. *Genpatsu to houshasen hibaku no kagaku to rinri.* Tokyo: Senshu University Press.

Shimazono, Susumu, Hiroko Goto, and Atsushi Sugita. 2016. *Kagaku fushin no jidai wo tou: Fukushima genpatsu saigai gono kagaku to shakai.* Tokyo: Gōdō Shuppan.

Shimura, Tsutomu, Ichiro Yamaguchi, Hiroshi Terada, Erick Robert Svendsen, and Naoki Kunugita. 2015. "Public Health Activities for Mitigation of Radiation Exposures and Risk Communication Challenges after the Fukushima Nuclear Accident." *Journal of Radiation Research* 56, no. 3: 422–29. https://doi.org/10.1093/jrr/rrv013.

Shineha, Ryuma, ed. 2021. *Saika wo meguru "kioku" to "katari."* Kyoto: Nakanishiya Shuppan.

Shinnami, Kyosuke. 2015. *Ushi to tsuchi: Fukushima, 3.11 sonogo.* Tokyo: Shūeisha.

Silko, Leslie Marmon. 1977. *Ceremony.* New York: Viking.

Simmel, Georg. 1950. "The Stranger." In *The Sociology of Georg Simmel.* Translated by Kurt H. Wolff, 402–8. New York: Free Press.

Slater, David H. 2013. "Urgent Ethnography." In *Japan Copes with Calamity: Ethnographies of the Earthquake, Tsunami and Nuclear Disasters of March 2011,* edited by Tom Gill, Brigitte Steger, and David H. Slater, 25–50. Bern: Peter Lang.

Slater, David H., and Rika Morioka with Haruka Danzuka. 2014. "Micro-Politics of Radiation: Young Mothers Looking for a Voice in Post-3.11 Fukushima." *Critical Asian Studies* 46, no. 3: 485–508.

Sōma History. 1969. *Ousoushi.* Fukushima: Sōma City.

———. 1983. *Sōma shi shi vol. 3.* Fukushima: Sōma City.

Son, Wonchel, Mitsuru Matsutani, Yusuke Sakaguchi, and Kayo Ushijima. 2015. *Owaranai hisai no jikan: Genpatsu jiko ga Fukushima-ken nakadorino oyako ni ataeru eikyo(sutoresu).* Fukuoka: Sekifuusha.

Star, Susan Leigh. 1999. "The Ethnography of Infrastructure." *American Behavioral Scientist* 43, no. 3: 377–91.

Stawkowski, Magdalena E. 2016. "'I Am a Radioactive Mutant': Emergent Biological Subjectivities at Kazakhstan's Semipalatinsk Nuclear Test Site." *America Ethnologist* 43, no. 1: 144–57. https://doi.org/10.1111/amet.12269.

Steinhauser, Georg, Alexander Brandl, and Thomas E. Johnson. 2014. "Comparison of the Chernobyl and Fukushima Nuclear Accidents: A Review of the Environmental Impacts." *Science of the Total Environment* 470–71: 800–17. https://doi.org/10.1016/j.scitotenv.2013.10.029.

Stengers, Isabelle. 2018. *Another Science is Possible: A Manifesto for Slow Science.* Stephen Muecke (trans.) Cambridge: Polity.

Sternsdorff-Cisterna, Nicolas. 2015. "Food after Fukushima: Risk and Scientific Citizenship in Japan." *American Anthropologist* 117, no. 3: 455–67.

———. 2020. *Food Safety after Fukushima: Scientific Citizenship and the Politics of Risk.* Honolulu: University of Hawai'i Press.

Stoler, Ann Laura. 2008. *Imperial Debris: Reflections on Ruins and Ruination.* Durham, NC: Duke University Press.

Stone, Philip R. 2012. "Dark Tourism and Significant Other Death: Towards a Model of Mortality Mediation." *Annals of Tourism Journal* 3, no. 3: 1565–87.

Sugano, Hisao. 2022. *Fukushima ni Ikiru: Atamaage kussezu jyunen higaisha no shōgen.* Tokyo: Shin Nippon Press.

Sutou, Shizuyo. 2016. "A Message to Fukushima: Nothing to Fear but Fear Itself." *Genes and Environment* 38: 12. https://doi.org/10.1186/s41021-016-0039-7.

Suzuki, Gen. 2021. "Communicating with Residents about 10 Years of Scientific Progress in Understanding Thyroid Cancer Risk in Children after the Fukushima Dai-Ichi Nuclear Power Station Accident." Supplement, *Journal of Radiation Research* 62: i7–14. https://doi.org/10.1093/jrr/rraa097.

Suzuki, Hiroshi. 2021. *Fukushima genpatsu saigai 10 nen wo hete.* Shinjyuku: Jichitai Kenkyu Sha.

Tahara, Shigesuke. 1979. "Tokyo denryoku Fukushima dai-ichi genshiryoku hatsudensho service hall." *Journal of the Atomic Energy Society of Japan* 21, no. 4: 48–52.

Taira, Yasuyuki, Yudai Inadomi, Shota Hirajou, Yasuhiro Fukumoto, Makiko Orita, Yumiko Yamada, and Noboru Takamura. 2019. "Eight Years Post-Fukushima: Is Forest Decontamination Still Necessary?" *Journal of Radiation Research* 60, no. 5: 714–16. https://doi.org/10.1093/jrr/rrz047.

Takagi, Ryusuke, Akihiko Sato, and Toshiyuki Kanai. 2021. *Genpatsu jiko hisaijichitai no saisei to kunō: Tomioka-machi jyunen no kiroku.* Tokyo: Daiichi Houki.

Takahashi, Hara, and Norichika Horie. 2021. *Shisha no chikara: Tsunami hisaichi "reiteki taiken" no Shisei-gaku.* Tokyo: Iwanami Shoten.

Takahashi, Satsuki. 2011. "Fourfold Disaster: Renovation and Restoration in Post-Tsunami Coastal Japan." *Anthropology News* 52, no. 7: 5, 11.

———. 2018. "The Future of 'Fukushima Future.'" *Japanese Journal of Cultural Anthropology* 83, no. 3: 441–58. https://doi.org/10.14890/jjcanth.83.3_441.

Takase, Jin, Kazushi Yoshida, and Wataru Kumagai. 2012. *Jinjya ha keikokusuru: Kodai kara tsutawaru tsunami no message.* Tokyo: Koudansha.

Takebayashi, Shun. 2001. *Ao no gunzou: Genshiryoku sousou no koro.* Tokyo: Denki Shimbun.

Takebayashi, Yoshitake, Hiroshi Hoshino, Yasuto Kunii, Shin-Ichi Niwa, and Masaharu Maeda. 2020. "Characteristics of Disaster-Related Suicide in Fukushima Prefecture after the Nuclear Accident." *Crisis: The Journal of Crisis Intervention and Suicide Prevention* 41, no. 6: 475–82. https://doi.org/10.1027/0227-5910/a000679.

Takezawa, Shoichiro. 2022. *Genpatsu hinanmin ha dou ikitekitaka: Hishousei no jinruigaku.* Tokyo: Toushindo.

TallBear, Kim. 2019. "Caretaking Relations, Not American Dreaming." *Kalfou: A Journal of Comparative and Relational Ethnic Studies* 6. no. 1: 24–41. https:// doi.org/10.15367/kf.v6i1.228.

Tatewaki, Isao, Noriyuki Kumazawa, and Eri Ariga. 2009. *Genshiryoku to chiiki shakai: Tōkai mura JOC rinkaijiko karano saisei 10 nenme no shougen.* Tokyo: Bunshindō.

Tatsuta, Kazuto. 2017. *Ichi-F: A Worker's Graphic Memoir of the Fukushima Nuclear Power Plant.* Translated by Stephen Paul. New York: Kodansha Comics.

Terada, Hiroaki, Haruyasu Nagai, Katsunori Tsuduki, Akiko Furuno, Masanao Kadowaki, and Toyokazu Kakefuda. 2020. "Refinement of Source Term and Atmospheric Dispersion Simulations of Radionuclides during the Fukushima Daiichi Nuclear Power Station Accident." *Journal of Environmental Radioactivity* 213 (March): 106104. https://doi.org/10.1016/j.jenvrad.2019.106104.

Terashima, Hideya. 2021. *Hisaichi no journalism: Higashinihon daishinsai 10 nen "yorisou" no imi wo motomete.* Tokyo: Akashi Shoten.

Thanks & Dream. 2017. *3.11 Hinansha no koe—Toujisha archive.* Osaka: Thanks & Dream.

Tierney, Kathleen. 2008. "Hurricane Katrina: Catastrophic Impacts and Alarming Lessons." In *Risking House and Home: Disasters, Cities, Public Policy,* edited by John M. Quigley and Larry A. Rosenthal, 119–38. Berkeley, CA: Berkeley Public Policy Press.

Toki, H., T. Wada, Y. Manabe, S. Hirota, T. Higuchi, I. Tanihata, K. Satoh, and M. Bando. 2020. "Relationship between Environmental Radiation and Radioactivity and Childhood Thyroid Cancer Found in Fukushima Health Management Survey." *Scientific Reports* 10, no. 1: 4074. https://doi.org/10.1038/s41598-020-60999-z.

Tokita, Tamaki. 2015. The Post-3/11 Quest for True Kizuna: Shi no Tsubute by Wagō Ryōichi and Kamisama 2011 by Kawakami Hiromi. *Asia-Pacific Journal: Japan Focus* 13, issue 7, no. 7. https://apjjf.org/2015/13/6/Tamaki-Tokita/4283 .html.

Tokyo Electric Company Fukushima Daiichi Nuclear Plant Operation. 2008. *Kyosei to Kyoshin: Chiiki to tomoni: Fukushima dai-ichi genshiryoku hatsudensho 45 nen no ayumi.* TEPCO.

Tokyo Electric Power Company (TEPCO). 1983. *Tokyo Denryoku 30 nen shi.* Tokyo: Tokyo Electric Power Company.

Tomioka Town and Fukushima University. 2017. *Kokyo wo omou mamoru tsunagu: Chiiki no daigaku to machi yakuba no kokoromi.* Fukushima: Nisshindo.

Torpey, John. 2001. "'Making Whole What Has Been Smashed': Reflections on Reparations." *Journal of Modern History* 73, no. 2: 333–58.

Toyama History. 1983. *Kindaishi,* vol. 4. Kanazawa: Toyama Prefecture.

Toyota, Masatoshi. 2008. *Genshiryoku hatsuden no rekishi to tenbou.* Tokyo: Tokyo Bool Press.

Tsing, Anna Lowenhaupt. 2015. *The Mushroom at the End of the World: On the Possibility of Life in Capitalist Ruins.* Princeton, NJ. Princeton University Press.

Tsubokura, Masaharu. 2016. "Exposure and Current Health Issues in Minamisoma." Supplement, *Annals of the ICRP* 45, no. 2: 129–34. https://doi.org/10.1177/0146645316666708.

———. 2018. "Review: Secondary Health Issues Associated with the Fukushima Daiichi Nuclear Accident, Based on the Experiences of Soma and Minamisoma Cities." *Journal of National Institute of Public Health* 67, no. 1: 71–83.

Tsubokura, Masaharu, Shigeaki Kato, Tomohiro Morita, Shuhei Nomura, Masahiro Kami, Kikugoro Sakaihara, Tatsuo Hanai, Tomoyoshi Oikawa, and Yukio Kanazawa. 2015. "Assessment of the Annual Additional Effective Doses amongst Minamisoma Children during the Second Year after the Fukushima Daiichi Nuclear Power Plant Disaster." *PLOS ONE* 10, no. 6: e0129114. https://doi.org/10.1371/journal.pone.0129114.

Tsubokura, Masaharu, Yuto Kitamura, and Megumi Yoshida. 2018. "Post-Fukushima Radiation Education for Japanese High School Students in Affected Areas and Its Positive Effects on Their Radiation Literacy." Supplement, *Journal of Radiation Research* 59: ii65–74. https://doi.org/10.1093/jrr/rry010.

Tsubokura, Masaharu, Michio Murakami, Shuhei Nomura, Tomohiro Morita, Yoshitaka Nishikawa, Claire Leppold, Shigeaki Kato, and Masahiro Kami. 2017. "Individual External Doses below the Lowest Reference Level of 1 MSv Per Year Five Years after the 2011 Fukushima Nuclear Accident among All Children in Soma City, Fukushima: A Retrospective Observational Study." *PLOS ONE* 12, no. 2: e0172305. https://doi.org/10.1371/journal.pone.0172305.

Tsubokura, Masaharu, Yosuke Onoue, Hiroyuki A. Torii, Saori Suda, Kohei Mori, Yoshitaka Nishikawa, Akihiko Ozaki, and Kazuko Uno. 2018. "Twitter Use in Scientific Communication Revealed by Visualization of Information Spreading by Influencers within Half a Year after the Fukushima Daiichi Nuclear Power Plant Accident." *PLOS ONE* 13, no. 9: e0203594. https://doi.org/10.1371/journal.pone.0203594.

Tsujiuchi, Takuya, and Tom Gill. 2022. *Fukushima genpatsu jiko hisaisha kunan to kibō no jinruigaku: Bundan to tairitsu wo norikoerutameni.* Tokyo: Akashi Shoten.

Tsujiuchi, Takuya, and Kazutaka Masuda. 2019. *Fukushima no iryo jinruigaku: Genpatsujiko shien no fieldwork.* Tokyo: Toumi Shobo.

Tuck, Eve. 2009. "Suspending Damage: A Letter to Communities." *Harvard Educational Review* 79, no. 3: 409–28. https://doi.org/10.17763/haer.79.3.n0016675661t3n15.

Turner, Victor Witter. 1988. *The Anthropology of Performance.* New York: PAJ Publications.

Ui, Jun. (1971) 2006. *Kougai genron gouhon new edition.* Tokyo: Aki Shobo.

United Nations Scientific Committee on the Effects of Atomic Radiation (UNSCEAR). 2021. *Sources, Effects and Risks of Ionizing Radiation, UNSCEAR 2020–21 Report: Report to the General Assembly, with Scientific Annexes B. Levels and Effects of Radiation Exposure Due to the Accident at the Fukushima Daiichi Nuclear Power Station: Implications of Information Published since the UNSCEAR*

*2013 Report.* UN. https://web.archive.org/web/20220521154755/https://www
.unscear.org/unscear/uploads/documents/unscear-reports/UNSCEAR_
2020_21_Report_Vol.II.pdf 43f39-en.

Utagawa, Keisuke. 2016. *Shinsaigo no fushigi na hanashi.* Tokyo: Asuka Shinsha.

Valaskivi, Katja, Anna Rantasila, Mikihito Tanaka, and Risto Kunelius. 2019.
*Traces of Fukushima: Global Events, Networked Media and Circulating Emotions.*
Singapore: Palgrave Macmillan.

Van Wyck, Peter C. 2005. *Signs of Danger: Waste, Trauma, and Nuclear Threats.*
Minneapolis: University of Minnesota Press.

Veteran Mothers' Society. 2014. *First Held in Minamisoma: Radiation and Health
Seminar by Dr. Tsubokura.* Fukushima: Veteran Mothers' Society.

Vetter, Kai. 2020. "The Nuclear Legacy Today of Fukushima." *Annual Review of
Nuclear and Particle Science* 70: 257–92. https://doi.org/10.1146/annurev-nucl-
101918-023715.

Virilio, Paul. 2010. *The University of Disaster.* Translated by Julie Rose. Cambridge:
Polity.

Voice of Fukushima. 2015. *Ima tsutaetai: The Voice of Fukushima.* Tokyo: Touho
Tsushin.

Von Hippel, Frank N. 2011. "The Radiological and Psychological Consequences of
the Fukushima Daiichi Accident." *Bulletin of the Atomic Scientists* 67, no. 5:
27–36. https://doi.org/10.1177/0096340211421588.

Wakamatsu, Jotaro. 2012. *Fukushima kakusai kimin: Machi ga meltdown shiteshi-
matta.* Tokyo: Coal Sack.

Wakamatsu, Jotaro, and Arthur Binard. 2016. *What Makes Us.* Tokyo: Seiryu.

Wakui, Masayuki. 2012. *Naze ippon no matsudakega ikinokottanoka: Kiseki to kibo
no matsu.* Tokyo: Soeisha.

Walker, Brett L. 2001. "Commercial Growth and Environmental Change in Early
Modern Japan: Hachinohe's Wild Boar Famine of 1749." *Journal of Asian Studies*
60, no. 2: 329–51. https://doi.org/10.2307/2659696.

———. 2005. *The Lost Wolves of Japan.* Seattle: University of Washington Press.

———. 2010. *Toxic Archipelago: A History of Industrial Disease in Japan.* Seattle:
University of Washington Press.

Walker, J. Samuel. 2000. *Permissible Dose: A History of Radiation Protection in the
Twentieth Century.* Berkeley: University of California Press.

Watanabe, Etsuji, Jyunko Endo, and Kousaku Yamada. 2016. *Houshasen hibaku no
souten: Fukushima genpatsu jiko no kenkōhigai ha nainoka.* Tokyo: Ryukufu
Suppan.

Watanabe, H. 2012. "3.11 Higashinihon daishinsai tokushū." *Kashima Rekishi Aikoukai Shi* 5, no. 27: 159–78.

Watanabe, Kazue. 2020. *Kikikaki Minamisōma.* Tokyo: Shin Nippon
Shuppansha.

Wenger, Dennis, James D. Skyes, Thomas D. Sebok, and Joan L. Neff. 1975. "It's a
Matter of Myths: An Empirical Examination of Individual Insight into Disaster
Response." *Mass Emergencies* 1: 33–46.

Wiemann, Anna. 2018. *Networks and Mobilization Processes: The Case of the Japanese Anti-Nuclear Movement after Fukushima.* Munich: Iudicium Verlag.

Wynne, Brian. 1996. "May the Sheep Safely Graze? A Reflexive View of the Expert-Lay Knowledge Divide." In *Risk, Environment and Modernity: Towards a New Ecology,* edited by Scott M. Lash, Bronislaw Szerszynski, and Brian Wynne, 44–83. London: SAGE.

Yagi, Ekou. 2021. "Suffering the Effects of Scientific Evidence." In *Legacies of Fukushima: 3.11 in Context,* edited by Kyle Cleveland, Scott Gabriel Knowles, and Ryuma Shineha, 169–80. Philadelphia: University of Pennsylvania Press.

Yamaguchi, Ichiro, Hiroshi Terada, Tsutomu Shimura, Toshihiko Yunokawa, and Akira Ushikawa. 2021. "Measures Taken to Ensure Radiation Safety of Food after the Accident at TEPCO's Fukushima Daiichi Nuclear Power Station—Summary of Measures Implemented over 10 Years." [Japanese]. *Journal of National Institute of Public Health* 70, no. 3: 273–87.

Yamaguchi, Tomomi, and Muto Ruiko. 2012. "Muto Ruiko and the Movement of Fukushima Residents to Pursue Criminal Charges against Tepco Executives and Government Officials." *Asia-Pacific Journal: Japan Focus* 10, issue 27, no. 2. https://apjjf.org/2012/10/27/Tomomi-Yamaguchi/3784/article.html.

Yamamoto, Kahoruko. 2017. "'Genpatsu hinan' wo meguru mondai no shosou to kadai." In *Genpatsu shinsai to hinan: Genshiryoku seisaku no tankan ha kanouka,* edited by Kouichi Hasegawa and Kahoruko Yamamoto, 60–92. Tokyo: Yuhikaku.

Yamamoto, Kana, Shuhei Nomura, Masaharu Tsubokura, Michio Murakami, Akihiko Ozaki, Claire Leppold, Toyoaki Sawano, et al. 2019. "Internal Exposure Risk Due to Radiocesium and the Consuming Behaviour of Local Foodstuffs among Pregnant Women in Minamisoma City near the Fukushima Nuclear Power Plant: A Retrospective Observational Study." *BMJ Open* 9, no. 7: e023654. https://doi.org/10.1136/bmjopen-2018-023654.

Yamaoka, Jyunichiro. 2012. *Houshanō wo seotte: Minamisōma shichō Sakurai Katsunobu to shimin no sentaku.* Tokyo: Asahi Shimbun Shūppan.

Yamashita, Yusuke, Takashi Ichimura, and Atsuhiko Sato. 2013. *Ningennaki fukko: Genpatsu hinan to kokumin no "furikai" wo megutte.* Tokyo: Akashi Shoten.

Yasui, Kiyotaka, Yuko Kimura, Kenji Kamiya, Rie Miyatani, Naohiro Tsuyama, Akira Sakai, Koji Yoshida, et al. 2017. "Academic Responses to Fukushima Disaster: Three New Radiation Disaster Curricula." Supplement, *Asia Pacific Journal of Public Health* 29, no. 2: 99S–109S. https://doi.org/10.1177/1010539516685400.

Yasui, Shojiro. 2018. "Lessons Learned from Radiation Protection for Emergency Response and Remediation/Decontamination Work Relating to the Fukushima Daiich Nuclear Power Plant Accident in 2011." *Journal of National Institute of Public Health* 67, no. 1: 84–92.

Yasumoto, Shinya, and Naoya Sekiya. 2018. "Why Uneasy Feelings to the Fukushima Prefecture Products Cannot Be Mitigated in Hokkaido and Kinki: Focusing on Agenda Setting by the Local Paper." [Japanese]. *Journal of Regional Safety* 33: 127–36.

Yazzie, K. Melanie, and Cutcha Risling Baldy. 2018. "Introduction: Indigenous Peoples and The Politics of Water." *Decolonization: Indigeneity, Education & Society* 7, no. 1: 1–18.

Yokemoto, Masafumi. 2013. *Genpatsu baisho wo tou: Aimaina sekinin, honrousareru hinansha.* Tokyo: Iwanami Shoten.

———. 2021. "Genshiryoku songai baishō to shudan soshō: 'Furusato no soushitsu' higai wo chushinni." *Gakujyutsu no doukō* 26, no. 3: 38–41.

Yokemoto, Masafumi, and Toshihiko Watanabe. 2015. *Genpatsu saigai ha naze fukintouna fukko wo motarasunoka: Fukushima jiko kara "ningen no fukko" chiiki saisei he.* Kyoto: Minerva Shobo.

Yoneyama, Lisa. 1999. Hiroshima Traces: Time, Space and the Dialectics of Memory. Berkeley: University of California Press.

Yoshida, Chia. 2018. *Sonogo no Fukushima: Genpatsu jiko wo ikiru hitobito.* Kyoto: Jinbunnshoin.

———. 2020. *Korui: Futaba-gun shouhoushitachi no 3.11.* Tokyo: Iwanami Shoten.

Yoshihara, Naoki. 2013. *"Genpatsusama no machi kara no dakkyaku": Ōkuma-machi kara kangaeru community no mirai.* Tokyo: Iwanami Shoten.

———. 2016. *Zetsubō to kibō: Fukushima hisaisha to community.* Tokyo: Sakuhinsha.

———. 2021. *Shinsai fukkō no chiiki shakaigaku: Ōkuma-machi no jyunen.* Tokyo: Hakusuisha.

Yoshikawa, Akihiro. 2021. *Imagining a Better Tomorrow: The Fukushima Daiichi Nuclear Power Station Disaster and Its Impact on Society.* Fukushima: AFW.

Yoshimura, Ryoichi. 2018. "Genpatsu jiko niokeru 'furusato soushitsu songai' no baishō." *Ritsumeikan Law* 2, no. 378: 223–48.

Yoshioka, Hitoshi. 2011. *Genshiryoku no shakaishi.* Tokyo: Asahi Shimbun.

Yotsumoto, Yukio, and Shunichi Takekawa. 2016. "The Social Structures of Victimization of Fukushima Residents Due to Radioactive Contamination from the 2011 Nuclear Disaster." In *Japan after 3/11: Global Perspectives on the Earthquake, Tsunami, and Fukushima Meltdown,* edited by Pradyumna P. Karan and Unryu Suganuma, 251–68. Lexington: University Press of Kentucky.

Yu, Miri. 2020. *Minamisoma Medley.* Tokyo: Daisanbunmeisha.

Yukawa, Hideki. (1943) 1976. *Meni mienai mono.* Kyoto: Kobun Sha.

Zhang, Hui, Zijun Mao, and Wei Zhang. 2015. "Design Charrette as Methodology for Post-Disaster Participatory Reconstruction: Observations from a Case Study in Fukushima, Japan." *Sustainability* 7: 6593–6609.

# INDEX

Note: The term "1F" can be found at the listing: "TEPCO ("1F" [*ichi efu*] Tokyo Electric Power Company Fukushima Dai-ichi Nuclear Power Plant)." The term "3.11" can be found as if alphabetized as "Three.11." Page numbers in *italics* denote illustrative material.

compensation: altar replacement, 190–92, *191*; Alternative Dispute Resolution (ADR) for settlement of items not specified in the guidelines, 72, 108; applying and maintaining, as significant burden, 72, 84, 211; authority of TEPCO to determine what constitutes harm, 72; average paid to individual residents, 265n12; categories and subcategories of, 68–69, 71–72, 265n11; civil actions to obtain additional, 108; claim filing, and supporting documents required for, 72; consultation events hosted by local governments, 82; death (*kanrenshi*), 263n22; denials by TEPCO, 72; development of 1F and payments to farmers and fishers, 161, 274–75nn30–32; each family member receiving, 104; evacuation zone designations determining amount, duration, and categories of, 68; free medical and dental care, 56; the Japanese state ruled not responsible for paying, 258n14; JOC criticality accident, 113, 268n15; loans by the state to TEPCO for, 72, 101, 265n13; the logic of correlation and failures of, 95–96; negotiation of longer periods for, 104; negotiation of more money for, 104; psychological damage category for, 68, 104, 204; as "reparation politics," 267n6; rumor-related damage (*fūhyō higai*), 91–93, 111, 113; status of "the exposed" and the right to, 65–66; total payments made by TEPCO, 72; for trees, 189. *See also* compensation exclusions and inequalities; temporary housing units (*kasetsu*); temporary leased housing (*kariage*); victimhood

Compensation Act, 71–72, 265n11

compensation exclusions and inequalities: anxiety not compensable, 94, 100; as harm, 96, 106; as harm, and international call to reconsider the structure of, 266n14; lifestyle and secondary health effects (the state and TEPCO failing to be accountable for), 94, 100, 137, 190, 192; reopening of evacuation zones ends compensation, 85, 101–2, 104–6, 134;

residents outside the evacuation zones, 48, 68, 69, 77–78, 232; residents who stayed instead of evacuating, 41; residents who voluntarily evacuated, 94–95. *See also* compensation inequalities, divisions in the community produced by

compensation inequalities, divisions in the community produced by: overview, 65–66, 70–71; arbitrariness of the zones, 41, 77–78, 94–95, 99–100, 232; and average annual income prior to 3.11, 68; the compensation economy and, 103–6; the compensation game and, 106–10, *108*; district-based identities exacerbated, 51; exposability of abandoned people (*kimin*), 95; friendships ending, 74, 77; job losses by residents whose address makes them ineligible for compensation, 69–70; judgments and critiques by those ineligible to receive compensation, 68, 69, 77, 81–82; reterritorialization of the evacuation zones and, 68, 77–78; surveillance of compensated evacuees by those ineligible, 82; "victims" defined by compensation policy vs. individual experiences of harm, 77. *See also* division and corrosion of community

contamination of Fukushima: accusations of being "pronuclear" of scientists who wish to debunk the stereotype of, 127, 132, 134, 137, 138, 142, 271n34; collective imagination as "land of contamination," 35, 152; confirmation bias and perception of, 142; as "contaminated community," 45–46, 52, 109; map showing levels of, 29, *30*; the media spectacle as fixing the conception of, 2, 3–4, 8, 35, 49, 93; as patchy and ever-changing, 43, 44; as selective cartography, 35, 46, 116, 261n8. *See also* discrimination; stigma of the TEPCO accident

convenience stores, 40, 42, 81

corrosive communities, 52

cosmic rays, 101, 119

COVID-19 pandemic: earthquake and tsunami during (2022), 236–37; and the

decontamination *(continued)*
recontamination, 180; and disorder, elimination of, 181; egalitarianism of, 179; *en* as ripped apart by, 171, 182; of forests, 185, 189; and forgetting, 154, 163–64; Futaba town, 156; geographic procession of, 171; geographic zones and state vs. municipal responsibility for (SDA vs. ICSA), 177, *178*; habitability improvements via, 181; initial radiation survey and inspection of property (agent meeting), 164, 165, 172, 174–77, *175*, 179; Ministry of the Environment as responsible for, 172, 276n7; as moving vs. removing contaminants, 185; paradox of, 165–66, 168–69; pollution remediation as language of, 173; process of, 173; psychological effects of, 180–82, 276n10; radiation background levels as focus of, 173–74; radiation level reductions overall, 179–80; radiation levels of property already lower than the standard, before decontamination, 174, 177, 180; radiation standard mandated to achieve, as changing, 85, 172, 174, 266n6, 276n6; renovating homes and delays in returning, 189; reopening of zone as signaled by, 169–70, 181–82; scientific uncertainty about how to achieve, 35, 85, 166, 173, 179–80; of Shinto shrine (Hiwashi), 198–99; as symbolic/ritual process, 170, 180–81; waiting to return until completion of, 169, 265n5; "white space" areas nearly impossible to achieve, 169; and willingness of evacuees to return, importance of, 169–70, 174, 181. *See also* decontaminated waste (nuclear waste); decontamination, and possessions of the owners; decontamination workers; demolitions; disaster recovery and reconstruction (*fukkyū-fukkō*); ecology of harm produced by decontamination policy
decontamination, and possessions of the owners: demolition of home and necessity of following the protocols of sorting through, 184–87; and the ecology of harm produced by decontamination

policy, 229; farmers' land and soil, 169, 185, 186–87; and homogenization of unique histories of the removed objects, 168, 179, 208; and loss of livelihoods, 169, 174, 177–79, 181–82, 184–87; as opportunity to throw away unnecessary things, 176–77; owners choosing to keep, 172, 176–77, 179; persimmon tree decision, 185, 205, 207, 229; sacrifice by owners as expectation, 174; as widespread destruction of residents' familiar environments, 184–85, 186
decontamination-support initiative for follow-up activities (TEPCO), 214; drones project to chase away boars, 214–16, 223, 229, 278n2
decontamination workers: choosing the work because they are ineligible for compensation, 70; convenience stores expanding due to, 40, 42; criticisms of, by residents, 176; economic exploitation of, 180; exploitation by corporate contractors, 180; free medical care given to, 60; health risks of radiation exposure for, 70, 81; lower socioeconomic backgrounds of, 60; number of, 180; photographs of property taken by, 174–76, *175*; preexisting medical issues of, 60; traffic due to, 40; younger adults in demand for, 60, 69–70
demersal fish, consumption of, 131
demolitions: for earthquake damage, 183–84; possessions of owner, necessity of following the sorting protocol for decontamination, 184–87; rebuilding homes, delays in, 189; rebuilt homes, discomfort of returnees with, 190; as recontaminating areas previously decontaminated, 180
dental care, 56
desire of residents to stay or return to coastal Fukushima: overview, 45–46, 247; approval not sought for, 24, 32, 45; Buddhist temple caretakers and need for community service, 16–17, 193–94; compensation denied for failure to evacuate, 41; as ignored in the opinions of experts, 45; "life goes on," 45, 67; as

dying well (*iishinikata*): and desire not to
burden those who remain, 247–48;
detachment and, 248; family graves, 183,
226–27, 246; home as inseparable in, 22,
170, 183; and intergenerational succession, commitment to, 247, 248; and
necessity to return to evacuated homes,
55, 84, 183, 187, 208; and yearning for
home held to be a biological impulse,
170. *See also* intergenerational succession; spirituality

earthquake of 3.11: aftershocks of, as continuing, 235, 238; demolition of houses
due to damage by, 183–84; narratives of
experience, 58; and risk awareness of
additional quakes, 98. *See also* deaths
due to 3.11 (*kanrenshi*)
earthquakes: of 2022, and tsunami, 236–37;
and attention to the state of 1F reactors,
235, 237; number of serious, since 3.11,
237; as reference point for disasters, 3,
257n8
ecology of harm produced by decontamination policy: overview, 215–16, 229; and
accidental collaboration into the production of nuclear waste, 209–10,
216–17, 226, 228–29; *en* severed by, 171,
182, 216–17, 229, 230–31, 232, 233; hunters coerced into indiscriminate killing
to survive, 222, 224–26, 229; and the
limits of the half-life politics of separating humans from the radioactive environment at all costs, 212–13, 223, 228,
229–33; multiplicities of loss from, 229;
structural inequality, 232. *See also*
decontamination, and possessions of
the owners; half-life politics (narrow
emphasis on radiation exposure as the
sole consequence of the TEPCO
accident)
Edelstein, Michael, 45, 52
elite panic, 2
*en* (Japanese folk theory of the interconnectedness of things, people, land, and
ancestors): and abduction, 216; addressing harmful rumor (*fūhyō*) and forgetting (*fūka*) by engaging with, 143; as a

divine force that brings things together,
195; with boars, 217–18, 219–20, 225–26;
bonds (*kizuna*) as much stronger form
of relationality, 275n34; decontamination as ripping apart, 171, 182; definitions and colloquial uses of, 11–12, 37;
the ecology of harm as severing, 171, 182,
216–17, 229, 230–31, 232, 233; and exploration of different lines, 99; as gray
zone, 20, 243; of Hiwashi shrine and
TEPCO volunteers, 199; as infrastructure of meaningfulness, 12, 37; as invisible thread (*mienai ito*), 54, 144, 216,
242; as kinship, 242, 280n7; loss of
livelihoods as loss of, 170–71;
Minamisōma and Nanto city recultivating their migration-based *en*, 278n17;
and mortuary practice, 126–27; origins
in the Buddhist philosophy, 11, 20;
relational population interested in
Fukushima Prefecture, 233–34; as the
social and cultural fabric, damaged by
the TEPCO accident, 209; as supernatural force, 53, 192; and synchronicity
(*kyouji*), 143, 144, 192, 233–34; tours of
1F and, 151; and transmigration of the
soul, 248; as weak connection (*yowai
tsunagari*), 216, 234, 242, 279n5; willingness to cultivate and sustain, 54. *See
also* ethnography, *en*-inflected
environmental disasters: characterized as
manifesting discrimination and division, 5, 25, 134; and the legal concept of
indirect damage (*kansetsu songai*),
92–93; pollution language of decontamination and, 173
ephemera of loss, 158
Erikson, Kai, 3, 95, 96, 170
ethnography: critical, of fallout, 23; disaster
ethnography, 4–5, 259n36; oikography,
260n46; as simultaneously a method
and a product of written work, 44; as
slow science, 259n36; "urgent ethnography," 45. *See also* anthropology; ethnography, *en*-inflected
ethnography, *en*-inflected: overview, 26–27;
belief in *en* by the residents as the sali-
ent fact, 216, 242; as coproducing a new

intersubjective narrative, 245; the ethnographer as servant of *en,* 244–48; as Japanese folk understanding of ethnography, 12; mediation of, 245; the nuclear ghost and, 28, 37–38, 244; participants finding *en* in, 242

—METHODOLOGY OF THE RESEARCHER: car lent for fieldwork, 34–35; familiarity with local geography, 38; Japanese scholarship and sources, 26; long-term participant observation, 17–18, 26–27, 243–48, 258n17, 259n36; the nuclear ghost (*houshanō obake*) and, 17–18, 24; referral process of finding interlocutors, 4

European Union (EU), guidelines for radioactive materials in foods, *88,* 89

evacuation: burglary and destruction by human intruders during absences, 201, 277n12; destruction of property by wildlife and pest during absences, 188–89, 212, 217, 218, 220, 224; and disintegration of families, 56–57; gas shortage adding difficulty to, 98; in-between disaster (*saikan*), 66–67, 74–75, 97–98, 242–43, 264n3; individuals unable to evacuate due to disabled or aged family members, 41; of Kashima district, 41; mandatory, 41; Minamisōma not enforcing, 42, 46–47; pets lost in, 58, 201; physical distance from 1F as continuing reference point for evacuees, 74–75, 77; scarcity of material goods and need for, 48; voluntary, 41, 48, 94–95. *See also* absence of humans (anthropause); evacuation zones; evacuees; TEPCO accident

evacuation zones: aging of the population in, 8; of Chornobyl, as permanent, 9; as ineffective in protecting people from airborne radiation, 51, 100; infrastructure cutoffs and, 8; map (2014), *39;* population in, 8; radiation monitoring of items transferred from exclusion zones, 158, 203–4; as roughly overlapping preexisting district boundaries, 51; rules for persons visiting exclusion zones, as varying, 200–201; size of

original order, 7; social tensions produced by, 51–52. *See also* evacuation zones, reopening; evacuation zones, reterritorialization of

evacuation zones, reopening: overview, 9; compensation as ending with, 85, 101–2, 104–6, 134; decontamination as signaling the state's intention for, 169–70, 181–82; delays in evacuee returns, 189–90; equated with the end of harm, 267n5; Futaba district, 273nn10,13; Futaba town, 280n5; Haramachi district, 103; and ideological division between those who desire to go back as soon as possible and those who do not wish to go back, 105–10; Iitate, 170; intention to open the highly contaminated sections, 200, 280n5; Katsurao, 280n5; Minamisōma, 171; Namie, 273n10, 277n11; Naraha, 273n13; numbers of returnees, 171, 212, 277n11; Odaka district, partial and full, 42, 43, 189, 194, 212, 265n5; Ōkuma town, 280n5; as ongoing, 238, 280n5; radiation exposure threshold in addition to known average dose, 101; radiation exposure thresholds, shifting, 101

evacuation zones, reterritorialization of: overview of changes made (through March 2012), 75, 263n26, 265n1; map (April 22, 2011), *76;* and Namie, 200; and population recovery in Minamisōma, 48–49; revision based on measurable contamination levels, 68, 74–75, *76;* special evacuation designation zone, 68; in the twenty-kilometer zone, 68, 264n4. *See also* evacuation zones, reopening

evacuees: ambivalence about luck of survival, 5; length of time before returning home, 85, 265n5; number of, 9, 58, 131; staying in touch with friends who left, 82–83; those who never came back, 57, 85, 98, 105; transportation problems, 54, 56, 79; voluntarily remaining evacuated to avoid radiation exposure, 43, 49, 57, 104. *See also* atomic livelihood; compensation; desire of residents to stay or

farmers' testing program (Miura), 90–91, 93, 119, 204; fish testing, 141; local testing centers, 204, 205, 277n15; quantities of sample required for testing, 277n14; rice testing and tracing, 90–91, 266n11; state standards for contamination, 87–89, *88*, 205–6

forests: cleaning of (*souji*), 198–99; compensation for loss of trees, 189; decontamination of, 185, 189; of Hiwashi (Shinto shrine), TEPCO volunteers cleaning, 198–99; as more susceptible to the concentration of contaminants, 47, 198, 259n35; as percentage of Fukushima Prefecture, 185; as percentage of Minamisōma, 47; Shinto shrines typically surrounded by, 197, 198; surviving pine trees, care for, 40–41, 262nn16–17; tsunami destruction of pine trees, 40–41, 262n16. *See also* persimmon trees and fruit

forgetting (*fūka*): addressed by individuals interested in cultivating *en*, 143; containment of contamination as facilitating, 154, 163–64; definition and spelling of, 141–42; disappearance of 1F from collective memory until the accident, 155, 163; ignoring radiation monitoring readings and desire for, 64; as intertwined with harmful rumor (*fūhyō*), 142–43, 154, 163–64; outsider pressure to remember and relive, 64–66

Fortun, Kim, 27, 202

*frecon baggu. See* decontaminated waste and *frecon baggu*

French Polynesia nuclear contamination, 4

Freudenburg, William, 52, 71, 109

*fukkyū-fukkō. See* disaster recovery and reconstruction

*Fukugō Saigai. See* 3.11 (*san ten ichi ichi*)

Fukushima nuclear disaster. *See* TEPCO accident

Fukushima Prefecture: overview, 7, 258n13; and relational population, 233–34. *See also* contamination of Fukushima; Minamisōma; roads; stigma of the TEPCO accident

Fukushima Robot Test Field (RTF), 236

Futaba district: overview, 145; boars as obstacle for returnees, 214; deaths due to 3.11, 156; evacuees' plans to return or not, 85, 156; map of, *146*; radiation contamination of, 85, 145, 156; reopening of, 273nn10,13; tsunami damage, 85, 145, 156

Futaba town: and development of 1F, 160–63, 274–75nn27–32; disaster heritage preservation in, 155–60; reopening of, 280n5

future generations. *See* intergenerational impacts; intergenerational succession

gambling, 81–82

gamma waves, 268n4

Gardner, James, 156–57, 158

Geiger counters: as measuring phenomena beyond the threshold of human experience, 14; Namie readings on, 202, 203; readings indicating contamination to be less than official reports, 58, 174, 177, 180; readings of interest to residents as historical comparison, 62–63; as sign of outsider status, 62; as sign of the radiation-centered narrative, 15–16, 17, 44, 65, 66; as "truth," 15

Ghana, 272n5

ghosts (*obake* or *yūrei*): Buddhist funeral rituals to guide the spirit away from becoming, 194; encountered by 2011 tsunami survivors, 14, 259n33; ghost-story tradition (*Kaidan*), 30–31, 261n1; Japanese conceptualization of, 13–14, 16, 64; "radioactive ghosts," 14. *See also* nuclear ghost (*houshanō obake*)

Gill, Tom, 37–38, 51, 57

Goffman, Erving, 17, 181

Gordon, Avery, 13, 64

gray zone of the TEPCO accident, 19–20; Kishō Kurokawa's conception of the culture of gray, 11, 19–20, 243, 260n40, 264n3; the nuclear ghost and, 20, 21, 29–30, 67

green pheasants, 211

Gunma Prefecture, 261n8

*hairo no jidai. See* decommissioning, age of

*hairo. See* decommissioning (*hairo*) project

Ibaraki Prefecture, radioactive materials released in, 261n8
Ibaraki Science Museum of Atomic Energy, 112–13, 114, *115*
*Ibaraki Shinbun* (newspaper), 275n35
identity, regional, 51
Iitate: booklet collaboration with Dr. Tsubokura, 136; contamination of, 51; corrosion of community in, 51; evacuation zone, 75; request to state to reopen without decontamination, 170; in Sōma domain, 47
in-between disaster (*saikan*), 66, 74–75, 97–98, 242–43, 264n3
Indigenous North American cultures: assessment of radiation exposures among, 205–6; nuclear colonialism of the southwest, 4, 21, 155, 232–33; and place, 17, 55
Indigenous people in Japan (Emishi), 36
Indigenous peoples: human and more-than-human kinship as idea among, 280n7; and nature, relation of harmony with, 277n6
indirect damage (*kansetsu songai*), 92–93
infrastructure: railroads, 40, 43, *146*, 262n13. *See also* roads
Innovation Coast Framework, 236
*inoshishi. See* boars
intergenerational impacts: nuclear risk as dislocated and deferred to the future, 27–28, 233, 248; persistent concerns of the Japanese public and Fukushima residents about adverse hereditary consequences, 9–10, 238–39, 245
intergenerational succession: of Buddhist temple caregiving, 193; difficulties of, as harm of contamination, 239; dying well and commitment to, 247, 248; dying well and the desire not to burden those who remain, 247–48; farmers and the soil, 186, 211, 238, 247; home as the nexus of, 22; and nature, coexistence with the supernatural entities of, 195; respect for the land as requisite for, 248; of Shinto shrine caregiving, 196; the TEPCO accident as unable to annihilate the desire and efforts for, 206, 245;

transmigration of the soul as dependent on commitment to, 248
Interim Storage Facility (decontaminated waste), 182
internal radiation exposure (*naibu hibaku*): and biological vs. physical half-life, 118, 268n3; definition of, 119; fears of local food contamination despite data showing no levels of, 136; mosquito bite question, 118–20, 121, 126, 134, 136, 138; sources of, 119, 268n5; whole body counter (WBC) data measuring, 128, 130–31, 132, 136, 268n2, 271nn27–28,33. *See also* foods, radiological contamination of
International Atomic Energy Agency (IAEA), 9
International Commission on Radiological Protection (ICRP): and the global conventional practice of containment, 230; "Reference Man" defined by, 270n23; standard for dose limits, 85, 101, 266n6
International Nuclear and Radiological Event Scale: Chornobyl, 9; the TEPCO accident, 9, 258n23
iodine-131, 130, 269n8, 271n27
Isewan typhoon (1959), 197
Ishido, Satoru, 99, 239, 241
Ishinomaki: Buddhist temples caretaking unclaimed cremated remains of tsunami victims from, 194; ghosts encountered by survivors of, 14, 259n33
Iwaki, 49, 153
Iwamoto, Yoshiteru, 37, 47, 206, 232
Iwate Prefecture, surviving pine tree in, 262n16

Japan Atomic Energy Agency (JAEA): establishment of, 111–12, 267n10, 275n35; JOC criticality accident, 113, 120, 268n15; reactors of, 111, 113, 267nn11–13
Japan Environmental Storage and Safety Corporation, 182. *See also* decontaminated waste (nuclear waste)
Japanese raccoon dogs, 193
job losses due to 3.11: and necessity to live off compensation, 81; and residents

media (continued)
  visualizing contamination, 168; frequently appearing evacuees in ("the sufferer" and "the exposed"), 4, 258n11; and hyperlocal conception of radiation exposure, 10; and hyperreality, 5; and hypervisualization of radiation, 61–62; and misinformation, 124, 125–26, 138; radio station providing recovery information, 117–20, 127; the safety myth perpetuated by, 112–13. See also documentary film
medical care: and the clinic as site of socializing, 56; dental care, 56; emergency care by Dr. Tsubokura (2011), 127; free care for reconstruction and decontamination workers, 60; free care to evacuees and evacuation zone residents, 56, 60–61. See also medical radiation exposure; entries at health effects
medical radiation exposure: average annual exposure in Japan, 269n6; CT scans, 119, 120, 123, 268n5; equivalent risks for, 123; mammography, 268n5; radiocontrast agents (e.g., iodine and barium), 268n5; scientific data derived from, 124; X-rays, 41–42, 268n4
Meiji-Sanriku tsunami (1896), 259n33
mercury poisoning (Minamata disease), 5, 173, 243
migration to Minamisōma: eighteenth century, 206–7, 278nn17–20; the Fukushima Robot Test Field (RTF) and hope to attract, 236
Minakata, Kumagusu, 11, 20
Minamata disease (mercury poisoning), 5, 173, 243
Minami-Migita, 40, 40–41, 262n15
Minamisōma: assertion as anti–nuclear energy city, 145; and boars, 211–12, 220, 223; chilly atmosphere of, 30–31, 66; decontamination of, and removal of waste, 171, 182; district-based identities of residents, 51; economic structure of, 47, 236; the evacuation not enforced in, 42, 46–47; and the Fukushima Robot Test Field (RTF), 236; geographic divisions of, 48; location in relation to

Tokyo, 46; map (2014), 39; Merger Policy (2006) and social tensions, 50–51; as periphery, 7–8, 11, 19, 259n29, 260n39; population after the TEPCO accident, 8, 35, 48, 171, 238; population prior to the TEPCO accident, 7, 47; remoteness of, 7, 36, 40, 49, 262nn12–13; reopening of, 171; size in square kilometers, 46; slogan established for rebirth of, 117; subsidies paid by the state for cooperating with nuclear power plant placements, 144–45, 272n6; subsidies, refusal of, 145; the two different cities confronted by residents in their everyday lives, 61; uniqueness among disaster-affected cities, 46–47; voluntary evacuation due to scarcity of goods, 41, 48. See also aging of population due to 3.11; division and corrosion of community; farmers and farming; forests; Futaba district; gray zone of the TEPCO accident; Haramachi district; Kashima district; Odaka district; wildlife and pests
missing persons, 2–3, 257n5
Miura, Hiroshi, 87, 90–91, 93, 110, 144, 204
Miyagi Prefecture: deaths due to 3.11, 259n33; Minamisōma as culturally closer to, 40; radioactive materials released in, 261n8; roads as tsunami wall, 262n14
Miyamoto company, 227
monkeys (macaques): absence of humans and increased presence of, 193; and the ecology of harm produced by decontamination policy, 229; previous coexistence with, 211; viewed as destructive pest, 188, 212, 218
mosquitoes, question about, 118–20, 121, 126, 134, 136, 138
mothers. See risk-avoidant people
Murakami, Haruki: overview, 5–6; After Dark (2002), 165; After the Quake (2007), 250; Colorless Tsukuru Tazaki and His Years of Pilgrimage (2013), 140; IQ84 (2011), 29, 61, 211; Kafka on the Shore (2005), 28, 97, 99; New Yorker interview (2018), 188; Norwegian Wood

nuclear ghost *(continued)*
Minamisōma as periphery, 11; namelessness and, 102; as the psychosocial effects
of radioactivity, 14; radiation-centered
approach (half-life politics) as incorrect
interpretation, 3–4, 13, 31–32, 35, 139,
192, 240–41; and reluctance of residents
to burden generations to follow, 247–
48; respect for the land as message of,
248; as a situated experience, 32; willingness to receive the message of, 99
nuclearity, 260n43
nuclear power: deaths and illnesses of
workers, 2, 9, 10, 113, 268n15; displacement and deferral of risk to future
generations, 233, 248; displacement of
risk to the periphery (development of
underdevelopment) model for placement of reactors, 162–63, 275n35; global
expansion of, 27–28, 236; humanity
held to be powerless against, 141; meltdown of, as achieving its intended goal
of "going green" by purging humans
from their built environment, 202–3;
municipal subsidies to encourage cooperation for (Three Electric Power Development Laws), 144–45, 272n6; number
of reactors in Japan and status of, 236;
proposed but never-built (phantom)
plants, regional effects of, 144, 272n5;
rehabilitation as "peaceful use of
atoms," 274n27, 275n36; uncontainable
risks of, 200; as unsustainable structural
threat, 248; uranium enrichment and
processing, 111–12, 268n15; uranium
mining, 4, 21, 155, 232–33. *See also*
antinuclear discourses and movements;
Chornobyl; decommissioning *(hairo)*
project; decontamination; Japan
Atomic Energy Agency (JAEA);
nuclear workers; safety myth *(anzen
shinwa)*; TEPCO ("1F" *[ichi efu]* Tokyo
Electric Power Company Fukushima
Dai-ichi Nuclear Power Plant); TEPCO
accident
nuclear things: interconnectedness and
solidarity in the face of, 20–21, 260n42;
technopolitical distributions of, 21,

260n43. *See also* half-life politics (narrow emphasis on radiation exposure as
the sole consequence of the TEPCO
accident)
nuclear uncanny. *See* disorientation
(nuclear uncanny)
nuclear waste: international law not recognizing the decontaminated waste from
the TEPCO accident as, 276n13. *See
also* decontaminated waste (nuclear
waste)
nuclear weapons: atmospheric radiation
exposures due to, 130, 131; the psychosocial impacts as disorientation, 15,
259n34; uranium enrichment, 111–12.
*See also* Hiroshima and Nagasaki
atomic bombings; nuclear power
nuclear workers: deaths and illnesses of, 2,
9, 10, 113, 268n15; decommissioning
*(hairo)* project, 141, 149, 153–54, 235–36,
237; disaster recovery and reconstruction, 42, 60, 241; dislocation of
TEPCO responsibility for the accident
onto, 154, 273n21; scientific data derived
from, 124; who are also TEPCO accident evacuees, 143, 153–54, 199. *See also*
decontamination workers

Ochi, Sae, 134–35, 137, 269n10
Odaka district: overview, 42–43; compensation given to evacuees from, 68, 104;
decontamination project begins operating in, 171–72, *172*, 176; delays in evacuee returns, 189–90; earthquake-
damaged houses, 43; evacuees returning
home to maintain farm, home, and
altars, 182–83; flower-planting by locals,
43; health benefits of returning home,
131; Merger Policy into Minamisōma,
social tensions due to, 50–51; monster
serpent legend of, 194–95, 277n3; as
nostalgic countryside *(inaka)*, 212;
population after 3.11, 43, 212; rebuilding
homes in, 189–90; renovating homes in,
189; reopenings, partial and full, 42, 43,
189, 194, 212, 265n5; and reterritorialization of evacuation zones, 68; in the
twenty-kilometer zone, 42, 59. *See also*

decontamination; demolitions; Hiwashi (Shinto shrine)

oikography, 260n46

Ōkuma town: and development of 1F, 160–63, 274–75nn27–32; reopening, 280n5; rules for persons entering exclusion zones, 201; in the Sōma domain, 47

1F. *See* TEPCO ("1F" [*ichi efu*] Tokyo Electric Power Company Fukushima Dai-ichi Nuclear Power Plant)

Ono, Masahiro, 31–32, 44, 124–25

Ortiz, Alfonzo, 17

Osseo-Asare, Abena Dove, 112, 272n5

Ōtomi, 132, *133*

outsiders (*yosomono*): compensation assumed to be received by all the residents by, 109; crimes committed by, 33; desire to bring residents back to the TEPCO accident, 64–66; familiarity with the local geography as a necessity for, 38; and judgments about irrationality of living with radiation, 34; as less careful about exposure, 131; "parachuting" research by, 37–38; radiation-centered focus of, 34, 44, 46; solidarity among residents about how to deal with, 34; as status of those lacking awareness of historical radiation readings, 63; as status of those who carried Geiger counters, 62; as status of those who would walk on Route 6, 33, 34, 38; 3.11 as bringing in large numbers of, 33; wariness of residents toward, 33–34, 46

Parmentier, Richard, 62, 245

Peirce, Charles Sanders, 216

perception of risk. *See* risk-avoidant people; risk perception

peripheries (*shū-en*): coexistence and interconnection in, 11; displacement of radiation threat to, 21, 162–63, 275n35; and emergence of new things and ideas, 259n29; Minamisōma as, 7–8, 11, 19, 259n29, 260n39; structural inequalities between the center and, 152–53, 155, 162–63

persimmon trees and fruit: cultural importance of, 206, 207; family decision to allow destruction of tree, 185, 205, 207, 229; fruit from Namie, as radioactive, 203–4, 205–6, 277n15; wildlife eating fruit, 203, 204, 207, 217, 218, 229

pests. *See* wildlife and pests

Peta, as unit of measure, 258n24

Petryna, Adriana, 10, 25

pets, lost in the evacuation, 58, 201; and need for pet-friendly evacuation centers, 201

place, 17, 55

Polleri, Maxime, 37–38, 135, 260n44

polonium, 119

potassium, 130

precarity of 3.11 life: and the decommissioning project, 235–36; decontamination jobs, 180; and desire for a return to previous life, 67, 84; as entanglement of uncertainty, 66–67; in-between disaster (*saikan*), 66–67, 74–75, 97–98, 242–43, 264n3; and return of farmers to their livelihood, 205; risk of additional nuclear accidents, 235, 237, 238; wildlife in relation of shared precarity with humans, 228

psychological damage as compensation category, 68, 104, 204

psychological effects of decontamination, 180–82, 276n10

psychological harm of evacuation: being unable to search for missing friends and family members, 3, 257n5; as compensation category, 68, 104, 204; inability to be at home, 60; loss of laughter and joy, 8; loss of pets, 58, 201; PTSD, 170; sleep difficulties, 57–58, 59, 61, 142, 183; suicides, 98. *See also* anxiety about radiation exposure; health effects of the TEPCO accident (lifestyle and secondary effects); psychosocial effects of the TEPCO accident

psychosocial effects of the TEPCO accident: overview of, 94–95; and food contamination, 89; the nuclear ghost and, 14. *See also* disorientation (nuclear uncanny); division and corrosion of community; health effects of the TEPCO accident (lifestyle and secondary effects); risk perception; science and policy, mistrust of

PTSD, 170
public opinion: generally supportive sentiment of Japanese people for those in Fukushima, 35; persistent concerns about adverse intergenerational hereditary consequences, 9–10, 238–39, 245; relational population, 233–34. *See also* media; outsiders (*yosomono*); stigma of the TEPCO accident
purity and danger, 20, 181, 208, 233

radiation-centered narrative. *See* half-life politics (narrow emphasis on radiation exposure as the sole consequence of the TEPCO accident)
radiation exposure: Cloud Chamber as visualizing, 114, *115*; cultural sensitivity to history of (*hibaku*), 18–19, 119, 122, 137, 140, 260nn37–38; defined for this text as chronic low-dose exposure, 19; exposable people, 95, 125, 232, 237; external (*gaibu hibaku*), defined, 119; fatal dose, 151; as hyperlocal, 10; initial reaction of fear of the unknown, 18; normalization of, as "radioactive performance," 270n15; as omnipresent and universal, 10, 21, 116, 122–23, 130; universalization and normalization by radiation safety officers, 122–23, 130, 132, 270n15; wildlife research shows resilience to, 231. *See also* health effects of chronic low-dose radiation exposure; intergenerational impacts; internal radiation exposure (*naibu hibaku*); medical radiation exposure; naturally occurring radiation exposure; radiation monitoring; risk communication; risk perception
radiation monitoring: *becquerel* (Bq) as unit of measure, 89; and boar-incineration nuclear waste, 228–29; and causal links to disease, difficulty of proving, 120–21; citizen-centered projects, 92–93, 124–25, 135, 141, 204; and disaster heritage preservation, 157–58; as historical vs. immediate interest for residents, 62–63, 64; ignoring, and the desire to forget the accident, 64; the

meaning of the numbers as uncertain, 64; Namie visits requiring, 201, 203–4; personal dosimeters, 62–63, 64, 148, 151; the radiation-centered narrative as reinforced by, 62–63, 64; radiation monitoring posts (MPs), 63–64, *63*, 264n2; refusal of testing for exposure to avoid being the state's experimental body, 10, 122; as sign of outsider status, 62; on tour bus of 1F, 148, 149–51, 273n17; for transferring items out of exclusion zones, 158, 203–4; whole body counter (WBC) data (Dr. Tsubokura), 128, 130–31, 132, 136, 268n2, 271nn27–28,33; wildlife as radiotracers, 219, 229–31. *See also* foods, testing for radiological contamination; Geiger counters
radiation monitoring posts (MPs), 63–64, *63*, 264n2
Radiation Safety Officer (RSO) training course (Columbia University), 122–24, 130, 132, 270n15. *See also* ALARA; risk communication
radioactive materials released (fallout) in the TEPCO accident: overview, 9; Chornobyl compared to, 9, 124, 269n8; concentrated in forests and bushy landscapes, 15, 47, 259n35; as crossing international boundaries, 130; data collection by Dr. Tsubokura (WBC results), 128, 130–31, 132, 136, 268n2, 271nn27–28,33; exact quantity as unknown, 270n24; isotopes calculated based on their ratio to cesium, 271n25; as less than exposures to nuclear weapons testing and Chornobyl, 131; Shinto shrines as vulnerable to, 198. *See also* cesium; decontamination; foods, radiological contamination of; radiation monitoring
radioactivity: *becquerel* (Bq) unit, 89; definition of, 266n10
radio station (Hibari FM) providing recovery information, 117; Dr. Tsubokura's information program, 117–20, 127
radon, 101, 119, 130
railroads, 40, 43, *146*, 262n13

recovery and reconstruction. *See* disaster recovery and reconstruction (*fukkyū-fukkō*)

relational population interested in Fukushima Prefecture, 233–34

religiosity among the Japanese, 192. *See also* Buddhism; Shintoism; spirituality

residents who stayed or returned. *See* desire of residents to stay or return to coastal Fukushima; evacuees

rice, testing and tracing, 90–91, 266n11

risk: displacement and deferral to future generations, 233, 248; displacement to the periphery (development of underdevelopment) model, 162–63, 275n35; as egalitarian, 179; individualization of risk as defining modern society, 138; of nuclear power, as uncontainable, 200. *See also* risk-avoidant people (e.g., mothers with young children); risk communication; risk perception

risk-avoidant people (e.g., mothers with young children): anxiety of, and avoidance of information, 94; citizen-based monitoring of radiation levels of foods by, 92–93, 124–25, 204; fear of local food despite WBC data showing no internal contamination, 136; and livelihood in a specific place, 137; respect for varying perceptions of risk, Dr. Tsubokura on importance of, 131, 137; tours of 1F reopening as TEPCO attempt to address, 140–41; voluntarily remaining evacuated to avoid radiation exposure, 43, 49, 57, 104. *See also* risk-avoidant people, labeled as "irrational"

risk-avoidant people, labeled as "irrational": overview, 97; construction of a "nuclear ghost" as irrational fear, 31–32, 44, 124–25; as "radioactive brain" (*housha nō*), 124–25, 270n18; as "radiophobia," 124; rumor-related damage (*fūhyō higai*) and, 92; as silencing residents, 100, 103; in top-down risk communication, 124–26, 130, 134, 270n18

risk communication: Chornobyl as reference point in, 126, 128, 130, 131; and dismissive/oppressive use of science,

126; every message hurts someone, 133, 134, 137; failure to regain the trust of the residents and public, 238–39; Fukushima as reference point in future, 126; inconsistencies in, mistrust of experts as emerging from, 134; and life information needed for informed decisions of residents, 134–35, 136–38; scientists' certainty of low risk and labeling of risk-avoidant people as "irrational," 124–26, 130, 134, 270n18; top-down model of, as ineffective, 125; trust issues blamed on the public in top-down model of, 125; universalization and normalization of radiation in, 122–23, 130, 132, 270n15; as violence, 137. *See also* risk communication strategy of Dr. Tsubokura

risk communication strategy of Dr. Tsubokura: accusations of Tsubokura being pronuclear and promoting the safety myth, 127, 132, 134, 137, 138, 271n34; action-oriented approach for reducing exposure, 130, 131; communication should be done by non-scientist care experts, 134; consistency of communications, 134; and the data showing lifestyle and secondary health issues to be the greatest health risk, 131–32, 135, 137–38, 271n29; and the data showing radiation levels to be not medically significant, 127, 128, 131, 132, 136; every message hurts someone, 133, 134, 137; and the need for specific local data, 128, 136–37; as never denying the harms of the TEPCO accident, 130; the risk as never zero, 128; and risk perception as varying among individuals, respect for, 131, 137; and the stigma of Fukushima, desire to combat, 132–33, 136, 142; youth education as focus, 136. *See also* Tsubokura, Masaharu

risk perception: disagreements over perceived risks as silencing residents, 57, 84, 85–86, 100, 103–4; and livelihood in a specific place, 137; postfallout individualization of, 138; respect for varying perceptions, Dr. Tsubokura on

risk perception *(continued)*
importance of, 131, 137. *See also* risk-
avoidant people (e.g., mothers with
young children); risk communication
roads: closures due to 3.11, and remoteness
of Minamisōma, 36; former Route 6
(Rikuzen Hamakaidō), 38, 40, 182; and
the *frecon baggu* wastescape, 171–72, *172*,
182, 184, 185, 208, 276n11; and the
imagination of distance, 37; and
improved access between central and
coastal Fukushima, 262n12; Kurokawa
on *en* and, 20; Sanrokusen road, 182,
184, 276n11; tsunami damage to, 40; as
tsunami wall, 40, 262n14. *See also*
Route 6 (Rokkoku)
Rokkoku. *See* Route 6 (Rokkoku)
Röntgen, Wilhelm Conrad, 123–24
Route 6, former (Rikuzen Hamakaidō), 38,
40, 182
Route 6 (Rokkoku): overview, 38, 40, 48;
barricaded at twenty-kilometer boundary, 15, 42, 59, 145; and cultural affinity
with Miyagi Prefecture, 40; and the
*frecon baggu* wastescape, 171–72, *172*,
182; maps of, *39*, *146*; the ocean as visible
from, 144; outsider status revealed by
those who would walk on, 33, 34, 38;
traffic on, 40, 145; the tsunami and
damage to, 40; as tsunami wall, 40
rumor-related damage (*fūhyō higai*) compensation, 91–93, 111, 113
rumor. *See* harmful rumor (*fūhyō*)

safety myth (*anzen shinwa*): accusations
against Dr. Tsubokura as promoting,
127, 132, 134, 137, 138, 271n34; accusations against Hiroshi Kainuma as
promoting, 142; the claim that Chornobyl was different from Fukushima, 100;
definition of, 162; the media as perpetu-
ating, 112–13; and the spectacle of
societal hierarchies, 162; and unconditional trust given by the locals, 161–62,
163
*saikan* (in-between disaster), 66, 74–75,
97–98, 242–43, 264n3
Sakurai, Katsunobu, 42, 145

Samuels, Richard J., 6, 61
Sanrokusen (road), 182, 184, 276n11
schools: bullying of evacuees in, 74, 105;
disaster heritage preservation project in,
156–58, *157*; radiation monitoring of, 64;
reopening, 48; subsidies to
Minamisōma for cooperating with
nuclear power used to pay for teachers,
272n6. *See also* children; college
science and policy, mistrust of: and arrogance of not paying respect to the power
of nature, 195; citizen-based radiological
testing and, 92–93, 124–25, 135, 141,
204; and the COVID-19 pandemic, 239,
240; and the ideological currents of
society, 121; and inconsistencies in risk
communication, 134, 239; loss of trust in
local food despite data showing safety
of, 92–93, 135–36; persistent belief in
adverse intergenerational hereditary
effects, 9–10, 238–39, 245; and the
scientific uncertainty about decontamination, 179–80; and the scientific
uncertainty about the dangers of radiation exposure, 121, 127, 238; and structural violence, 121; thyroid cancer and,
121, 126; top-down model of risk communication as blaming the public for
trust issues, 125. *See also* risk
communication
scientific and policy uncertainty: and
decontamination protocols, 35, 85, 166,
173, 179–80; and decontamination
standards to achieve, 85, 266n6, 276n6;
health effects of chronic low-dose
radiation exposure, 31, 43–44, 262n19;
and mistrust of science and policy by
the residents and evacuees, 121, 127,
179–80, 238; as source of social anxiety,
121; "too much" radiation, 123
scientific modernization, critiques of, 37
scientific research in radiation exposure:
collective bodies often exposed involuntarily, 121–22, 124; critiquing the
anthropocentric framing of radiation
safety, 230; exposure of scientists, 123–
24; the residents of coastal Fukushima
as nonconsensual experimental bodies,

stigma of the TEPCO accident: bitterness of Fukushima residents about, 70; bullying of evacuees and residents outside Minamisōma, 74, 81, 105, 142; on compensation recipients, 68, 69, 81–82; Dr. Tsubokura's desire to combat, 132–33, 136, 142; family disintegration and, 57; and foodstuffs produced in the area, 87, 90, 91–93, 136; and historical experience of Hiroshima and Nagasaki survivors, 18; judgments by scientists, 45; judgments made about those living with radiation, 6, 34; relocation of businesses to avoid Fukushima address, 83; rumor-related damage (*fūhyō higai*) compensation, 91–93, 111, 113; silence about and/or concealment of Fukushima origins due to, 7, 81, 98; as structural inequality, 152–53, 232; and thyroid cancer risk, 269n10; travelers from Fukushima experiencing, 6–7, 81; tsunami survivors as subject to, 81; warnings against entering Fukushima, 35. *See also* contamination of Fukushima; discrimination

stress: as health effect of the evacuation, 59–60, 108–9; wildlife showing less stress due to absence of humans, 231

strontium-90, 130, 271n27

structural inequality: asymmetrical distribution of risks, 24; between the center and periphery, 152–53, 155, 162–63; and ecology of harm, 232; radioactive colonialism, 20–21, 155, 232–33; and the stigma of Fukushima, 152–53, 232

structural violence of half-life politics: overview, 102–3; agency of individuals constrained by, 102; definition of, 102; and loss of livelihood, 233; mistrust of science and policy and, 121; as not the fault of the individual, 102; risk communication and, 137; social anxiety due to scientific uncertainty, 121

suffering: disaster capitalism and, 5, 180; ethnography and the search for, 4–5; of Hiroshima and Nagasaki atomic bomb victims (*hibakusha*), 18–19, 119, 137, 140; interpretive struggle about 3.11 and, 5

supernatural. *See* disorientation (nuclear uncanny); *en*; ghosts (*obake* or *yūrei*); nuclear ghost (*houshanō obake*); spirituality

surfing, 40, 262n15

the surreal (*cho-genjitsuteki*): coexisting with the real in the gray zone, 61; environmental disasters as, 5–6; and half-life politics, 25; the nuclear ghost experience as, 155; radiation and, 17; social life of, 110

Sv (*sievert*) unit, 89

synchronicity (*kyouji*): Carl Jung's conception of, 144; and persons engaging with *en* (*kyoujisha*), 143, 144, 192, 233–34, 242. *See also en*

System for Prediction of Environmental Emergency Dose Information (SPEEDI), 200, 277n9

Tamaichi (monk), 195, 198

Tamura city, 170

technocratic and biomedical framing. *See* half-life politics (narrow emphasis on radiation exposure as the sole consequence of the TEPCO accident)

tellurium-132, 130

temporalities of the ordinary, disruption of, 66–67, 264n3

temporary housing units (*kasetsu*): overview, 50, 263nn32–3; and desire of evacuees to remain close to their evacuated homes, 50, 59; in Haramachi district, 42; and Kashima residents' sense of inequality, 41; relief organizations bringing goods to, 78–80; resentment by other evacuees due to privileges given to residents of, 79, 80; social issues due to mixture of tsunami and radiation-related evacuees, 50, 51

temporary leased housing (*kariage*): overview of, 50, 263n29; confidentiality of location of, 78–79; garden plots planted, 86–87, *86*, 204–5; as generally preferred by evacuees, 78, 80; noise issues, 83; relief organization distributions not available to residents of, 78–79, 80

The California Series in Public Anthropology emphasizes the anthropologist's role as an engaged intellectual. It continues anthropology's commitment to being an ethnographic witness, to describing, in human terms, how life is lived beyond the borders of many readers' experiences. But it also adds a commitment, through ethnography, to reframing the terms of public debate—transforming received, accepted understandings of social issues with new insights, new framings.

*Series Editor: Ieva Jusionyte (Brown University)*

*Founding Editor: Robert Borofsky (Hawaii Pacific University)*

*Advisory Board: Catherine Besteman (Colby College), Philippe Bourgois (UCLA), Jason De León (UCLA), Laurence Ralph (Princeton University), and Nancy Scheper-Hughes (UC Berkeley)*

1. *Twice Dead: Organ Transplants and the Reinvention of Death,* by Margaret Lock

2. *Birthing the Nation: Strategies of Palestinian Women in Israel,* by Rhoda Ann Kanaaneh (with a foreword by Hanan Ashrawi)

3. *Annihilating Difference: The Anthropology of Genocide,* edited by Alexander Laban Hinton (with a foreword by Kenneth Roth)

4. *Pathologies of Power: Health, Human Rights, and the New War on the Poor,* by Paul Farmer (with a foreword by Amartya Sen)

5. *Buddha Is Hiding: Refugees, Citizenship, the New America,* by Aihwa Ong

6. *Chechnya: Life in a War-Torn Society,* by Valery Tishkov (with a foreword by Mikhail S. Gorbachev)

7. *Total Confinement: Madness and Reason in the Maximum Security Prison,* by Lorna A. Rhodes

8. *Paradise in Ashes: A Guatemalan Journey of Courage, Terror, and Hope,* by Beatriz Manz (with a foreword by Aryeh Neier)

9. *Laughter Out of Place: Race, Class, Violence, and Sexuality in a Rio Shantytown,* by Donna M. Goldstein

10. *Shadows of War: Violence, Power, and International Profiteering in the Twenty-First Century,* by Carolyn Nordstrom

Founded in 1893,
UNIVERSITY OF CALIFORNIA PRESS
publishes bold, progressive books and journals
on topics in the arts, humanities, social sciences,
and natural sciences—with a focus on social
justice issues—that inspire thought and action
among readers worldwide.

The UC PRESS FOUNDATION
raises funds to uphold the press's vital role
as an independent, nonprofit publisher, and
receives philanthropic support from a wide
range of individuals and institutions—and from
committed readers like you. To learn more, visit
ucpress.edu/supportus.

www.ingramcontent.com/pod-product-compliance
Ingram Content Group UK Ltd.
Pitfield, Milton Keynes, MK11 3LW, UK
UKHW041926120125
453522UK00004B/237